Regulation of
Enzyme Synthesis
and Activity in
Higher Plants

Annual Proceedings of the Phytochemical Society

1. **Biosynthetic Pathways in Higher Plants**
   Edited by J. B. PRIDHAM AND T. SWAIN

2. **Comparative Phytochemistry**
   Edited by T. SWAIN

3. **Terpenoids in Plants**
   Edited by J. B. PRIDHAM

4. **Plant Cell Organelles**
   Edited by J. B. PRIDHAM

5. **Perspectives in Phytochemistry**
   Edited by J. B. HARBORNE AND T. SWAIN

6. **Phytochemical Phylogeny**
   Edited by J. B. HARBORNE

7. **Aspects of Terpenoid Chemistry and Biochemistry**
   Edited by T. W. GOODWIN

8. **Phytochemical Ecology**
   Edited by J. B. HARBORNE

9. **Biosynthesis and its Control in Plants**
   Edited by B. V. MILBORROW

10. **Plant Carbohydrate Biochemistry**
    Edited by J. B. PRIDHAM

11. **The Chemistry and Biochemistry of Plant Proteins**
    Edited by J. B. HARBORNE AND C. F. VAN SUMERE

12. **Recent Advances in the Chemistry and Biochemistry of Plant Lipids**
    Edited by T. GALLIARD AND E. I. MERCER

13. **Biochemical Aspects of Plant-Parasite Relationships**
    Edited by J. FRIEND AND D. R. THRELFALL

ANNUAL PROCEEDINGS OF THE PHYTOCHEMICAL SOCIETY   NUMBER 14

Regulation of Enzyme Synthesis and Activity in Higher Plants

PROCEEDINGS OF THE
PHYTOCHEMICAL SOCIETY SYMPOSIUM
OXFORD, APRIL, 1976

Edited by

H. SMITH

*Department of Physiology and Environmental Studies,
University of Nottingham, England*

1977

ACADEMIC PRESS

LONDON   NEW YORK   SAN FRANCISCO

A Subsidiary of Harcourt Brace Jovanovich, Publishers

ACADEMIC PRESS INC. (LONDON) LTD.
24/28 Oval Road,
London NW1

United States Edition published by
ACADEMIC PRESS INC.
111 Fifth Avenue
New York, New York 10003

Library of Congress Catalog Card Number: 77-71837
ISBN: 0-12-650850-X

PRINTED IN GREAT BRITAIN BY
WILLIAM CLOWES & SONS, LIMITED
LONDON, BECCLES AND COLCHESTER

Contributors

P. A. AITCHISON, *Department of Botany, University of Edinburgh, Edinburgh EH8 9YL Scotland* (p. 63).

M. J. AL-AZZAWI, *Biology Building, University of Sussex, Falmer, Brighton BN1 9QG England* (p. 329).

E. ELLEN BILLETT, *University of Nottingham, School of Agriculture, Sutton Bonington, Loughborough LE12 5RD England* (p. 93).

J. BUC, *Laboratoire de Physiologie Cellulaire Végétale du C.N.R.S., Université d'Aix-Marseille, 70 Route Léon-Lachamp, F–13288 Marseille-Cedex 2 France* (p. 155).

N. CARFANTAN, *Physiologie des Organes Végétaux après récolte, Station du froid, 4 ter rte des Gardes 92190 Meudon France* (p. 197).

J. DAUSSANT, *Physiologie des Organes Végétaux après récolte. Station du froid, 4 ter rte Gardes 92190 Meudon France* (p. 197).

D. D. DAVIES, *School of Biological Sciences, University of East Anglia, University Plain, Norwich NR 88C England* (p. 41).

P. D. G. DEAN, *Department of Biochemistry, University of Liverpool, P.O. Box 147, Liverpool L69 3BX England* (p. 271).

J. L. FIELDING, *Biology Building, University of Sussex, Falmer, Brighton BN1 9QG England* (p. 329).

A. B. GILES, *University of Nottingham, School of Agriculture, Sutton Bonington, Loughborough LE12 5RD England* (p. 93).

J. L. HALL, *Biology Building, University of Sussex, Falmer, Brighton BN1 9QG England* (p. 329).

M. J. HARVEY, *Department of Biochemistry, University of Liverpool, P.O. Box 147, Liverpool L69 3BX England* (p. 271).

D. T-H. HO, *Washington University, St Louis, Missouri 63130 U.S.A.* (p. 83).

C. B. JOHNSON, *University of Nottingham, School of Agriculture, Sutton Bonington, Loughborough LE12 5RD England* (p. 225).

C. LAURIERE, *Physiologie des Organes Végétaux après récolte, Station du froid, 4 ter rte des Gardes 92190 Meudon France* (p. 197).

R. M. LEECH, *Department of Biology, University of York, Heslington, York YO1 5DD England* (p. 289).

A. J. MACLEOD, *Department of Botany, University of Edinburgh, Edinburgh EH8 9YL Scotland* (p. 63).

J.-C. MEUNIER, *Laboratoire de Physiologie Cellulair Végétale du C.N.R.S., Université d'Aix-Marseille, 70 Route Léon-Lachamp, F–13288 Marseille-Cedex 2 France* (p. 155).

B. J. MIFLIN, *Rothampsted Experimental Station, Harpenden, England* (p. 23).

J. NARI, *Laboratoire de Physiologie Cellulaire Végétale du C.N.R.S., Université d'Aix-Marseille, 70 Route Léon-Lachamp, F–13288 Marseille-Cedex 2 France* (p. 155).

D. RHODES, *University of Manchester, Oxford Road, Manchester M13 9PL England* (p. 1).

M. J. C. RHODES, *ARC Food Research Institute, Colney Lane, Norwich NR 70F England* (p. 245).

J. RICARD, *Laboratoire de Physiologie Cellulaire Végétale du C.N.R.S., Université d'Aix-Marseille, 70 Route Léon-Lachamp, F–13288 Marseille-Cedex 2 France* (p. 155).

J. G. SCANDALIOS, *Department of Genetics, Box 5487, North Carolina State University, Raleigh, North Carolina 27607 U.S.A.* (p. 129).

A. SKAKOUN, *Physiologie des Organes Végétaux après récolte, Station du froid, 4 ter rte des Gardes 92190 Meudon France* (p. 197).

H. SMITH, *University of Nottingham, School of Agriculture, Sutton Bonington, Loughborough LE 12 5RD England* (p. 93).

G. R. STEWART, *University of Manchester, Oxford Road, Manchester M13 9PL England* (p. 1).

J. E. VARNER, *Washington University, St Louis, Missouri 63130 U.S.A.* (p. 83).

W. WALLACE, *University of Adelaide, Waite Agricultural Research Institute, Department of Agricultural Biochemistry and Soil Science, Glen Osmond, South Australia 5064 Australia* (p. 177).

M. M. YEOMAN, *Department of Botany, University of Edinburgh, Edinburgh EH8 9Y1 Scotland* (p. 63).

Preface

The shape, function and activities of all organisms are determined by the genetic information encoded within the nuclear and organellar DNA. The responsible agents for the expression of genetic information are the ultimate gene products, the enzymes and structural proteins. Through the organized regulation of the formation, degradation and activity of enzymes and structural proteins, both in space and time, the cellular and organic structure of the organism is built up and its physiological properties are developed. A clear understanding of the development and behaviour of multi-cellular organisms is thus dependent on a detailed knowledge of the mechanisms underlying the regulation of enzyme synthesis and activity.

In recent years, outstanding advances have been made in this area, at least in respect of micro-organisms and some animals. The situation regarding plants, however, has been less encouraging. To the developmental enzymologist, plants simultaneously offer challenging theoretical opportunities, and daunting technical problems. It was in order to bring these opportunities and problems together, and thus perhaps to provoke new interest and provide some solutions, that the Symposium upon which this book is based was organized.

The concept of the Symposium, which was held in both the Botany School, and Pembroke College, Oxford in April 1976, under the auspices of the Phytochemical Society, was to spend the first half of the available time outlining what is known of enzyme control in plants and describing some of the areas where much remains to be learnt, with the second half devoted to the principal technical problems and suggestions of how they may be overcome.

Thus, ostensibly at least, Chapters 1 to 9 describe various aspects of the regulation of enzyme synthesis, degradation and activity in higher plants as affected by internal and external factors, whilst Chapters 10 to 15 cover the technical problems and their solutions. In fact, as was expected, the actual Chapters are in all cases broader than the original conception and thus an excellent coverage has been achieved; many of the "theoretical" chapters include technical suggestions of value, and the "technical" chapters reveal a number of theoretical issues of considerable importance.

As programme organizer, I was delighted and gratified at the success of the Symposium; as Editor I was similarly pleased that the authors were not more than usually dilatory in submitting their manuscripts. I would like to record here my debt to the members of the Phytochemical Society Committee (Professor J. C. Friend and Professor T. A. Swain) and others (Dr C. B.

Johnson and Dr E. E. Billett) who advised on the Symposium programme, to Dr V. S. Butt who was a most efficient local organizer, to Professor F. R. Whatley, for permission to use the Botany School for the Symposium and to several colleagues in my own Department who have helped in various ways. I am also grateful to the staff of Academic Press for their expert assistance in preparing this volume for publication.

June, 1977   HARRY SMITH

Contents

Contributors v
Preface vii

CHAPTER 1

Control of Enzyme Levels in the Regulation
of Nitrogen Assimilation

G. R. Stewart and D. Rhodes

I. Introduction 1
II. Do plants have substrate-inducible and end product repressible enzymes? 1
III. Substrate induction of nitrate reductase 7
IV. End product repression of the enzymes of nitrogen assimilation . . 11
V. Speculations 17
Acknowledgements 19
References 19

CHAPTER 2

Modification Controls in Time and Space

B. J. Miflin

I. Introduction 23
II. The one-dimensional approach 25
    A. Correlation between the test tube and the plant 26
    B. Evolutionary aspects 29
III. The dimension of space 31
    A. Microscale 31
    B. Macroscale 37
IV. The dimension of time 37
V. Conclusion 38
References 38

CHAPTER 3

Control of pH and Glycolysis

D. D. Davies

I. Introduction 41
II. Conceptual difficulties 42
    A. Linear steady-state treatment of glycolysis 42
    B. Non-linear approximations 45
    C. Identification of control points 46

III.   Experimental difficulties   47
       A. Collection of data   47
       B. Enzyme assays   48
       C. The concentration of metabolites   51
IV.   The control of glycolysis   51
       A. Aldolase   51
       B. Adenine nucleotides   53
 V.   The control of pH   56
       A. Interaction between glycolysis and the control of pH   .   .   .   56
       B. Unidirectional inhibition   59
       C. The equilibrium reaction catalysed by malic enzyme   .   .   .   60
References   61

CHAPTER 4

Regulation of Enzyme Levels During the Cell Cycle

M. M. Yeoman, P. A. Aitchison and A. J. MacLeod

  I.   Introduction   63
 II.   Synchrony in plants and plant parts   64
       A. Naturally synchronous systems   64
       B. Synchronized systems   65
       C. Significance of synchrony in the intact plant   68
III.   Patterns of enzyme activity   69
       A. In developing pollen   69
       B. In cell suspension cultures   70
       C. In the artichoke explant system   71
IV.   Factors underlying periodic changes in enzyme levels   74
 V.   Significance of periodic changes in enzyme levels   76
       A. Linear reading and gene dosage   76
       B. Programmed development   77
       C. General effects of perturbation   78
VI.   Conclusion   79
Acknowledgements   80
References   80

CHAPTER 5

Hormonal Control of Enzyme Activity in Higher Plants

J. E. Varner and D. T-H. Ho

   I.   Introduction   83
  II.   The cereal grain aleurone layer response to gibberellins, abscisic acid and
        ethylene   84
 III.   The soybean hypocotyl response to auxin and cytokinins   .   .   .   88
 IV.   The pea seedling response to auxin   88
  V.   Response of *Rhoeo* leaf sections to auxin and abscisic acid .   .   .   89
 VI.   The oat internode response to gibberellins   89
VII.   Cytokinin-dependent increases in nitrate reductase in *Agrostemma githago*
        embryos   90
References   91

CHAPTER 6

The Photocontrol of Gene Expression in Higher Plants

H. Smith, E. Ellen Billett and A. B. Giles

I.	Introduction.	93
	A. The Photoregulation of plant development	93
	B. Sites for the regulation of gene expression	95
	C. The perception of light	96
II.	General aspects of the effects of light on enzyme levels	97
	A. The range of enzymes controlled by light	97
	B. Techniques	100
III.	The photoregulation of phenylpropanoid synthesis enzymes	102
	A. The photocontrol of PAL activity	102
	B. The evidence for inactive PAL	106
	C. A high molecular weight PAL inhibitor from gherkins	107
	D. Summary	110
IV.	Regulation by specific enzyme inhibitors	111
V.	Photocontrol of other enzymes	111
	A. Ascorbic acid oxidase in mustard	111
	B. Nitrate reductase in mustard	111
	C. Phosphoenolpyruvate carboxylase in sugar cane	113
	D. Lipoxygenase in mustard	113
	E. Peroxidase in squash membrane preparations	115
	F. Invertase in radish	115
	G. Conclusion	116
VI.	Photocontrol of transcription and translation	116
	A. Technical problems	116
	B. Photocontrol of the pattern of protein synthesis	117
	C. Photocontrol of polyribosome levels	118
	D. Photocontrol of messenger-RNA levels	121
	E. Does light regulate transcription?	123
VII.	General comments	124
	Acknowledgements	124
	References.	125

CHAPTER 7

**Isozymes: Genetic and Biochemical Regulation
of Alcohol Dehydrogenase**

J. G. Scandalios

I.	Introduction	129
II.	Terminology	130
	A. Isozymes	130
III.	Alcohol dehydrogenase as a model system	131
	A. The enzyme	131
	B. Temporal control of ADH expression	133
	C. Spatial distribution of ADH	134
	D. Genetic Control of ADH	138
	E. Purification of maize ADH	141
	F. Properties of ADH isozymes	143

    G. Regulation of ADH activity 146
References 153

CHAPTER 8

Conformation Changes and Modulation of Enzyme Catalysis

J. Ricard, J. Nari, J. Buc and J.-C. Meunier

    I. Regulation of monomeric enzymes 156
  II. Unexpected effects of subunit interactions 166
 III. Regulation of enzyme activity through protein-protein interactions . 170
 IV. Conclusions 174
References 174

CHAPTER 9

Proteolytic Inactivation of Enzymes

W. Wallace

    I. Introduction 177
  II. Occurrence and main properties of inactivating enzymes . . . 178
    A. Tryptophan synthase inactivation in yeast 178
    B. Group specific proteases in animal tissues 178
    C. Nitrate reductase inactivating enzyme in the maize root . . 180
    D. Other examples of specific protease action in higher plants . . 183
 III. Characterization of inactivating enzymes 183
    A. Active site inhibitors 183
    B. Estimation of protease activity 184
    C. Esterase activity 187
 IV. Specificity of inactivating enzymes 187
  V. Evidence for inhibitors of inactivating enzymes . . . 189
 VI. Intracellular compartmentation of the inactivating enzymes . . 190
VII. Control of inactivating enzymes during cell extraction . . 192
Acknowledgement 193
References 193

CHAPTER 10

Immunochemical Approaches to Questions Concerning
Enzyme Regulation in Plants

J. Daussant, C. Laurière, N. Carfantan and A. Skakoun

    I. Introduction 198
  II. Principles of immunochemical methods based on the specific precipitation between antigens and antibodies 198
    A. Immune sera: biological reagents for identifying protein . . 198
    B. Antigen–antibody specific precipitation 199
    C. Techniques of precipitation in gel 199
    D. Remarks 200
    E. Principles of absorption techniques 201
    F. Preliminary investigations on the anti-enzyme immune serum . 202

III. Detection of "inactive" enzymes 202
  A. Search for inactive enzymes using techniques of specific precipitation in gels 203
  B. Search for inactive enzymes using the techniques of competitive absorption 206
  C. Remarks 206
IV. Multiple forms of enzymes at different physiological stages . . 207
  A. No enzymatic characterization reaction is available . . . 208
  B. Enzymatic characterization reaction on the gels is available . . 210
V. Identification of phenomena involved in the appearance and disappearance of enzyme activity and in the modification of enzyme physico-chemical properties 213
  A. Search for the cause leading to the disappearance of α-amylase of developing wheat seeds upon maturation 216
  B. Indirect or direct proof for enzyme synthesis 218
  C. Changes in physico-chemical properties of wheat β-amylase during germination 219
VI. Changes in the specific activity of enzymes 220
VII. Studies on enzyme structure and localization of enzyme synthesis using anti-enzyme subunits immune sera 221
VIII. Conclusion 221
References 222

CHAPTER 11

The Use of Density Labelling Techniques in Investigations into the Control of Enzyme Levels

C. B. Johnson

I. Introduction 225
II. The methodology of density labelling 225
  A. Density labels 226
  B. Solutes for density gradient centrifugation 228
  C. Centrifugation conditions 232
III. Density labelling as an experimental technique 233
  A. For the demonstration of de novo protein synthesis . . . 233
  B. For the measurement of enzyme turnover 236
  C. For investigations into the control of enzyme activity . . . 238
  D. Measurement of incorporation of density labels into amino acid pools 240
IV. Conclusions 242
Acknowledgements 242
References 242

CHAPTER 12

The Extraction and Purification of Enzymes from Plant Tissues

M. J. C. Rhodes

I. Introduction 245

II. General problems of extraction and stabilization of enzymes from higher plant tissues 247
    A. General considerations 247
    B. Problems due to phenolic compounds and tannins . . . 248
    C. Some factors affecting the stability of isolated enzymes . . 256
III. Techniques for purification of enzymes from higher plant tissues . 258
    A. Solubilization of membrane bound enzymes 259
    B. Concentration of enzyme extracts 261
    C. Desalting of enzyme extracts 262
    D. Chromatographic methods 262
    E. Electrophoretic procedures 265
IV. Criteria for assessing the purity of isolated enzymes . . . 266
V. Conclusions 267
References 267

CHAPTER 13

Affinity Chromatography

P. D. G. Dean and M. J. Harvey

I. Introduction 271
II. General principles 271
III. Selection of ligand 274
IV. Synthesis of affinity adsorbents 277
V. Elution procedures 281
Acknowledgements 286
References 286

CHAPTER 14

Subcellular Fractionation Techniques in Enzyme Distribution Studies

R. M. Leech

I. Isolation of subcellular components 290
    A. Choice of plant material 292
    B. Tissue homogenization methods 295
    C. Cellular fractionation procedures 296
II. Non-aqueous methods of subcellular fractionation 310
III. Other methods of cellular fractionation 311
    A. Thin layer counter-current distribution 311
    B. The use of Sephadex columns for particle separation . . . 312
    C. Electrophoretic methods of separation 312
IV. Characterization of subcellular components 313
    A. Assessment of contamination and organelle damage . . . 313
    B. Assessment of organelle heterogeneity in enriched fractions of one organelle 320
    C. Assessment of enzyme activity 322
References 323

CHAPTER 15

**Microscopic Cytochemistry in Enzyme Localization
and Development**

J. L. Hall, M. J. Al-Azzawi and J. L. Fielding

I. Introduction 329
II. Range of methods 332
   A. Simultaneous capture mechanisms 334
   B. The substrate film methods 335
   C. Autoradiographic procedures 335
   D. Immunochemical methods 336
III. Factors affecting localization 337
   A. Preservation of enzyme activity and cell structure . . . 337
   B. Effects of aldehyde fixation on enzyme activity 340
   C. Precision of the localization procedure 343
   D. Specificity of the reaction 345
   E. Controls 346
IV. Applications 346
   A. Peroxidase localization and differentiation in roots . . . 347
   B. Enzymic changes during leaf abscission 349
   C. Enzymic changes in washed storage tissue discs 353
References 361

Author Index 365
Subject Index 381

CHAPTER 1

Control of Enzyme Levels in the Regulation of Nitrogen Assimilation

G. R. STEWART AND D. RHODES

Department of Botany, The University, Manchester, England

I.	Introduction .	.	.	.	.	.	.	.	.	.	.	1
II.	Do Plants Have Substrate-inducible and End Product Repressible Enzymes?	.	.	.	.	.	.	.	.	.	.	1
III.	Substrate Induction of Nitrate Reductase	.	.	.	.	.	.	.	7
IV.	End Product Repression of the Enzymes of Nitrogen Assimilation .	.	11
V.	Speculations .	.	.	.	.	.	.	.	.	.	.	17
	Acknowledgements	.	.	.	.	.	.	.	.	.	19
	References .	.	.	.	.	.	.	.	.	.	.	19

I. Introduction

"The really critical shortcoming at present is the lack of experimentally induced and repressed enzymes". (Oaks and Bidwell, 1970)

In this contribution we will begin by considering the occurrence of substrate induction and end product repression as components of the regulatory strategies employed by higher plants. We will then consider some of the possible reasons why the enzymes of nitrogen assimilation are subject to stringent regulation by mechanisms which for the most part control enzyme level. At the outset it should be stressed that, here, we are using the terms induction and repression in a phenomenological sense; meaning that the level of an enzyme is observed to increase or decrease in response to a particular metabolite. No mechanistic basis is implied in the use of these terms and for the most part we will not be concerned with the precise molecular mechanisms underlying a change in enzyme level.

II. Do Plants Have Substrate-inducible and End Product Repressible Enzymes?

The alteration of enzyme levels in response to exogenously supplied substrates and products is a common phenomenon in many microorganisms;

in *Escherichia coli* the total number of enzymes whose formation is controlled in this way has been "conservatively" estimated at 100 (Mandelstam, 1971) and in species of *Pseudomonas* the number of inducible enzymes may be in excess of 300 (Stanier *et al.*, 1965). Similarly, the levels of many enzymes, in animal systems, change in response to alterations in dietary intake (Schimke and Doyle, 1970). There are, however, relatively few examples of this type of adaptive enzyme in higher plants. By contrast the levels of many plant enzymes change in response to external factors such as light and exogenously applied plant hormones, and during development (see, e.g. Filner *et al.*, 1969).

It is interesting and it may be of some physiological significance that the best documented examples of substrate inducible enzymes in higher plants are those of nitrate reduction (see, e.g. Beevers and Hageman, 1969; Hewitt, 1975). The addition of nitrate characteristically results in an increase in the activity of nitrate and nitrite reductase. The reduction of nitrate to nitrite is not necessary for the formation of nitrite reductase since nitrate will induce this enzyme in the absence of a catalytically active nitrate reductase (Chroboczek–Kelker and Filner, 1972; Stewart, 1972a). In general nitrite seems to be less effective than nitrate as an inducer of nitrate reductase (Ferguson, 1969; Joy, 1969) and in some plants is less effective in inducing nitrite reductase (Ferguson, 1969; Stewart, 1972a). However in *Phaseolus vulgaris* cotyledons nitrite is more effective as an inducer of nitrate reductase and the suggestion has been made that the enzyme is product, rather than substrate inducible (Lips *et al.*, 1973). Other studies also suggest a requirement for nitrate is not obligatory, in that the enzyme can be induced by a wide range of compounds including cytokinins (Borriss, 1967), growth retardants (Knypl, 1974) and even chloramphenicol (Shen, 1972). Evidently this absence of a nitrate-requirement for the induction of nitrate reductase brings into question its status as a substrate–inducible enzyme.

Another possible candidate for a substrate–inducible enzyme is glutamate dehydrogenase. Until recently this enzyme was assumed to have a major role in the assimilation of ammonia by higher plants (see Miflin and Lea, 1976) and increases in activity have been observed in plants supplied with exogenous ammonia (Kretovich *et al.*, 1969; Gamborg and Shyluk, 1970; Kanamori *et al.*, 1972; Barash *et al.*, 1973; Shepard and Thurman, 1973). Similar increases occur however in response to high concentrations of some amino acids (Soulen and Olsen, 1969; Bayley *et al.*, 1972; Sahulka, 1972). This may however involve changes in different isoenzymes of glutamate dehydrogenase since, in oat leaves ammonia promotes the *de novo* synthesis of a specific isoenzyme (Barash *et al.*, 1975). It may be that there are two functionally distinct glutamate dehydrogenases, an assimilatory isoenzyme induced by ammonia and a catabolic isoenzyme induced by amino acids. This would resemble the situation in fungi, but here the assimilatory enzyme is NADP-linked and the catabolic is NAD-linked (see Sanwal and Lata, 1961).

The other much-quoted example of a substrate–inducible enzyme in higher plants is thymidine kinase, but in this case it is clear that endogenous factors, other than substrate availability, are important in its regulation (Hotta and Stern, 1965).

The lack of substrate-inducible enzymes in higher plants may of course simply reflect their nutritional limitations. Many of the inducible enzymes in microorganisms are those of carbohydrate degradation; most higher plants have presumably little need to degrade exogenous carbohydrates but an examination of saprophytic higher plants (e.g. *Orebranche majus*) might prove to be interesting.

Again, as with substrate-inducible enzymes there are relatively few examples of product-repressible enzymes in higher plants. Invertase can be repressed by the reaction products of its activity (Sacher and Glasziou, 1962) but also by a large number of other sugars (Glasziou and Waldron, 1964). Phosphatase activity is derepressed 25-fold in phosphate-deficient *Spirodela oligorrhiza* (Reid and Bieleski, 1970) and the increase is specific to an externally located alkaline phosphatase (Bieleski, 1974). The increases, which occur during germination, in the activity of phytase, and isocitrate lyase and malate synthetase, can be prevented by exogenous phosphate (Bianchetti and Sartirana, 1967) and glucose (Longo and Longo, 1970), respectively.

The enzymes of nitrate reduction are in some plants subject to end product repression. Nitrate reductase induction can be inhibited by ammonia (Joy, 1969; Smith and Thompson, 1971; Orebamjo and Stewart, 1975a), amino acids (Filner, 1966; Stewart, 1972a; Radin, 1975) and amides (Stewart, 1972a; Oaks, 1974; Radin, 1975). The extent to which some of these compounds, particularly amino acids, can be regarded as end products of nitrate assimilation is debatable. The general occurrence of this phenomenon is doubtful (see, e.g. Ingle *et al.*, 1966; Schrader and Hageman, 1967) but it may be significant that it can be readily demonstrated in cell-cultures (Filner, 1966; Oaks, 1974), roots (Smith and Thompson, 1971; Stewart *et al.*, 1974; Radin, 1975) and morphologically somewhat "simple" plants (Stewart, 1972a). It may be that these systems are less "buffered" and accumulate potential repressors at the site at which control over nitrate reductase is exerted. In most cases the repressive action of these end products cannot be accounted for by an inhibition of inducer uptake (Heimer and Filner, 1970; Stewart, 1972b; Oaks, 1974) which implies their effect is on the enzyme or its induction. Studies of what is now regarded as the primary ammonia-assimilating enzyme in higher plants (Lea and Miflin, 1974), glutamine synthetase have shown it to be repressed by glutamine (Rhodes *et al.*, 1975) and a similar response has been reported, earlier, for wheat embryos (Rijven, 1961).

This control, by end products, over the enzymes of inorganic nitrogen assimilation contrasts with higher plant biosynthetic enzymes. Almost every biosynthetic pathway which has been examined in bacteria is apparently regulated by the "dynamic duo" of feedback inhibition and end product repression (Savageau, 1972). In higher plants only control by feedback inhibition has been reported. Thus, studies of the biosynthetic enzymes for the aromatic amino acids (Belser *et al.*, 1971; Widholm, 1971; Chu and Widholm, 1972), the branched-chain amino acids (Oaks, 1965; Miflin and Cave, 1972) and arginine (Shargool, 1973) failed to demonstrate end product repression of any of the enzymes examined.

There are several plausible explanations for this apparent absence of end product repression. It could result from a difficulty in experimentally altering the endogenous repressor pools (Oaks and Bidwell, 1970). The initial enzyme levels may already represent the fully repressed state, so that the addition of exogenous end products has no effect. Certainly, this is the explanation for the failure to detect the phenomenon in many microorganisms (see, e.g. Donachie, 1964; Sercarz and Gorini, 1964) and it is only following starvation of auxotrophic mutants for a specific end product that derepression of the appropriate biosynthetic enzymes can be demonstrated. Another possibility is that exogenously supplied end products do not reach the site at which control is exerted; that is, there is compartmentation and exogenous end products do not enter the endogenous repressor pool. This would imply however that this repressor pool is distinct from the pool of end product which

TABLE I

Variation in specific activity of acetohydroxyacid synthetase and the branched chain aminotransferase in *Lemna minor*

Medium	Specific activity (nmoles/h/mg protein)				μmoles/gfw		
	Aceto-hydroxy[a] acid synthetase	Branched chain amino transferase[b]					
		Isoleu	Leu	Val	Isoleu	Leu	Val
NO$_3$ (5 mM)	72	530	300	310	0·08	0·20	0·05
NH$_4$ (5 mM)	75	500	300	330	0·35	0·4	0·15
Amino acid mixture	70	480	290	280	0.60	0·8	0.7
NO$_3$ + Isoleu + Val[c]	77	510	310	350	1·3	0·15	1·4
NO$_3$ + Isoleu + Leu + Val	90	540	300	310	1·6	1·25	1·75
NO$_3$ + Isoleu	67	500	320	330	1·8	0·10	0·04

Plants of *Lemna minor* grown for 3-subcultures on the above media. Amino acid mixture was that given in Rhodes *et al.* (1975).

[a] Acetohydroxy acid synthetase determined according to Miflin and Cave (1972).

[b] Branched chain aminotransferase determined according to Aki and Ichihara (1970). Extracts for enzyme determination were desalted on G25 prior to assay.

[c] Isoleucine, Leucine and Valine supplied at 50 μM concentrations.

acts as a feedback inhibitor since the addition of end product results, in many cases, in a growth inhibition (see, e.g. Miflin, 1973), which can be ascribed to a failure of the feedback regulation system to cope with an excess of a single end product.

Alternatively it might be that the patterns of control in higher plants are different from those in bacteria and consequently the "wrong" signals are being used. This possibility is worth considering since in several fungi (Carsiotis and Lacy, 1965; Guerzoni, 1972; Schurch *et al.*, 1974) and *Euglena gracilis* (Lara and Mills, 1973) a phenomenon known as cross-pathway regulation has been reported. For example histidine starvation of *Neurospora crassa* results in a depression of not only the histidine biosynthetic enzymes but also those of arginine and tryptophane biosynthesis (Carsiotis *et al.*, 1970). Similarly the discovery that the glutamine synthetase of *Klebsiella aerogenes* (see, e.g. Magasanik *et al.*, 1974) and the NADP-glutamate dehydrogenase of *Saccharomyces cerevisiae* (Dubois *et al.*, 1973) participate in the regulation of other enzymes of nitrogen metabolism, indicate that it may be necessary to consider as repressors, metabolites other than the obvious end products of a particular pathway.

Ideally the isolation of auxotrophic mutants would provide the most suitable system to examine the relationship between biosynthetic enzymes and the endogenous end products. However the application of this approach to higher plants in general, is limited by the characteristics of their genetic systems since they are generally diploid or even polyploid. Examination of the partially auxotrophic mutants of "haploid" *Nicotiana* calluses (Carlson, 1970) might be useful with regard to end product repression.

With some of these considerations in mind an investigation of the relationship between the level of biosynthetic enzymes and the tissue concentration of end products was carried out by growing plants of *Lemna minor* on a range of nitrogen sources. Intermediates of the arginine pathway were used in an attempt to increase the naturally present end product pools; in this way it was hoped to avoid the possibility of compartmentation which might arise from supplying the end product exogenously. Plants were also grown on a mixture of amino acids to test for the possibility of "alternative" signals. The results in Table I show the variation in specific activity of acetohydroxyacid synthetase and the branched chain amino transferase in plants grown on different media. It can be seen that there is no obvious relationship between the activity of these enzymes and the tissue concentrations of isoleucine, leucine and valine. In fact exogenously supplied isoleucine, leucine and valine promote an increase in specific activity of acetohydroxyacid synthetase. Essentially similar results were obtained for three enzymes of the arginine pathway, acetylornithine-glutamate acetyl transferase, ornithine carbamyl transferase and arginosuccinate lyase (Table II). Here, exogenously supplied arginine promotes a 2–3 fold increase in the specific activity of ornithine carbamyl transferase and arginosuccinate lyase. Similar increases in activity, in response to exogenously supplied end products have been reported for acetohydroxyacid synthetase (Miflin and Cave, 1972) and arginosuccinate lyase (Shargool, 1973). Although not decisive these results tend to confirm previous reports that end product repression is not a component in the control of amino acid biosynthesis in higher plants. There is however much evidence from bacterial systems that its function is not to limit the synthesis of intermediary metabolites when an excess is available but rather that it is a mechanism for economizing the cell's energy resources (Davis, 1961; Mandelstam, 1971). Studies of mutants in which allosteric control is absent (desensitized mutants) indicate that repression alone does not function as an efficient control in restricting the overproduction of end products (see, e.g. Moyed, 1961). The question then arises as to why higher plants do not employ substrate induction and end product repression as adaptive devices for economizing their energy resources and as a corollary do they employ alternative regulatory strategies for this purpose? Part of this alternative strategy may be that in a highly differentiated system this economy is achieved through the division of biochemical function among different cell types, the control of enzyme formation, both spatially and temporally, being associated with the mechanisms underlying differentiation. Oaks and Bidwell (1970) have made the intriguing suggestion that many natural enzyme inductions are in fact the result of altered compartmentation of substrate and end products. This being so there would of course be many higher plant enzymes subject to induction

TABLE II

Variation in the specific activity of the enzymes of arginine biosynthesis in *Lemma minor*

Medium	Specific activity (nmoles/h/mg protein)			
	Acetyl ornithine-glutamate acetyl transferase	Ornithine carbamyl transferase	Argino-succinate lyase	Arginine μmoles/gfw
NO$_3$ (5 mM)	7	39	4	0·10
NH$_4$ (5 mM)	8	42	5	0·15
Amino acid mixture	8	44	6	1·5
NH$_4$ + 1 mM Citralline	8	72	5	0·8
NH$_4$ + 1 mM Glutamate	9	57	5	1·8
NH$_4$ + 1 mM Arginine	10	115	12	3·1

Plants of *Lemna minor* grown for 3-subcultures on the above media. Amino acid mixture was that given in Rhodes *et al.* (1975). Enzyme activities were determined as described in Cybis and Davis (1975).

and end product repression, but the strategy of control would differ from that employed by bacteria in that the critical determinant would be the regulation of intra- and intercellular transport. In relation to this it is striking that the blue-green algae appear to have even fewer adaptive enzymes than higher plants (see Carr, 1973) and a peculiar feature of their physiology is the excretion into the growth medium of amino acids and peptides (Fogg, 1952). There may well be some interesting evolutionary implications regarding the similarities between them and higher plants with respect to control over enzyme level and in the differences between them with respect to control over metabolite transport.

While this survey is by no means comprehensive it does illustrate the paucity of higher plant enzymes which are controlled through induction, derepression and repression. The most obvious exceptions to this would seem to be the enzymes of nitrogen assimilation where both positive and negative control elements regulate the level of these enzymes.

III. SUBSTRATE INDUCTION OF NITRATE REDUCTASE

The molecular mechanisms underlying the increase in nitrate reductase activity during induction are by no means well established. There are several possible mechanisms and these can be categorized as follows:

1. Increase in rate of enzyme synthesis;
2. Decrease in rate of enzyme degradation;
3. Chemical modification of the enzyme, resulting in an increased activity;
4. Physical modification of the enzyme, again resulting in increased activity;
5. Changes in a soluble activator or inhibitor which associates with the enzyme to increase or decrease its activity, respectively.

Most studies of the induction process indicate a requirement for RNA and protein synthesis, in the sense that induction of the enzyme is blocked by inhibitors of RNA and protein synthesis (see Hewitt, 1975). Direct evidence for *de novo* synthesis is available for the induction of the enzyme in tobacco callus cells (Zielke and Filner, 1971). However the exact mechanism was not determined, but clearly changes in the rate of enzyme synthesis and or degradation are implicated. Kinetic analysis is often employed to obtain information on the relative importance of degradation and synthesis (see, e.g. Schimke, 1969; Yagil, 1975). We have attempted to fit the data shown in Fig. 1 to two types of equation which describe the kinetics of the transition from non-induced to induced state. The first of these is widely used in studies of animal systems (Segal and Kim, 1963; Berlin and Schimke, 1965) and is the prototype of post-transcriptional control. This model assumes unstable protein is coded for by pre-existing stable RNA and consequently the rate of enzyme synthesis is time-independent and leads to exponential formation and decay of the enzyme:

$$\frac{\mathrm{d}[E]}{\mathrm{d}t} = k_f - k_d[E]$$

where k_f and k_d are the constants for formation and decay of the enzyme respectively; and $[E]$ is the enzyme concentration. When this equation is applied (using the procedure of Segal and Kim, 1963) to the results in Fig. 1 a half-life of 12·6 h is obtained for decay following transfer to nitrate-free medium or if 6-methyl purine is used to block RNA synthesis and a half-life of 9·9 h when protein synthesis is blocked with cycloheximide. In contrast a half-life of 3·4 h is obtained for the ascending phase. According to the steady-state model, half-lives derived from the induction phase or from the decay phase should be the same, unless enzyme degradation is increased during the decay phase. Clearly the data obtained here suggest the reverse, namely nitrate induction actually increases the rate of degradation. This would then imply a very marked increase in the rate of enzyme synthesis in the induced state compared to the non-induced.

The second equation is based on a model which assumes an unstable protein is synthesized from unstable RNA (Yagil, 1975), which increases

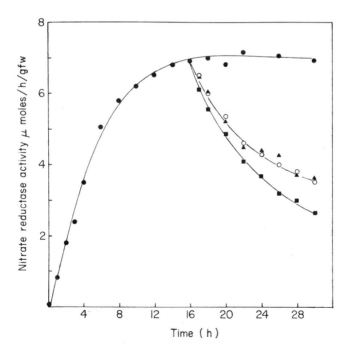

FIG. 1. Time course of nitrate reductase induction and decay. 1 mM asparagine plants transferred to 5 mM NO_3 at zero time. At 16 h: 250 $\mu g/ml$ 6-methyl purine added (O); 2 $\mu g/ml$ cycloheximide (■); or transfer back to 1 mM asparagine (▲).

during induction:

$$\frac{d[M]}{dt} = k_f{}^m - k_d{}^m[M] : \frac{d[E]}{dt} = k_f{}^e - k_d{}^e[E]$$

where $k_f{}^m$, $k_d{}^m$ are the constants for mRNA and $k_f{}^e$, $k_d{}^e$ those for enzyme. Again in this equation the half-time for enzyme accumulation should equal the half-time for enzyme disappearance. Neither of these steady-state models seem to provide an adequate description of the kinetics of nitrate reductase induction. The possibility that nitrate increases the rate of degradation is interesting but will require direct determination of the rates of degradation under different conditions.

The role of nitrate in the induction process is uncertain although in *Neurospora crassa* it has been suggested that it promotes translation of the m-RNA rather than acting at the transcription level (Sorger and Davies, 1973). In a number of higher plants it is clear that nitrate is not an obligate-requirement for induction. Knypl and Ferguson (1975) have shown that incubation of cucumber cotyledons at pH 3·0 with a wide range of compounds leads to marked increases in nitrate reductase activity (see Table III). Some of these compounds, in particular citrate, are more effective than nitrate in inducing the formation of the enzyme. A similar phenomenon is observed with some unicellular algae where growth on certain nitrogen sources or under conditions of nitrogen starvation results in the appearance of nitrate reductase (Morris and Syrett, 1963; Syrett and Hipkin, 1973; Rigano and Violante, 1973; Rigano *et al.*, 1974). In *Platymonas tetrathele* substantial levels of nitrate reductase are present in cells grown on some amino acids and in nitrogen-starved cells (Table IV). As with the observations of Knypl and Ferguson (1975), cells grown on some nitrogen sources contain higher activities of the enzyme than nitrate-grown cells (Rigano and Violante, 1973).

These results with cucumber cotyledons and unicellular algae are in marked contrast to those obtained for some higher plants where very high concentrations (100 mM) of nitrate are required to induce maximum levels of nitrate reductase (see, e.g. Ferrari and Varner, 1969; Lips *et al.*, 1973; Radin, 1974). Interpretation of these results is complicated by the possibility of separate inducer, substrate and storage pools of nitrate (Heimer and Filner, 1971; Ferrari *et al.*, 1973), but they suggest the possibility that nitrate may not be acting as an inducer *sensu stricta*.

The relationship between nitrate and nitrate reductase level has been investigated by growing plants of *Lemna minor* in continuous-flow culture at different nitrate concentrations. In this way low external nitrate concentrations can be maintained, avoiding the problem of nitrate depletion which occurs during growth in batch culture. It is clear from the results in Fig. 2 that high levels of nitrate reductase can be maintained at low external nitrate concentrations, but that this is partly the result of a highly efficient nitrate-uptake system which maintains the tissue concentration of nitrate.

TABLE III

Non-specific induction of nitrate reductase in cucumber cotyledons incubated at pH 3·0. None of the compounds tested, apart from nitrate, induced nitrate reductase when supplied at pH 6·0

Compound	Nitrate reductase as % activity on 10 mM KNO_3
Water	0
KNO_3, 10 mM	100
KCl, 10 mM	1·4
NaCl, 10 mM	0
Phosphate, 5 mM	0
Phosphate, 10 mM	16·3
TRIS, 10 mM	0
HEPES, 10 mM	45·4
Acetate, 10 mM	37·2
L-Ascorbate, 10 mM	69·9
L-Aspartate, 10 mM	52·0
Citrate, 10 mM	126·4
p-Coumarate, 5 mM	7·1
Fumarate, 10 mM	19·3
L-Glutamate, 10 mM	51·5
α-Ketoglutarate, 10 mM	8·2
D, L-Malate, 10 mM	29·4
Oxalate, 10 mM	45·4
Succinate, 10 mM	39·0
D(+)-Tartrate, 10 mM	75·5

After Knypl and Ferguson (1975).

The K_i for induction is high, internal concentration of 5×10^{-4} M giving a two-fold increase of the basal enzyme activity (a value of 0·9 nmoles/M/Mg being used as the basal level). Of course this assumes a uniform distribution of nitrate within the plants and since we are dealing with a highly compartmentalized system it is probably an unwarranted assumption. If nitrate is acting as an inducer in the strict sense then the affinity for the inducer is rather low.

There are, then, a number of observations which raise problems regarding both the molecular mechanisms underlying the induction process and the role of nitrate in this process. Degradation of the enzyme would seem to be an important component in the regulatory mechanism. Specific degrading enzymes have been reported in maize (Wallace, 1973, 1974) and in rice (Kavam et al., 1974) and an adaptive degradation system has been implicated in the dark inactivation of the barley enzyme (Travis et al., 1969). It is possible that nitrate could act by antagonizing, in some way, this degradation system and that under certain conditions (e.g. cucumber cotyledons) other factors can substitute or render this system inoperative.

TABLE IV

Nitrate reductase activity of *Platymonas tetrathele* grown on different sources of nitrogen

Nitrogen source (5 mM)	Nitrate reductase (nmoles/min/10^8 cells)
Nitrate	21·2
Ammonia	2·64
Glutamate	3·86
Serine	5·40
Aspartate	2·80
Proline	3·0
Citrulline	1·7
Ornithine	1·7
Phenylalanine	7·2
Tryptophane	5·2
Uric acid	63·0
Ammonia to ($-$N) 8 h	10·5

Enzyme determinations were as described by Syrett and Hipkin (1973).

IV. END PRODUCT REPRESSION OF THE ENZYMES OF NITROGEN ASSIMILATION

It is perhaps somewhat surprising that end product repression is utilized in the control of the enzymes of nitrogen assimilation when there are relatively few other instances of this form of control in higher plants. The occurrence of this stringent control over the level of these enzymes may be related to the potential pathways of ammonia assimilation (see Fig. 3). In *Lemna minor* there is the enzymic potential for assimilation via either the glutamine or glutamate pathways (Table V). In nitrate-grown plants assimilation of ammonia, derived from the reduction of nitrate, occurs almost exclusively via the glutamine pathway (Stewart and Rhodes, 1976). The two pathways of ammonia assimilation differ in certain important respects: the glutamine pathway is regarded as being the high affinity and high energy requiring pathway; while the glutamate pathway is a low affinity, low energy requiring pathway (Brown *et al.*, 1974). In nitrate grown *Lemna*, ammonia assimilation occurs almost exclusively via the glutamine pathway and there is therefore a requirement for NADH (nitrate reductase), ATP (glutamine synthetase) and reduced ferredoxin (nitrite reductase, glutamate synthase). Previous studies have shown that as the tissue glutamine concentration increases there is a reduction in glutamine synthetase level and a concurrent increase in that of glutamate dehydrogenase (Rhodes *et al.*, 1976), indicating a switch from assimilation via the glutamine to the glutamate pathway. This

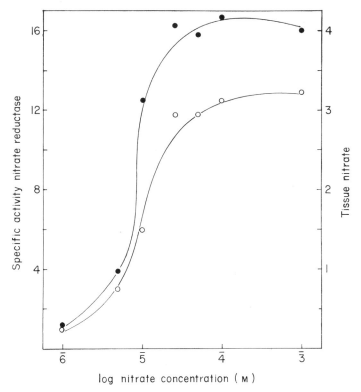

FIG. 2. Relationship between nitrate reductase level and nitrate concentration. Plants were grown in continuous-flow culture at various nitrate concentrations, flow rates were adjusted to maintain external NO_3 in culture vessel between 40–60% of initial concentration. Nitrate reductase specific activity, nmoles/m/mg protein (●). Tissue nitrate concentration μmoles/gfw (○).

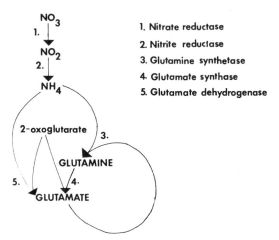

FIG. 3. Pathways of nitrogen assimilation in *Lemna minor*.

presumably would serve to limit the overproduction of glutamine at the expense of ATP and glutamate.

The concurrent regulation of glutamine synthetase and glutamate dehydrogenase has prompted us to re-examine the end product control of nitrate reductase. Previous studies had shown that ammonia, amino acids and amides could inhibit the induction of nitrate and nitrite reductases. Although it is possible to regard amino acids as end products of nitrate assimilation it is difficult to envisage any natural situation in which they would accumulate sufficiently to function as endogenous repressors. A number of the amino acids which inhibit nitrate reductase induction are potent inhibitors of growth. This inhibition of growth does not arise solely from their repressive effects on nitrate assimilation since they inhibit the growth of ammonia- and asparagine-assimilating plants. This contrasts with the situation in tobacco cell-cultures where threonine, for example, inhibits only the growth of nitrate-assimilating cells (Heimer and Filner, 1970). The inhibition of growth by some amino acids arises from the failure of the feedback inhibition mechanisms to cope with an excess of a single end product. For example leucine, isoleucine and valine are potent inhibitors of growth when supplied singly but in certain combinations such as isoleucine and valine or a mixture of all three, have little effect on growth (Table VI). Under conditions where they no longer inhibit growth there appears to be little repression of nitrate reductase. Similar results have been obtained with certain other amino acids and it is evident that they do not exert a direct effect on nitrate reductase formation. This cannot however be the explanation for repression by ammonia and the amides (Stewart, 1972b) since these compounds are not inhibitors of growth.

TABLE V

Variation in the specific activities of nitrate reductase (NR), glutamine synthetase (GS), glutamate dehydrogenase (GDH), glutamate synthase (GOGAT) in *Lemna minor*

Nitrogen source (5 mM)	Specific activity (nmoles/min/mgm protein)			
	NR	GS	GDH	GOGAT
NO_3	15·8	98	95	240
NH_4	0	56	160	175
Glutamate	0·5	70	160	100
Glutamine	0	41	223	132
NO_3 + Glutamine	3·2	40	220	115
NH_4 + Glutamate	0	33	225	105

Enzyme activities were determined as described by Rhodes *et al.* (1976).

TABLE VI

Variation in specific activity of nitrate reductase, and growth rate of *Lemna minor* on nitrate medium supplemented with combinations of isoleucine, leucine and valine

Medium	Specific activity nitrate reductase (nmoles/min/mgm)	growth as % control
NO$_3$	16·3	100
NO$_3$ + Isoleucine	12·6	68
NO$_3$ + Leucine	5·8	10
NO$_3$ + Valine	5·3	5
NO$_3$ + Isoleucine + Valine	13·9	90
NO$_3$ + Isoleucine + Leucine + Valine	15·8	85

NO$_3$ was supplied at 5 mM, and amino acids at 50 μM concentrations. Nitrate reductase was determined as described by Stewart (1972a).

The relationship between nitrate reductase activity and various soluble nitrogenous compounds has been examined by growing plants on media containing nitrate and various additional (non-inhibitory) nitrogen sources. Plants were adapted to these media through successive sub-cultures, so that "steady-state" enzyme and metabolite concentrations were established. This approach avoids many of the problems encountered in induction experiments, where there are marked changes in metabolite concentrations during the transition from one steady-state to another. The results in Fig. 4 show that as the tissue concentration of glutamine increases there is a corresponding decrease in the specific activity of nitrate reductase. No such relationship was evident for the tissue concentrations of ammonia, asparagine or total amino acids and nitrate reductase level. In general the tissue nitrate concentration increased in proportion to the extent of enzyme repression, indicating repression was not associated with exclusion of the inducer.

The relationship between nitrate reductase activity and that of glutamine synthetase is shown in Fig. 5; there is parallel repression of the two enzymes under these conditions. It would seem then, in *Lemna*, that there is a common element of control over these three enzymes of nitrogen assimilation, with glutamine repressing the formation of nitrate reductase and glutamine synthetase but promoting that of glutamate dehydrogenase. This situation is in certain respects similar to that reported in *Klebsiella aerogenes* (Magasanik *et al.*, 1974), *Saccharomyces cerevisiae* (Dubois *et al.*, 1973) and *Aspergillus nidulans* (Pateman and Kinghorn, 1975) where there is again a common control element for enzymes of ammonia metabolism.

The repression of nitrate reductase in *Lemna minor* appears then to be associated with a switch from ammonia assimilation via glutamine to assimilation via glutamate. It may therefore be important as a means of reducing competition between nitrate reductase and glutamate dehydro-

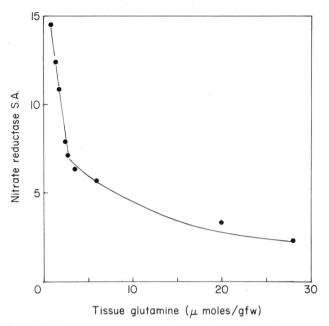

Fig. 4. Relationship between nitrate reductase level and tissue glutamine concentration. Plants were adapted to 5 mM NO_3 supplemented with amino acids and ammonia (see Rhodes *et al.*, 1976 for details).

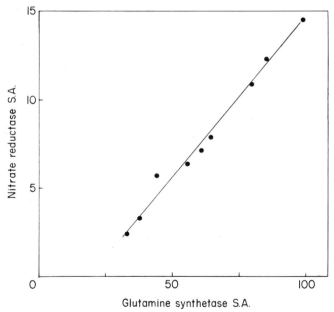

Fig. 5. Relationship between nitrate reductase level and glutamine synthetase level. See Fig. 4 for details. Enzyme activities are given as nmoles/min/mg protein.

genase for available reductant. A similar function has been ascribed to the rapid reversible inactivation of nitrate reductase which occurs on the addition of ammonia to nitrate-assimilating plants (Orebamjo and Stewart, 1975b).

The product–substrate relationships in the glutamine pathway are complex: glutamine being the substrate of glutamate synthase and the product of glutamine synthetase; and glutamate the substrate of glutamine synthetase and product of glutamate synthase. Herein may be the significance of the stringent control over glutamine synthetase. Under conditions where there is a high potential for glutamine synthesis, and a restriction on ATP and glutamate synthesis a rapid loss of glutamine synthetase occurs (Rhodes and Stewart, 1977). This is shown in Fig. 6; in this experiment ammonia-adapted

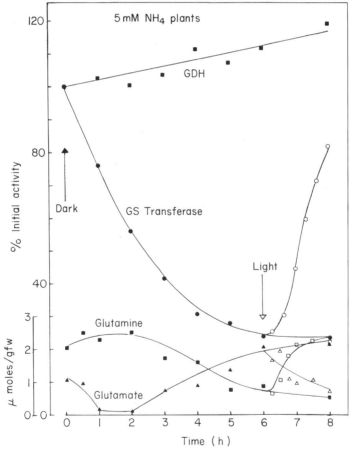

FIG. 6. Inactivation of glutamine synthetase in *Lemna minor*. 5 mM NH_4 Plants transferred from continuous light to darkness at zero time and after 6 h back to light. Glutamine synthetase (●). Glutamine (■). Glutamate (▲). Closed symbols — dark treatment; open symbols — light treatment.

plants of *Lemna* were placed in darkness. There is a rapid reduction in glutamine synthetase level and initially an increase in glutamine and a marked decrease in glutamate. Following the inactivation of glutamine synthetase, the glutamate concentration increases and that of glutamine decreases. Following transfer back to the light, the level of glutamine synthetase increases rapidly and accompanying this is a re-adjustment in the glutamate and glutamine concentrations. These changes in glutamine synthetase involve some form of reversible inactivation (Rhodes and Stewart, 1977); *in vitro* reactivation is temperature dependent and requires ATP, glutamate, magnesium and NADPH. Reversible inactivation serves as a means of limiting the overproduction of glutamine at the expense of ATP and glutamate.

Clearly the stringent control exercised over the enzymes of nitrogen assimilation is related to the complex substrate–product relationships of the assimilation pathways and to the necessity to conserve energy and reducing power. The rapid changes which occur in nitrate reductase levels in the dark (Travis *et al.*, 1969; Steer, 1973) may be related to the relationship between nitrate assimilation and the pathways of ammonia assimilation. It is interesting that in a higher plant such as *L. minor* the strategies of control employed in regulating nitrogen assimilation are similar to those found in microorganisms (see, e.g. Shapiro and Stadtman, 1970; Sims, 1975) albeit that we do not understand the precise molecular mechanisms employed in higher plants.

V. SPECULATIONS

Rather than attempt to draw any firm conclusions regarding the mechanisms controlling enzyme levels it may be worth while to consider some of the problems of regulation posed in higher plants and to speculate on the possible strategies employed by them in the regulation of their metabolism. The presence of a complex membrane system and membrane-bound organelles in a higher plant cell has some important implications as regards the control of enzyme levels, particularly since this design allows the spatial separation of enzymes from one another and from regulatory metabolites. The possible spatial separation of the site of enzyme activity, the site of enzyme synthesis and the pools of regulatory metabolites raises questions relating to what constitutes an effective regulatory strategy and which metabolites are effective regulatory signals. Some of the enzymes of nitrogen assimilation are spatially separated from one another and it may be significant that in *Lemna* they share a common control element, glutamine. The elegant studies of Cybis and Davis (1975) on the regulation of arginine biosynthesis in *Neurospora* illustrate very clearly the problems of regulation when an enzyme is spatially separated from an end product and how such problems are solved. Similar studies with higher plants might prove interesting in relation to the lack of end product repression of their biosynthetic enzymes. In relation to the possible separation of end product and enzyme it is interesting that the inactivation of glutamine synthetase in *Lemna* is potentiated by a decrease in ATP, glutamate and possibly magnesium (Rhodes and Stewart, 1977), rather than an accumulation of the end product, glutamine. This substrate-mediated control may be particularly relevant in higher plants where transport of end products from cells acting as sources to cells acting as sinks may occur (see, e.g. Pate, 1973).

Compartmentation itself may of course be a regulatory strategy. Here however control over intracellular and intercellular transport rather than enzyme level would be important. Control of metabolic compartmentation might of course explain why so few higher plants respond to exogenous substrates and end products.

Reversible inactivation of the type described here for glutamine synthetase has been reported for a number of photosynthetic enzymes (see, e.g. Latzko *et al.*, 1970; Hatch and Slack, 1969; Muller *et al.*, 1969) and would seem to be a regulatory device particularly appropriate for higher plants. In a mature non-dividing tissue removal, by growth, of a redundant enzyme is not possible as it is in unicellular systems. The importance of protein degradation as a regulatory device in this situation has been recently reviewed (Huffaker and Peterson, 1974; Holzer *et al.*, 1975). Reversible inactivation is an alternative means of achieving the same end result, namely cessation of enzyme activity, and does not have the disadvantage inherent in irreversible inactivation since it does not need to be coupled with a potential for resynthesis of the enzyme. This may be important in a differentiated cell where the potential for resynthesis may not be present (see, e.g. Mandelstam, 1971; Ashworth, 1971). The inactivated form of glutamine synthetase in *Lemna* is stable and can be reactivated *in vivo* (see Fig. 7) and *in vitro* (Rhodes and Stewart, 1977) for up to at least 72 h following inactivation. Reversible inactivation is then a flexible form of control and it may be significant that it is important in the regulation of carbon and nitrogen assimilation. There are certain similarities in the physiological conditions which bring about inactivation of these two groups of enzymes and it may be that this is an example of a "cascade" pattern of control. It will be of interest to establish the precise details of the mechanisms or possibly mechanism controlling the inactivation–reactivation of these enzymes.

Reversible inactivation also affords the possibility that inactivation may be potentiated by metabolites which need not be those which potentiate activation. This characteristic may be important in relation to the integration of the biochemical functions of different cell types. Studies of the kind carried out by Pate and his co-workers (see, e.g. Pate, 1973) illustrate very clearly the complex division of biochemical functions and the inter-relationships that exist between various organs and how these alter during development. At present we know little of the mechanisms which enable some

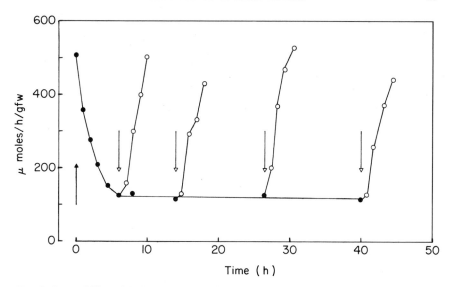

FIG. 7. Reversibility of dark-inactivation of glutamine synthetase. 5 mM NH₄ Plants transferred to dark at zero time and returned to light at times indicated by arrows. (Enzyme determined by an *in vivo* assay — Rhodes *et al.*, 1975.)

tissues to function as sources (i.e. centres of overproduction) of specific
metabolites and at the same time as sinks for other metabolites.

It is evident that higher plants exhibit many of the regulatory mechanisms
demonstrated for bacteria. Although they are not such amenable experimental
systems as bacteria, they possess characteristics which pose problems of
regulation not encountered in bacteria and these present challenging areas
of research, even in the absence of substrate-inducible and end product
repressible enzymes.

ACKNOWLEDGEMENTS

We would like to thank our colleagues, Janice Coulson, T. O. Orebamjo, G. A.
Rendon and A. Taylor, who have at different times participated in the studies
described here.

REFERENCES

Aki, K. and Ichihara, A. (1970). *In* "Methods in Enzymology" (Colowick, S. P. and
    Kaplan, N. O., eds), Vol. 17A, p. 811. Academic Press, New York and London.
Ashworth, J. M. (1971). *Symp. Soc. exp. Biol.* **25**, 27.
Barash, I., Sadon, T. and Mor, H. (1973). *Nature* **244**, 150.
Barash, I., Mor, H. and Sadon, T. (1975). *Pl. Physiol.* **56**, 856.
Bayley, J. M., King, J. and Gamborg, O. L. (1972). *Planta* **105**, 15.
Beevers, L. and Hageman, R. H. (1969). *A. Rev. Pl. Physiol.* **20**, 495.

Belser, W. L., Murphy, J. B., Delmer, D. P. and Mills, S. E. (1971). *Biochim. biophys. Acta* **237**, 1.

Berlin, C. M. and Schimke, R. T. (1965). *Molec. Pharmacol.* **1**, 149.

Bianchetti, R. and Sartirana, M. L. (1967). *Biochim. biophys. Acta* **145**, 485.

Bieleski, R. L. (1974). *Bull. R. Soc. N.Z.* **12**, 165.

Borriss, H. (1967). *Wiss. Z. Univ. Rostock Math naturw. Reihe.* **16**, 629.

Brown, C. M., MacDonald-Brown, D. S. and Meers, J. L. (1974). *Adv. microbial Physiol.* **11**, 1.

Carlson, P. S. (1970). *Science* **168**, 487.

Carr, N. G. (1973). *In* "The Biology of Blue-Green Algae" (Carr, N. G. and Whitton, B. A., eds), Botanical monographs No. 9, p. 39. Blackwell Scientific Publications, Oxford.

Carsiotis, M. and Lacy, A. M. (1965). *J. Bacteriol.* **89**, 1472.

Carsiotis, M., Jones, R. F., Lacy, A. M., Cleary, T. J. and Frankhauser, D. B. (1970). *J. Bacteriol.* **104**, 98.

Chroboczek-Kelker, H. and Filner, P. (1972). *Biochim. biophys. Acta* **252**, 69.

Chu, M. and Widholm, J. M. (1972). *Physiologia. Pl.* **26**, 24.

Cybis, J. and Davis, R. H. (1975). *J. Bacteriol.* **123**, 196.

Davis, B. D. (1961). *Cold Spring Harb. Symp. quant. Biol.* **26**, 1.

Donachie, W. D. (1964). *Biochim. biophys. Acta* **82**, 284.

Dubois, E., Grenson, M. and Wiame, J. M. (1973). *Biochem. biophys. Res. Commun.* **4**, 967.

Ferguson, A. R. (1969). *Planta* **88**, 353.

Ferrari, T. E. and Varner, J. E. (1969). *Pl. Physiol.* **44**, 85.

Ferrari, T. E., Yoder, O. C. and Filner, P. (1973). *Pl. Physiol.* **51**, 423.

Filner, P. (1966). *Biochim. biophys. Acta* **118**, 299.

Filner, P., Wray, J. L. and Varner, J. E. (1969). *Science, N. Y.* **165**, 358.

Fogg, G. E. (1952). *Proc. R. Soc. B.* **139**, 372.

Gamborg, O. L. and Shyluk, J. P. (1970). *Pl. Physiol.* **45**, 598.

Glasziou, K. T. and Waldron, J. C. (1964). *Aust. J. biol. Sci.* **17**, 609.

Guerzoni, M. E. (1972). *Arch. Mikrobiol.* **86**, 57.

Hatch, M. D. and Slack, C. R. (1969). *Biochem. J.* **112**, 549.

Heimer, Y. M. and Filner, P. (1970). *Biochim. biophys. Acta* **215**, 152.

Heimer, Y. M. and Filner, P. (1971). *Biochim. biophys. Acta* **230**, 362.

Hewitt, E. J. (1975). *A. Rev. Pl. Physiol.* **26**, 73.

Holzer, H., Betz, H. and Ebner, E. (1975). *In* "Current Topics in Cellular Regulation" (Horecker, B. L. and Stadtman, E. R., eds), Vol. 9, p. 103. Academic Press, London and New York.

Hotta, Y. and Stern, H. (1965). *J. Cell Biol.* **25**, 99.

Huffaker, R. C. and Peterson, L. W. (1974). *A. Rev. Pl. Physiol.* **25**, 363.

Ingle, J., Joy, K. W. and Hageman, R. H. (1966). *Biochem. J.* **100**, 577.

Joy, K. W. (1969). *Pl. Physiol.* **44**, 849.

Kanamori, T., Konishi, S. and Takahashi, E. (1972). *Physiologia. Pl.* **26**, 1.

Kavam, S. K., Gandhi, A. P., Sawkney, S. K. and Naik, M. S. (1974). *Biochim. biophys. Acta* **350**, 162.

Knypl, J. S. (1974). *Bull. R. Soc. N.Z.* **12**, 71.

Knypl, J. S. and Ferguson, A. R. (1975). *Z. Pfl. physiol.* **74**, 434.

Kretovich, V. L., Karyakina, T. I. and Tkemaladze, G. S. (1969). *Izv. Akad. Nank SSSR, Ser. Biol.* **5**, 749.

Lara, J. C. and Mills, S. E. (1973). *Abstr. A. Meet. Am. Soc. Microbiol.* **5**, 141.

Latzko, E., Garnier, V. R. and Gibbs, M. (1970). *Biochem. biophys. Res. Commun.* **39**, 1140.

Lea, P. J. and Miflin, B. J. (1974). *Nature, Lond.* **251**, 614.
Lips, S. H., Kaplan, D. and Roth-Bejerano, N. (1973). *Europ. J. Biochem.* **37**, 589.
Longo, C. P. and Longo, G. P. (1970). *Pl. Physiol.* **45**, 249.
Magasanik, B., Prival, J. M., Brenchley, J. E., Tyler, B. M., De Leo, A. B., Streicher, S. L., Bender, R. A. and Paris, C. G. (1974). *In* "Current Topics in Cellular Regulation" (Horecker, B. L. and Stadtman, E. R., eds), Vol. 8, p. 119. Academic Press, London and New York.
Mandelstam, J. (1971). *Symp. Soc. exp. Biol.* **25**, 1.
Miflin, B. J. (1973). *In* "Biosynthesis and its Control in Plants". (Milborrow, B. V., ed), p. 49. Academic Press, London and New York.
Miflin, B. J. and Cave, P. R. (1972). *J. exp. Bot.* **23**, 511.
Miflin, B. J. and Lea, P. J. (1976). (in press).
Morris, I. and Syrett, P. J. (1963). *Arch. Mikrobiol.* **47**, 32.
Moyed, H. S. (1961). *Cold Spring Harb. Symp. quant. Biol.* **26**, 323.
Muller, B., Ziegler, H. and Ziegler, I. (1969). *Europ. J. Biochem.* **9**, 101.
Oaks, A. (1965). *Pl. Physiol.* **40**, 142.
Oaks, A. (1974). *Biochim. biophys. Acta* **372**, 112.
Oaks, A. and Bidwell, R. G. S. (1970). *A. Rev. Pl. Physiol.* **21**, 43.
Orebamjo, T. O. and Stewart, G. R. (1974). *Planta* **117**, 1.
Orebamjo, T. O. and Stewart, G. R. (1975a). *Planta* **122**, 27.
Orebamjo, T. O. and Stewart, G. R. (1975b). *Planta* **122**, 37.
Pate, J. S. (1973). *Soil. Biol. Biochem.* **5**, 109.
Pateman, J. A. and Kinghorn, J. R. (1975). *In* "Filamentous Fungi" (Smith, J. E. and Berry, D., eds), No. 12. p. 159. Edward Arnold, London.
Radin, J. W. (1974). *Pl. Physiol.* **53**, 458.
Radin, J. W. (1975). *Pl. Physiol.* **55**, 178.
Reid, M. S. and Bieleski, R. L. (1970). *Planta* **94**, 273.
Rhodes, D. and Stewart, G. R. (1977). *Planta* (in press).
Rhodes, D., Rendon, G. A. and Stewart, G. R. (1975). *Planta* **125**, 201.
Rhodes, D., Rendon, G. A. and Stewart, G. R. (1976). *Planta* **129**, 203.
Rigano, C. and Violante, U. (1973). *Arch. Mikrobiol.* **90**, 27.
Rigano, C., Aliotta, G. and Violante, U. (1974). *Arch. Mikrobiol.* **99**, 81.
Rijven, A. H. G. C. (1961). *Biochim. biophys. Acta.* **52**, 213.
Sacher, J. A. and Glasziou, K. T. (1962). *Biochem. biophys. Res. Commun.* **8**, 280.
Sahulka, J. (1972). *Biol. Pl.* **14**, 308.
Sanwal, B. D. and Lata, M. (1961). *Can. J. Microbiol.* **7**, 319.
Savageau, M. A. (1972). *In* "Current Topics in Cellular Regulation" (Horecker, B. L. and Stadtman, E. R., eds), Vol. 6, p. 64. Academic Press, London and New York.
Schimke, R. T. (1969). *In* "Current Tropics in Cellular Regulation" (Horecker, B. L. and Stadtman, E. R., eds), Vol. 1, p. 77. Academic Press, London and New York.
Schimke, R. T. and Doyle, D. (1970). *A. Rev. Biochem.* **39**, 929.
Schrader, L. E. and Hageman, R. H. (1967). *Pl. Physiol.* **42**, 1750.
Schurch, A., Miozzari, J. and Hutter, R. (1974). *J. Bact.* **117**, 1131.
Segal, H. L. and Kim, Y. S. (1963). *Proc. natn. Acad. Sci. U.S.A.* **50**, 912.
Sercarz, E. E. and Gorini, L. (1964). *J. molec. Biol.* **8**, 254.
Shapiro, B. M. and Stadtman, E. R. (1970). *A. Rev. Microbiol.* **24**, 501.
Shargool, P. D. (1973). *Pl. Physiol.* **52**, 68.
Shen, T. C. (1972). *Pl. Physiol.* **49**, 546.
Shepard, D. V. and Thurman, D. A. (1973). *Phytochemistry* **12**, 1937.
Sims, A. P. (1975).

Smith, F. W. and Thompson, J. R. (1971). *Pl. Physiol.* **48**, 219.

Sorger, G. J. and Davies, J. (1973). *Biochem. J.* **134**, 673.

Soulen, T. K. and Olsen, L. C. (1969). *Planta* **86**, 205.

Stanier, R. Y., Hegeman, G. D. and Ornston, L. N. (1965). p. 227. Coll. int. CNRS, Marseille.

Steer, B. T. (1973). *Pl. Physiol.* **51**, 744.

Stewart, G. R. (1972a). *J. exp. Bot.* **23**, 171.

Stewart, G. R. (1972b). *Symp. Biol. Hung.* **13**, 127.

Stewart, G. R. and Rhodes, D. (1976). *FEBS Letts.* **64**, 296.

Stewart, G. R., Lee, J. A., Orebamjo, T. O. and Havill, D. C. (1974). *In* "Mechanisms of Regulation of Plant Growth" (Bieleski, R. L., Ferguson, A. R. and Cresswell, M. M., eds), Vol. 12, p. 41. Roy. Soc. New Zealand.

Syrett, P. J. and Hipkin, C. R. (1973). *Planta* **111**, 57.

Travis, R. L., Jordan, W. R. and Huffacker, R. C. (1969). *Pl. Physiol.* **44**, 1150.

Wallace, W. (1973). *Pl. Physiol.* **52**, 197.

Wallace, W. (1974). *Biochim. biophys. Acta* **341**, 265.

Widholm, J. M. (1971). *Physiologia. Pl.* **25**, 75.

Yagil, G. (1975). *In* "Current Topics in Cellular Regulation" (Horecker, B. L. and (Stedtman, E. R., eds), Vol. 9, p. 183. Academic Press, London and New York.

Zielke, R. H. and Filner, P. (1971). *J. biol. Chem.* **246**, 1772.

CHAPTER 2

Modification Controls in Time and Space

B. J. MIFLIN

Department of Biochemistry Rothamsted Experimental Station Harpenden, England

I. Introduction	23
II. The One-dimensional Approach	25
A. Correlation Between the Test Tube and the Plant	26
B. Evolutionary Aspects	29
III. The Dimension of Space	31
A. Microscale	31
B. Macroscale	37
IV. The Dimension of Time	37
V. Conclusion	38
References	38

I. INTRODUCTION

The title of this book emphasizes that the discussion should be about problems and techniques; not about answers. Consequently, this paper is an attempt to raise difficulties and pose questions; in some cases I have also tried to provide possible answers. Because of the fragmentary nature of our knowledge, these often involve unsupported assumptions, so they should be viewed as a stimulus to further experimentation rather than the definite article.

Modification controls are defined as those that change the rate of activity of a given amount of enzyme as distinct from controls that change the levels of an enzyme (Stewart and Rhodes, p. 7, this volume). They include:

Allosteric feedback—in which some metabolite, often the end product of a pathway, modifies the level of activity of a key enzyme by binding to that enzyme at a site other than the active site; this modification can be positive as well as negative. (For a general review see Umbarger, 1969.)

Energy charge—this is really a special case of the previous classification and involves the response of the enzyme to the ratio of ATP to ADP + AMP (Atkinson, 1966).

Metal ion effects—many enzymes are activated and/or inactivated by metal ions; consequently, the level of ions in the environment of the enzyme, which is in part under the control of the organism, can be used to control the rate of enzyme activity. These effects often involve changes in the polymeric structure of the enzyme related to changes in activity. This has been shown for malate dehydrogenase in plants (Jeffries *et al.*, 1969, Blackwood and Miflin, 1976).

pH—since enzymes usually have a defined pH optimum, changes in the pH of the enzymes' environment will modify activity and the consequent rate of flux through a given pathway (e.g. see Davies, 1973, and p. 56, this volume).

As further experimentation is done on isolated enzymes the apparent complexity of all the various controls that might operate *in vivo* increases. In most cases it is possible to construct plausible advantages that can be gained by the organism from these complex controls; fundamentally these revolve around the ability to coordinate its many various pathways to increase efficiency and minimize waste. However, the explanations may often indicate more about the ingenuity (or lack of) of the scientist than the organism. In trying to distinguish "real" effects from test tube artefacts, two particularly difficult problems are:

1. to distinguish whether the given concentration of modifier used to elicit a response in a test tube is likely to be in the range of concentrations found in the environment of the enzyme in the cell. Although it is very easy to conclude that the response of malate dehydrogenase to 0.5 M NH_4^+ (Blackwood and Miflin, 1976) is unlikely to be of physiological significance, it is much harder to know which range of concentration of intermediary metabolites is relevant to the *in vivo* situation;

2. to recognize whether a control mechanism is of value to an organism in its present normal environment or whether it is a relic of evolution no longer appropriate to present conditions; for example, a given modifier may affect an enzyme to a considerable extent over a concentration range known to exist *in vivo* but still not be of significance since the enzyme is in great excess and is not the limiting factor in the flux of compounds through the overall pathway (Fig. 1).

A further block to a complete understanding is the general tendency to consider controls operating in one dimension—that of the test tube. This produces the type of static picture that is represented for examples in Figs. 2 and 4;—these are useful as an initial description but they ignore most of the factors operating in an intact organism. This approach may be satisfactory when applied to bacteria where the population of cells are uniform and the internal compartmentation is minimal so that the biochemist is exchanging the bacterial cell membrane for the glass wall of the test tube; it is not so for

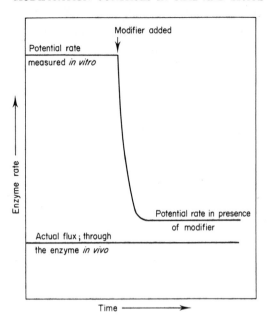

Fig. 1. Lack of correlation between *in vitro* and *in vivo* rates of activity in relation to control mechanisms.

plants. Plant subcellular structure is complex and a given cell contains a considerable number of compartments; the pathway of interest may occur in only one. Within a plant or plant organ considerable differences in function occur between plant cells so that only some may be carrying out the pathway of interest. The life pattern of a plant cell is different from a bacterial one in that it has a short, rapid growth phase and a long, stationary or mature phase; there is no reason to suppose that the control mechanisms operating are static through this time. I would now like to discuss some aspects of modification controls as they are affected by these different aspects. In doing this, examples from nitrogen metabolism will be chosen partly for personal preference and partly because controls on carbohydrate metabolism are likely to be discussed at length elsewhere (Davies, p. 51, this volume. Ap Rees, 1977).

II. THE ONE-DIMENSIONAL APPROACH

Many detailed reviews and papers exist (Priess and Kosuge, 1970; papers in Milborrow, 1973; Miflin, 1976) that outline the effects of the individual modification controls on a whole range of enzymes, and no further general cataloguing will be attempted. However, I would like to consider some of the problems raised in the Introduction.

A. CORRELATION BETWEEN THE TEST TUBE AND THE PLANT

The perennial doubt that plagues any biochemist interested in the whole organism is whether his observations are a true representation of *in vivo* metabolism or an artefact of isolation. Various attempts have been made to relate control mechanisms of isolated enzymes with events occurring in intact plants or plant organs. Oaks (1965a) has shown that leucine can inhibit its own synthesis from labelled acetate in maize-root tips and that the same is true for several other amino acids. She has followed this with a study of isopropyl malate synthase, the first enzyme unique to leucine biosynthesis (see Fig. 2), and has shown that its activity is inhibited by leucine (Oaks, 1965b). These studies and the growth inhibition studies discussed below may be criticized on the ground that they depend on supplying amino acids from the outside. Fletcher and Beevers (1971) measured the effects of cycloheximide, which inhibits protein synthesis and thus causes the internal pool of amino acids to increase, on the synthesis of various amino acids from ^{14}C precursors. They concluded that these results are consistent with the operation of feedback control mechanisms. Oaks and Johnson (1972) have criticized these experiments but in a later paper Fletcher (1975) has repeated his findings. Unfortunately, in any experiment with inhibitors there is always the possibility that the results are due to either non-specific or indirect effects. Miflin (1969, 1971) following the parallel provided by the work of Gladstone (1939) and Umbarger and Brown (1955) with bacteria, has related the inhibition of growth by leucine, isoleucine and valine to feedback control of isolated acetolactate synthase (Fig. 2) and shown that this inhibition can probably be explained in terms of feedback control mechanisms. Figure 3a

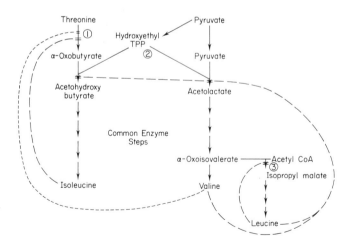

Fig. 2. Pathway of branched chain amino acid biosynthesis and its regulation. Enzymes (1) threonine deaminase, (2) acetolactate synthase, (3) isopropyl malate synthase. Key: ----- positive control — — — negative control.

shows a summary of the results; the cooperative effect of leucine and valine on the growth of the seedlings can be explained by their inhibition of acetolactate synthase and the consequent starving of the organism for isoleucine. Addition of small amounts of isoleucine can also partially overcome inhibition caused by large concentrations of valine even though the plant theoretically should be starved of leucine. The probable explanation is that the seedling's requirement for leucine can be served via transamination of

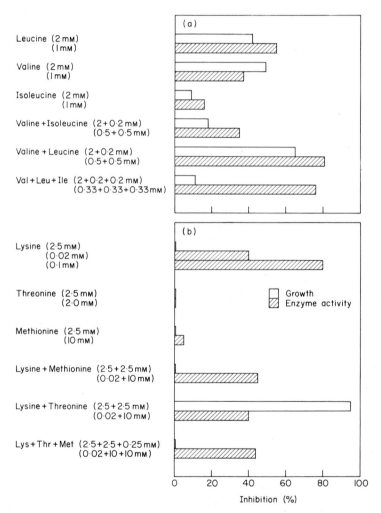

FIG. 3. The effect of combinations of various amino acids on growth and on isolated enzymes. (a) The effect of branched chain amino acids on the growth of barley seedlings and isolated barley acetolactate synthase (from Miflin, 1969, 1971 and Miflin and Cave, 1972); (b) The effect of lysine, methionine and threonine on the growth of maize seedlings and isolated maize aspartate kinase (from Green and Phillips, 1974 and Cheshire and Miflin, 1975).

valine to α-oxoisovalerate. This result indicates some lack of correspondence between growth inhibition and isolated enzyme studies.

Subsequent work has shown that synergistic growth inhibition caused by two amino acids may not correspond at all to a cooperative effect on a single enzyme and that care should be taken in interpreting growth inhibition patterns. Green and Phillips (1974) have shown that combinations of lysine plus threonine inhibit growth of maize seedlings in a synergistic manner (Fig. 3b). This inhibition is relieved by methionine indicating that lysine plus threonine cause the plant to be starved of methionine. Green and Phillips (1974) have concluded that this is most likely due to cooperative feedback control of aspartate kinase (Fig. 4). However, aspartate kinase has been isolated and characterized from maize and does not appear to be controlled in this way. Bryan *et al.* (1970) showed that it is only subject to feedback control by lysine and that threonine is without significant effect. Cheshire and Miflin (1975), using different assay techniques avoiding the use of hydroxylamine and high salt concentration which could have caused artefacts, essentially confirmed the previous findings of Bryan *et al.* (1970) (Fig. 3b). Despite various efforts, neither group has been able to find a significant inhibitory effect of threonine either alone or with lysine on aspartate kinase of maize. The probable explanation of the growth inhibitory effects is that aspartate kinase is not the only feedback sensitive enzyme in the aspartate pathway (Fig. 4); homoserine dehydrogenase (Bryan, 1969) and dihydrodipicolinate synthase (Cheshire and Miflin, 1975) are also affected. Growth inhibition by lysine and threonine could also result from combined effects of lysine on aspartate kinase and threonine on homoserine dehydrogenase. It is probable that maize aspartate kinase is not totally inhibited in the presence of 1 mM external lysine and that sufficient activity remains to provide the organisms with its threonine and methionine requirements. Conceivably, threonine inhibition of homoserine dehydrogenase is competitive with respect to aspartate semialdehyde; thus it will be more effective in the presence of lysine, which will lower the level of aspartate semialdehyde, than in its absence. The combined effect of lysine inhibition of aspartate kinase and more effective threonine inhibition of homoserine dehydrogenase could be sufficient to starve the plant of methionine. Further evidence of this view comes from some preliminary experiments with labelled acetate (reported by Miflin, 1975) which show that although lysine synthesis from acetate is inhibited in the presence of added lysine, there is no evidence for a cooperative effect of lysine and threonine on lysine synthesis. Threonine on its own increases lysine synthesis which can be explained by inhibition of homoserine dehydrogenase diverting more aspartate semialdehyde to lysine formation. The results do not support the idea that aspartate kinase is subject to cooperative feedback control by lysine and threonine. Henke *et al.* (1974) have shown that lysine and threonine supplied independently inhibit the growth of *Mimulus cardinalis* seedlings, and supplied together are

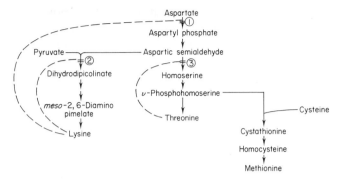

FIG. 4. Biosynthetic pathway of the aspartate family of amino acids and its control in maize. Enzymes (1) aspartate kinase (2) dihydrodipicolinate synthase (3) homoserine dehydrogenase. Key: ----- negative feedback control.

lethal. The growth inhibition can be relieved by methionine; correlated with the growth inhibition is an effect on protein synthesis and on incorporation of ^{14}C—aspartate into the soluble fraction of amino acids derived from it (Henke and Wilson, 1974). Unfortunately, incorporation into the protein amino acids was not measured. The authors have interpreted their results in terms of a possible cooperative control mechanism working at the level of aspartate kinase, although a sequential inhibition of methionine synthesis, as proposed above for maize, would also be possible.

Further complications in the interpretation of the inhibitory effects of amino acids are that they can cause inhibition of other processes, for example, nitrate uptake in which threonine is particularly effective (Filner, 1969) and methionine appears to be a general inhibitor of amino acid uptake (King and Hirji, 1975).

B. EVOLUTIONARY ASPECTS

Studies on the control of amino acid biosynthesis in bacteria have revealed a wide variety of controls for any given key enzyme involving both allosteric feedback inhibition and induction and repression (Umbarger, 1969) with considerable differences even between closely related species. In contrast, blue-green bacteria, green algae and higher plants appear to have few, if any repressive controls on the synthesis of enzymes of amino acid biosynthesis (Miflin, 1976) and rely on allosteric feedback mechanisms. Miflin and Cave (1972) attempted to determine the degree of species variation in the control of acetolactate synthase from a range of plants and found no significant differences. Subsequently, Stewart (personal communication), has found, using an *in vivo* assay, that the same regulatory pattern exists in *Lemna*. Any suggestion that this represents a lack of species variation in the control of all amino acid biosynthetic enzymes is unfounded since recent studies of aspartate kinase have shown wide variation (Table I). Following the initial finding by Bryan *et al.* (1970) of lysine inhibition of maize aspartate kinase, Wong and Dennis (1973) claimed that wheat aspartate kinase was cooperatively inhibited by lysine plus threonine although lysine at low concentrations was an effective regulator in its own right. Interpretations of Wong and Dennis' results are somewhat difficult due to the low level of activity in their preparation. In contrast, Aarnes and Rognes (1974) have found that the pea enzyme is completely unaffected by lysine and is sensitive to threonine; subsequently other plants have been found to have enzymes with similar control (Aarnes, 1974). Evidence for a cooperative effect of lysine plus threonine have also been found for cucumber. Recent studies at Rothamsted on barley aspartate kinase have shown that its regulation appears to be different from that of all previously described aspartate kinases in that it is cooperatively inhibited by lysine and methionine (Fig. 5), (Shewry and Miflin, 1977). The exact significance of the evolution of diverse controls for aspartate kinase, compared with the single control mechanism for acetolactate synthase is not apparent at present.

A recent finding of evolutionary interest has shown that both blue-green bacteria and higher plants share a pathway of tyrosine biosynthesis which is different from the one commonly present in bacteria (Jensen and Pierson, 1975). This observation coupled with the presence of many of the enzymes of amino acid biosynthesis in the plastids (see later) and the predominance of end product inhibition controls suggests that plants may have acquired many of their amino acid biosynthetic pathways via the blue-green bacteria.

TABLE I

Control of aspartate kinase from different plants by amino acids

Species	Amino acid(s) most effective in inhibiting aspartate kinase	Reference
Zea mays	Lysine	Bryan *et al.* (1970)
		Cheshire and Miflin (1975)
Pisum sativum	Threonine	Aarnes and Rognes (1974)
Helianthus annuus	Lysine	Aarnes (1974)
Sinapis alba	Threonine	Aarnes (1974)
Raphanus sativus	Lysine	Aarnes (1974)
Cucumis sativus	Lysine + Threonine	Aarnes (1974)
Chlorella pyrenoidosa	Threonine	Aarnes (1974)
Anacystis nidulans	Threonine	Aarnes (1974)
Triticum aestivum	Lysine or	
	Lysine + Threonine	Wong and Dennis (1973)
Hordeum vulgare	Lysine and	
	Lysine + Methionine	Shewry and Miflin (1976)

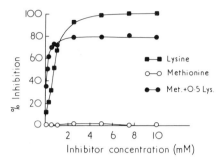

Fɪɢ. 5. Inhibition of barley aspartate kinase by lysine and methionine (from Shewry and Miflin, 1977).

III. THE DIMENSION OF SPACE

A. MICROSCALE

Let us consider the enzyme transported from the test tube back into the cell from which it was obtained. This enzyme may find itself within one or more of the various compartments of the cell: nucleus, cytoplasm, plastid, mitochondrion, microbody, dictyosome, or bound into any of the membranes. I propose to consider some of the enzymes involved in amino acid biosynthesis as examples. It has been found that several of these are located in the plastids, for example, nitrite reductase (Miflin, 1974a and references quoted therein), glutamine synthetase (Santarius and Stocking, 1969; Haystead 1973; O'Neal and Joy 1973a; Miflin 1974a), glutamate synthase (Lea and Miflin, 1974), acetolactate synthase (Miflin, 1974a), tryptophan synthase (Grosse and Eickhoff, 1974), homoserine dehydrogenase (Bryan and Miflin; Matthews and Bryan; unpublished results), diaminopimelate decarboxylase (Mazelis *et al.*, 1976) and several transaminases (Kirk and Leech, 1972; Miflin 1974a) (see also Leech, p. 289, this volume). Similarly, isolated intact chloroplasts have been shown capable of reducing nitrite to amino nitrogen (Malaghaes *et al.*, 1974; Miflin 1974b) and incorporating amino acids into protein (Blair and Ellis, 1973). Extrapolating from these results it is possible to envisage the pathway of nitrogen metabolism within the chloroplast as shown in Fig. 6. It must be emphasized that only a few of the enzymes involved in the synthesis of the individual amino acids at the ends of the various families have so far been shown to be present. One way to test directly the hypothesis contained in Fig. 6 is to feed $^{15}NO_2{}^-$ and/or ^{14}C-labelled oxoacids to intact choloroplasts in the light and measure their incorporation into the various amino acids; such experiments are in progress at Rothamsted. Proposals for regulating this pathway are usually in terms of positive effects of the initial substrate (nitrate or nitrite), on nitrate and nitrite reductase, and negative effects of amino acids on these enzymes and on glutamine synthetase and enzymes operating subsequent to ammonia assimilation.

I would now like to consider this in more detail and suggest that the picture is more complex, and that controls other than allosteric (or repression) feedback controls are of major importance. Glutamine synthetase (and not glutamate dehydrogenase) plays the major role in ammonia assimilation in chloroplasts (Miflin and Lea, 1976) and its properties are thus of key significance to the pathway shown in Fig. 6. The enzyme is probably present in the cytoplasm as well as in the chloroplast, and if enzymes from these two sources are not identical, then the information available on total isolated

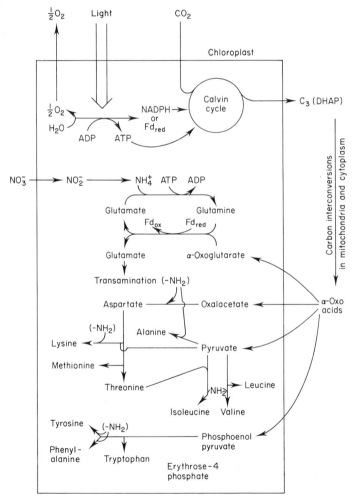

FIG. 6. Proposed pathway of nitrogen metabolism in plastids (with acknowledgement to schemes in Heber, 1974; Kirk and Leech, 1972; Lea and Miflin, 1974).

glutamine synthetase will reflect the average of both isoenzymes and obscure the precise properties of the chloroplast enzyme. However, in the absence of proof of this assertion and for the purposes of the following discussion I will assume that the results obtained by O'Neal and Joy (1973b, 1974, 1975) and other workers, apply to the chloroplast enzyme. These workers have shown that glutamine synthetase responds to magnesium concentration and pH as shown in Figs 7 and 8. The level of activity can go from zero to a maximum over a $\triangle$ [Mg^{++}] of 10 mM or less depending on the ATP concentration. Similarly, at a fixed [Mg^{++}] of 20 mM a change in pH from 6·8 to 7·6 increases activity from around 40% to 100%. What relevance have these properties to the environment of the chloroplast? Several authors have shown that the [Mg^{++}] in the chloroplast stroma changes in response to light, by values of 10–15 mM (e.g. see Walker 1973; Hind et al., 1974). Although neither the absolute values of Mg^{++} concentration, nor Mg^{++} activity are reported by these workers, a value for the total chloroplast of 16 mM has been calculated by Nobel (1969); thus the values are in a range where any change may be extremely effective in controlling the rate of glutamine synthetase activity.

Coincident with this increase of magnesium ions in the stroma there is a loss of hydrogen ions so that the pH increases from around 7·1 to 8·0 (Werdan et al., 1975). Although the response of glutamine synthetase to [Mg^{++}] and pH is affected by the [ATP], it is probable that the changes occurring upon illumination are sufficient to change substantially, the activity of the enzyme. This proposed activation is similar to that suggested for ribulose diphosphate carboxylase (Walker 1973; Werdan et al., 1975). There might be some significance in this similarity since it could be an advantage to the organism to have the key enzymes of CO_2 and NH_3 assimilation controlled in a similar manner. Glutamine synthetase has one further significant control and that is by energy charge. Figure 9 shows the results of O'Neal and Joy (1975) and also the changes occurring during illumination based on the results of Keys (1968). Despite the difficulties in determining the adenylate concentration in the chloroplast, the results suggest that the changes in energy charge could cause at least a two-fold change in the rate of glutamine synthetase. In contrast to the above controls in which there is a reasonable correlation between in vivo conditions and test tube observations, the feedback control of glutamine synthetase by amino acids requires high levels (> 10 mM) to obtain small effects ($< 40\%$ inhibition in the presence of Mg^{++}).

Movement of metabolites may also be important in the regulation of the enzymes responsible for the synthesis of the carbon skeletons of the amino acids since it is only the end product amino acids that are actually present in the plastid that will regulate the activity of the plastid enzymes. This compartmentation effect shows why it is impossible to correlate the levels of total amino acid pools of the tissue with the degree of regulation; for example, Widholm (1972) has shown that, although anthranilate synthase

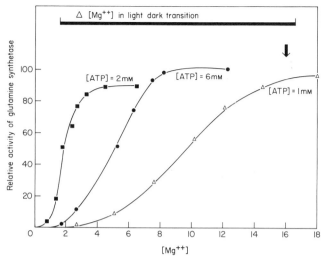

FIG. 7. Changes in the Mg^{++} concentration of the chloroplast stroma in light and dark and the effect of Mg^{++} on isolated glutamine synthetase at different concentrations of ATP. The horizontal bar denotes the change in $[Mg^{++}]$ in the stroma of the chloroplast during the light/dark transition but the absolute values are not given (Hind *et al.*, 1974). The arrow shows the average $[Mg^{++}]$ of chloroplasts calculated by Nobel (1969). The results for glutamine synthetase are from O'Neal and Joy (1974).

△—△ 1 mM ATP; ■—■ 2 mM ATP; ●—● 6 mM ATP.

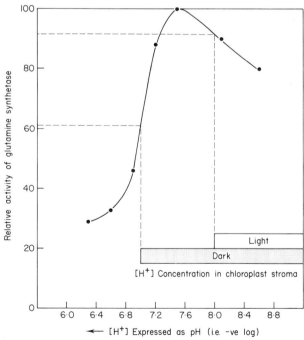

FIG. 8. The pH of the chloroplast stroma in light and dark and the effect of pH on glutamine synthetase (redrawn from data of Werdan *et al.*, 1975 and O'Neal and Joy, 1973b).

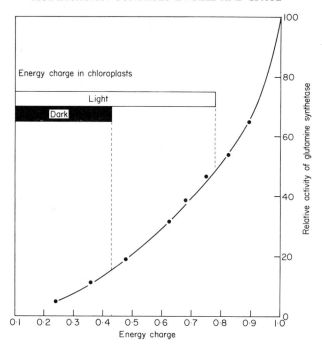

Fɪɢ. 9. The energy charge in chloroplasts in the light and dark and the effect of energy charge on glutamine synthetase (data redrawn from Keys, 1968 and O'Neal and Joy, 1975).

from carrot tissue cultures is 50% inhibited *in vitro* by 3·3 μM tryptophan, the average tryptophan pool size in the cells is 80 μM. Unfortunately, the data for the size of the tryptophan pool within the carrot plastids is not available, but is unlikely to approach this value. Sceptics may argue that this is the usual excuse that biochemists use when they cannot match their *in vitro* data to the whole organism and it may be that the control mechanism is a myth. Fortunately, Widholm has selected a mutant of carrots containing a tryptophan synthase much less sensitive to feedback control by tryptophan. Cells containing this enzyme have an average tryptophan pool of 27 times higher than that of the normal cells.

Some idea of the concentration of other amino acid modifiers within the chloroplast can be derived from the data of Aach and Heber (1967). Table II shows how these compare with the levels of amino acids known to cause 50% inhibition of the various enzymes. Despite all assumptions there is a reasonable correspondence between most of the values suggesting that the controls could be relevant to the *in vivo* situation. It should be noted that there is as yet no direct evidence that aspartate kinase is in the plastids although it is a reasonable assumption since all the intermediates between it and homoserine dehydrogenase are unstable and probably enzyme bound. It is possible that the control of these pathways within the plastid is also complex since the rate of amino acid synthesis will depend upon the rate of transport of carbon precursors into the chloroplast and the rate of amino acid transport out of the chloroplast. Although not a great deal is known about transport mechanisms of amino acid and oxoacids and diverse opinions are found, it is conceivable that they are under independent control (Heber, 1974). The significance of pH desensitization of allosteric control, for example, of acetolactate synthase (Miflin, 1971) in relation to the above-mentioned pH changes within the plastid is also uncertain.

TABLE II

Correlation of feedback regulation by amino acids with levels of amino acids found in chloroplasts

Enzyme	Amino acid	Levels (mM)		
		To give 50% inhibition	In chloroplasts of Spinacia[a]	Vicia[a]
Homoserine	Thr	$0.2 - 2.0^b$	0·3	0·2
dehydrogenase	Thr	$\leqslant 5.0^c$		
Acetolactate	Leu	0.5^d	0·4	2·6
synthase	Val	2.3^d	0·6	0·2
	Leu + Val	$0.03 + 0.03^d$		
Aspartate	Lys	0.7^e	0·3	0·1
kinase	Lys	0.3^f		
	Thr	0.5^c	0·3	0·2
	Lys + Thr	$0.2 + 2.0^f$		
	Lys + Met	$0.4 + 0.2^e$		

[a] Aach and Heber (1967). The values have been calculated from amounts/mg dry weight assuming that the chloroplast contains 10% dry matter.
[b] Matthews et al. (1975).
[c] Aarnes and Rognes (1974).
[d] Miflin (1971).
[e] Shewry and Miflin (1977).
[f] Wong and Dennis (1973).

In summary it is possible to propose a model for the control of nitrogen flow in plastids in which factors additional to substrate promotion and product feedback inhibition are concerned. As with most models, it must be remembered that it is based on various assumptions and extrapolations. The basis of this model is that the first part of the process is controlled by light— when the light is turned off, nitrogen flow does not occur. This is in accordance with the results of Canvin and Atkins (1974) using ^{15}N-labelled substrates in whole leaves. If this is to occur, mechanisms must exist to turn off reactions in the dark. One consequence of the dark is that there would be no production of reduced ferredoxin and this may be sufficient to prevent flow through nitrite reductase and glutamate synthase. However, the levels of ATP still remain higher than ADP even in the dark (Heber, 1974), and light/dark transitions may regulate the nitrogen flow through regulation of glutamine synthetase as described above. So far, suggested mechanisms of feedback regulation of glutamine synthetase do not appear significant, but over a slightly longer time (hours rather than minutes), the inactivation of the enzyme may be important (Stewart and Rhodes, p. 17, this volume). The second part of the pathway, i.e. flow of nitrogen from glutamate into various amino acids, is more probably controlled by end product feedback inhibition operating on the key enzymes responsible for the production of the carbon skeletons of amino acids. However, the role of light in the transport of effectors including H^+ ions should not be overlooked.

B. MACROSCALE

The discussion so far has centred on the control of the total enzyme extracted from a given tissue. Within the real world of the plant, the position that the cells find themselves may be of considerable importance. Conversely, the regulatory controls on enzyme activity allow the cell to adjust its metabolism to the local environment. Oaks (1966) has made a detailed study of leucine biosynthesis in maize-root tips. From kinetic experiments she has concluded that leucine delivered in the transport stream mixes with endogenously synthesized leucine and expands that pool. When a critical concentration is reached (i.e. 10^{-7} M) the endogenous synthesis is turned off. Other related effects of position within the roots are, the effect of glucose on asparagine synthesis (Oaks, 1975) and changes in the level of inactivators of nitrate reductase (Wallace, 1974). These observations may be as much related to the age of the cells as to their position.

One system where cells of the same age are concerned and in which regulation may differ is in the leaves of so-called C4 plants. The development of this type of carbon fixation has led to considerable differentiation within the leaf, particularly with respect to the plastids, and a partition of different aspects of carbon metabolism between the vascular bundle sheath tissue and the mesophyll cells. Studies on nitrogen metabolism also show differences in which the enzymes of nitrate reduction are in the mesophyll (Mellor and Tregunna, 1971; Harel et al., 1976) but those for ammonia assimilation (glutamine synthetase and glutamate synthase) are present in both (Harel et al., 1976). It is conceivable that the regulatory properties of the enzyme may vary between cells or specifically between the two types of plastid. If this is so, how valid are studies carried out on the mixed population of enzyme derived from a whole leaf of maize (or other C4 plants)?

IV. THE DIMENSION OF TIME

A bacterial cell in favourable circumstances has a simple growth pattern with rapid division. A plant cell in contrast, has a long period in an active but stationary phase; in most whole-plant organs there is a small region of cell division, a zone of cell expansion and a zone of mature tissue. Is it correct to assume that the feedback controls are identical as the cell passes through these various phases?

I have already considered this question to some extent insofar as position affects age. Further studies of Oaks et al. (1970) on the regulation of proline biosynthesis suggest that this becomes less pronounced as roots mature. Bryan and his co-workers (Matthews et al., 1975; Dicamelli and Bryan, 1975) have looked at the problem in more detail with respect to homoserine dehydrogenase from etiolated shoots, roots and light-grown tissues of maize seedlings. As the seedling grow, the extracted enzyme becomes progressively less sensitive to threonine. Similar changes were observed between meristematic and mature regions of the root and shoot. When excised shoots are cultured at 28°C for three days, the extractable level of the enzyme drops and desensitization occurs; these effects were not observed when the shoots were kept at 5–7°C. Various control experiments that were carried out support the contention that this is an in vivo effect.

Although more experimental results are required, particularly to see if the phenomenon can be observed in other monocotyledons and in dicotyledons, there is the possibility that the progressive desensitisation of regulatory controls during maturation is an important facet of biochemical control mechanisms in higher plants and may reflect the role of mature cells as site of production (sources) for the apical regions (sinks).

CONCLUSION

Studies in higher plant systems have indicated that biosynthetic enzymes are subject to modification controls in a similar manner to bacterial enzymes. Further work has emphasized the complexities that must be considered in trying to relate these test tube controls to the biochemistry of a whole plant or even to a whole cell. The control of nitrogen metabolism within a plastid provides a good example of some of these complexities, but it is by no means the only one. Similar complexities exist in the control of carbohydrate metabolism and in balancing respiratory and synthetic pathways. However, it is possible to construct hypotheses indicating the relevance of certain of the modification controls observed in vitro to the in vivo situation. The difficulties in attempting to do this are considerable and the temptation to draw facile conclusions needs to be resisted.

REFERENCES

Aach, H. J. and Heber, U. (1967). Z. Pfl. physiol. 57, 317.
Aarnes, H. (1974). Physiologia Pl. 32, 400.
Aarnes, H. and Rognes, S. E. (1974). Phytochemistry 13, 2717.

Ap Rees, T. (1977). *Symp. Soc. exp. Biol.* **31**.
Atkinson, D. E. (1966). *A. Rev. Biochem.* **36**, 85.
Baldry, C. and Coombs, J. (1973). *Z. Pfl. physiol.* **69**, 213.
Blackwood, G. C. and Miflin, B. J. (1976). *Pl. Sci. Letts.* **7**, 435.
Blair, G. E. and Ellis, R. J. (1973). *Biochim. biophys. Acta* **319**, 223.
Bryan, J. K. (1969). *Biochim. biophys. Acta* **171**, 205.
Bryan, P. A., Cawley, R. D., Brunner, C. E. and Bryan, J. K. (1970). *Biochem. biophys. Res. Commun.* **41**, 1211.
Canvin, D. T. and Atkins, C. A. (1974). *Planta* **116**, 207.
Cheshire, R. M. and Miflin, B. J. (1975). *Phytochemistry* **14**, 695.
Davies, D. D. (1973). *In* "Biosynthesis and its Control in Plants" (Milborrow, B. V., ed.), Academic Press, London and New York.
Dicamelli, C. A. and Bryan, J. K. (1975). *Pl. Physiol.* **55**, 999.
Filner, P. (1969). *Devl Biol. Suppl.* **3**, 206.
Fletcher, J. S. (1975). *Pl. Physiol.* **56**, 450.
Fletcher, J. S. and Beevers, H. (1971). *Pl. Physiol.* **48**, 261.
Gladstone, G. P. (1939). *Br. J. exp. Path.* **20**, 189.
Green, C. E. and Phillips, R. L. (1974). *Crop Sci.* **14**, 827.
Grosse, W. and Eickhoff, F. (1974). *Planta* **118**, 25.
Harel, E., Lea, P. J. and Miflin, B. J. (1976). *Pl. Physiol. Suppl.* **57**, 212.
Haystead, A. (1973). *Planta* **111**, 271.
Heber, U. (1974). *A. Rev. Pl. Physiol.* **25**, 393.
Heber, U. and Santarius, K. A. (1970). *Z. Naturf.* **256**, 718.
Henke, R. R. and Wilson, K. G. (1974). *Planta* **121**, 155.
Henke, R. R., Wilson, K. G. and McClure, J. W. (1974). *Planta* **116**, 333.
Hind, G., Nakatini, H. Y. and Izawa, S. (1974). *Proc. natn. Acad. Sci. U.S.A.* **71**, 1484.
Jeffries, R. L., Laycock, D., Stewart, G. R. and Sims, A. P. (1969). *In* "Ecological Aspects of the Mineral Nutrition of Plants" (Rorison, I. H., ed.), pp. 292–307. Blackwell Scientific Publications, Oxford.
Jensen, R. A. and Pierson, D. L. (1975). *Nature* **254**, 667.
Keys, A. J. (1968). *Biochem. J.* **108**, 1.
King, J. and Hirji, R. (1975). *Can. J. Bot.* **53**, 2088.
Kirk, P. R. and Leech, R. M. (1972). *Pl. Physiol.* **50**, 228.
Lea, P. J. and Miflin, B. J. (1974). *Nature* **251**, 614.
Magalhaes, A. C., Neyra, C. A. and Hageman, R. H. (1974). *Pl. Physiol.* **53**, 411.
Matthews, B. F., Gurman, A. W. and Bryan, J. K. (1975). *Pl. Physiol.* **55**, 991.
Mazelis, M., Miflin, B. J. and Pratt, H. M. (1976). *FEBS Letts.* **64**, 197.
Mellor, G. E. and Tregunna, E. B. (1971). *Can. J. Bot.* **49**, 137.
Miflin, B. J. (1969). *J. exp. Bot.* **20**, 810.
Miflin, B. J. (1971). *Archs Biochem. Biophys.* **146**, 542.
Miflin, B. J. (1973) *In* "Biosynthesis and its Control in Plants" (Milborrow, B. V., ed.), pp. 49–68. Academic Press, London.
Miflin, B. J. (1974a). *Pl. Physiol.* **54**, 550.
Miflin, B. J. (1974b). *Planta* **116**, 187.
Miflin, B. J. (1975). *In* "Fertilizer Use and Protein Production" pp. 53–74. Proc. 11th Coll. Int. Potash Institute, Berne, Switzerland.
Miflin, B. J. (1976). *In* "Proceedings of the International Workshop on Genetic Improvement of Seed Proteins" pp. 135–155. NRC, Washington, USA.
Miflin, B. J. and Cave, P. J. (1972). *J. exp. Bot.* **23**, 511.
Miflin, B. J. and Lea, P. J. (1976). *Phytochemistry* **15**, 873.
Milborrow, B. V. (ed.) (1973). *In* "Biosynthesis and its Control in Plants" Academic Press, London and New York.

Nobel, P. S. (1969). *Biochim. biophys. Acta* **172**, 134.

Oaks, A. (1965a). *Pl. Physiol.* **40**, 149.

Oaks, A. (1965b). *Biochim. biophys. Acta* **111**, 79.

Oaks, A. (1966). *Pl. Physiol.* **41**, 173.

Oaks, A. (1975). *Biochem. Physiol. Pfl.* **168**, 371.

Oaks, A. and Johnson, F. J. (1972). *Pl. Physiol.* **50**, 788.

Oaks, A., Mitchell, D. J., Barnard, R. A. and Johnson, F. J. (1970). *Can. J. Bot.* **48**, 2249.

O'Neal, D. and Joy, K. W. (1973a). *Nature New Biol.* **246**, 61.

O'Neal, D. and Joy, K. W. (1973b). *Archs Biochem. Biophys.* **159**, 113.

O'Neal, D. and Joy, K. W. (1974). *Pl. Physiol.* **54**, 773.

O'Neal, D. and Joy, K. W. (1975). *Pl. Physiol.* **55**, 968.

Priess, J. and Kosuge, T. (1970). *A. Rev. Pl. Physiol.* **21**, 433.

Santarius, K. A. and Stocking, C. R. (1969). *Z. Naturf.* **246**, 1170.

Shewry, P. R. and Miflin, B. J. (1977). *Pl. Physiol.* **59**, 69–73.

Umbarger, H. E. (1969). *A. Rev. Biochem.* **38**, 323.

Umbarger, H. E. and Brown, B. (1955). *J. Bacteriol.* **70**, 241.

Walker, D. A. (1973). *New Phytol.* **72**, 209.

Wallace, W. (1974). *Pl. Physiol.* **55**, 774.

Werdan, K., Heldt, H. W. and Milovancev, M. (1975). *Biochim. biophys. Acta* **396**, 276.

Widholm, J. M. (1972). *Biochim. biophys. Acta* **279**, 48.

Wong, K. F. and Dennis, D. T. (1973). *Pl. Physiol.* **51**, 322.

CHAPTER 3

Control of pH and Glycolysis

D. D. DAVIES

School of Biological Sciences, University of East Anglia, Norwich, England

I. Introduction	.	41
II. Conceptual Difficulties	.	42
A. Linear Steady-state Treatment of Glycolysis	.	42
B. Non-linear Approximations	.	45
C. Identification of Control Points	.	46
III. Experimental Difficulties	.	47
A. Collection of Data	.	47
B. Enzyme Assays	.	48
C. The Concentration of Metabolites	.	51
IV. The Control of Glycolysis	.	51
A. Aldolase	.	51
B. Adenine Nucleotides	.	53
V. The Control of pH	.	56
A. Interaction between Glycolysis and the Control of pH	.	56
B. Unidirectional Inhibition	.	59
C. The Equilibrium Reaction catalysed by Malic Enzyme	.	60
References	.	61

I. Introduction

This paper discusses metabolic control. Biological control systems resemble carefully engineered control systems and provide fertile ground for the application of teleology—a method of thinking about biological problems which has been praised by Krebs (1954) and used by most biochemists. Given one or two enzymes with unusual kinetics, one or two crossover points and a little imagination, anyone can produce ideas or theories about metabolic control. Most of the theories are unlikely to withstand critical experimentation, some may be interesting, but few are likely to have physiological significance. For example, at the previous symposium on metabolic control, I proposed that the nitrogen status of a plant could be reflected by the ratio pyridoxamine phosphate/pyridoxal phosphate and coordination between the enzymes of carbohydrate and nitrogen metabolism could be achieved if the enzyme activities were regulated by either pyridoxal phosphate or pyridoxamine phosphate. A low ratio pyridoxamine phosphate/ pyridoxal phosphate reflects a low nitrogen status and the inhibition of the enzymes of glycolysis by pyridoxal phosphate could be physiologically significant. However, we know that pyridoxal phosphate forms a Schiff's base with lysine residues which are involved in binding the phosphate group of sugar phosphates. Thus the fact that pyridoxal phosphate inhibits the enzymes of glycolysis may be a consequence of the chemistry of the active sites of glycolytic enzymes, rather than a mechanism for coordination of nitrogen and carbohydrate metabolism. However, theories die hard and biochemists are no better at discarding untenable theories than are other scientists, but as Max Planck noted, scientists are mortal and their untenable theories die with them.

Clearly enough was enough, because the organiser of this symposium has invited me to discuss the difficulties and problems associated with the study of the control of carbohydrate metabolism.

II. CONCEPTUAL DIFFICULTIES

The usual biochemical approach to a complex problem is to apply Occum's razor in the hope that a complex problem has a simple solution. This approach can be seen in studies of the control of metabolism in which changes in flux through a system are examined in relation to changes in substrate concentration. Krebs (1957) proposed that control points may be identified by changes in concentration of metabolic intermediates, arguing that if the flux increases, the substrate concentration of the pacemaker will decrease. In a number of cases (Hohorst et al., 1959), when glycolysis is stimulated all the intermediates of glycolysis increase in concentration so that the Krebs principle fails to identify the pacemaker. The explanation is probably that reactions upstream from the metabolites being analysed are stimulated and produce a flooding effect downstream. In principle, a single pacemaker is sufficient for the control of a metabolic sequence. However, it is clear that the control of glycolysis is more complex, involving several rate limiting reactions and exhibiting systemic properties.

A. LINEAR STEADY-STATE TREATMENT OF GLYCOLYSIS

Biochemical reactions are highly non-linear but a number of workers have attempted a linear analysis of control systems. Higgins (1965), Kacser and Burns (1973), Heinrich and Rapoport (1974) have defined a number of systems parameters and although their definitions are slightly different the parameters' "elasticity" and "sensitivity" defined by Kacser and Burns (1973) can be equated with the parameters "effector strength" and "control strength" as defined by Heinrich and Rapoport (1974).

The "elasticity coefficient" (ε) or "effector strength" (χ) is the relative change in the rate of a reaction produced by a relative change in the concentration of substrates (inner effectors) or compounds not directly controlled by the enzyme (outer effectors)

$$^{v}\varepsilon_{s} = (\mathrm{d}v/v)/(\mathrm{d}S/S) \tag{1}$$

Where v is the velocity of the *isolated* enzyme at the concentration of substrates, products and effectors found *in vivo*.

There are as many elasticity coefficients as there are metabolites and effectors which interact with the enzyme and each coefficient depends on the concentrations of the other reactants forming a control matrix. It is of interest to note that the substrate concentration range most sensitive to the action of an allosteric effector is that of its lowest concentration and not the concentration where sigmoidicity is greatest.

"Sensitivity" (Z) or "control strength" (C) is the extent to which each enzyme contributes to the overall control of the system:

$$(\mathrm{d}F/F)/(\mathrm{d}E/E) = Z \tag{2}$$

Here F is the flux through the system and E the concentration of the enzyme whose "sensitivity" is being considered. Equation (2) could be used to measure sensitivity coefficients in cell-free preparations. If the flux is measured and the activity of the enzyme whose sensitivity is being measured is increased by small increments, we would expect the result shown in Fig. 1. Increasing

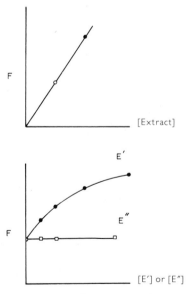

FIG. 1. Relationship between flux through a multi-enzyme system and increments of a single enzyme Upper: concentration of cell-free extract Lower: amount of additional enzyme (E) added to extract.

amounts of enzyme E'' have no effect on the flux and Z for this enzyme will be zero. The increase in flux on adding enzyme E' varies with the amount of E' added, and we need the response to an infinitesimally small change in E'. We have not yet obtained such data, but Table I presents some values from an experiment by Wu and Racker (1959) which was not designed to measure sensitivities and the increments of enzyme are far too large. Nevertheless, the results (Table I) indicate multisite control involving hexokinase, phosphofructokinase and triose phosphate dehydrogenase.

TABLE I

Glycolysis in extracts of ascites tumour cells with additions of purified glycolytic enzymes

Enzyme added	Activity in extract (units)	Activity added (units)	Lactate formed (μmoles/20 min)
None	—	—	2·5
Hexokinase	0·057	0·5	3·8
Glucose-6-p isomerase	1·22	2	2·5
Phosphofructokinase	0·12	0·5	3·4
Aldolase	0·41	0·8	2·7
Triose phosphate dehydrogenase	2·9	3	3·1

Recalculated from data of Wu and Racker, 1959. Experiments were carried out in a final volume of 0·6 ml containing extract (2·4 mg) tris buffer pH 7·4 (17 mM), MgCl$_2$ (6 mM), ATP (3·4 mM), NAD (1·7 mM), K phosphate buffer pH 7.4 (17 mM), KCl (10 mM), glucose (14 mM). Test tubes were incubated at 30° for 20 min. 1 ml of 20% CuSO$_4$ was added to each test tube to stop the reaction.

Heinrich and Rapoport (1974) have discussed the principles involved in the determination of *in vivo* sensitivities and Rapoport *et al.* (1974) have used these methods to determine the sensitivities of some glycolytic enzymes present in red blood cells. There are as many sensitivity coefficients as there are enzymes in a metabolic sequence and according to the summation theorem of Kacser and Burns (1973), the sum of all the sensitivity coefficients of a metabolic sequence is equal to unity. It should be noted that the sensitivity coefficient is a systemic property depending not only on the one enzyme but on all others. Thus, many enzymes contribute to control of the flux and no single enzyme will have a coefficient of one so that no single reaction can be strictly a master reaction—though in special cases the sensitivity coefficient may be close to unity and in such cases the concept of a "master reaction" will be valid. If a reaction is at equilibrium the sensitivity coefficient will be zero. However, if there is a flux through a system, no reaction can be at equilibrium. If a reaction is close to equilibrium its sensitivity coefficient is not necessarily small. Kacser and Burns (1973) have shown that for the metabolic sequence

$$S_1 \overset{E_1}{\rightleftharpoons} S_2 \overset{E_2}{\rightleftharpoons} S_3 \overset{E_3}{\rightleftharpoons} S_4$$

$$Z_1 : Z_2 : Z_3 : \ldots \equiv 1 - \rho_1 : \rho_1 (1 - \rho_2) : \rho_1 \rho_2 (1 - \rho_3) : \ldots \qquad (3)$$

where ρ = mass action ratio/equilibrium constant = disequilibrium ratio.

For the contrived situation shown in Table II, the step nearest to equilibrium (largest ρ) is not the least sensitive to control (smallest z). These considerations demonstrate that the widely held view, that reactions close to equilibrium are unimportant in the control of flux, may in certain circumstances be invalid. On the other hand, Rapoport et al. (1974) applied the concept of sensitivity to the study of the glycolytic flux in erythrocytes and assumed that "only non-equilibrium steps are important for the regulation of the chain". They found that the sensitivity coefficient of hexokinase is greater than that of phosphofructokinase but argued, (despite its smaller sensitivity coefficient), that most changes in the flux produced by effectors are caused by interaction with phosphofructokinase.

TABLE II

Step N	1	2	3
Assumed ρ	0·9	0·01	0·1
Calculated Z ratios	0·1	0·89	0·008

After Kacser and Burns, 1973.

B. NON-LINEAR APPROXIMATIONS

Biochemical systems are highly non-linear. For example, allosteric enzymes catalyse reactions of a high order with respect to substrates and/or effectors. The Hill equation is a widely used approximation for many non-Michaelis reactions. However, the Hill equation is not mathematically simple enough and Savageau (1969) has presented a non-linear approximation that utilizes a multidimensional power law function to describe the general rate law.

$$\text{Rate} = a[S]^b \qquad (4)$$

Kohen et al. (1973) have studied metabolic control in single cells by introducing metabolites by microelectrophoresis and using extremely sensitive techniques to measure NAD oxidation–reduction transients. The results obtained indicate that the utilization of glucose-6-phosphate in glycolysis can be described by the power formula (eq 4) where $a = 1·1$ and $b = 0·7$. This relationship appears to hold with concentrations of glucose-6-phosphate up to

10 mM. At concentrations above this, deviations occur which may be due to shunting of glucose-6-phosphate into the pentose phosphate pathway.

The extent to which the power law—which could dramatically reduce the number of individual assays necessary for the analysis of control—will be used, will probably be restricted by the mathematical limitations of experimental biochemists.

C. IDENTIFICATION OF CONTROL POINTS

Bücher and Rüssmann (1964) discussed the control of glycolysis in terms of an analogy in which a series of reservoirs are connected by waterfalls— the reservoirs correspond to reactions close to equilibrium and the water-falls to reactions far from equilibrium where control is effected by floodgates. The flow of water through the system can be increased by lowering the floodgates producing an increased gradient at the reservoirs and a reduced gradient at the waterfall. Thus, when the flux through a metabolic system is increased, reactions close to equilibrium move away from equilibrium and reactions far from equilibrium move towards equilibrium. This analysis suggests that control points will be located at quasi-irreversible reactions and if the flux increases, the reaction will move towards equilibrium, with the substrate of the control reaction decreasing in concentration whilst the product increases in concentration. Thus control points can be identified by measuring changes in the concentration of metabolites when the flux through the system is increased.

Data obtained in this way are usually presented as a crossover plot (Fig. 2) based on the crossover theorem of Chance (Chance et al., 1958).

The advantages of the method lie in the visual presentation of data, but this most frequently used method of identifying control points has frequently been misunderstood and misinterpreted. The following considerations should be noted.

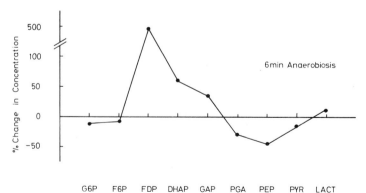

FIG. 2. Effect of anaerobiosis on the concentration of the glycolytic intermediates in washed carrot discs.

1. Control points are usually located at quasi-irreversible reactions and such reactions are by definition not affected by products. Thus, as noted by Rolleston (1972) there is no logical requirement to use the concentration of a product to detect a control point.

2. If a reaction involving a cofactor is close to equilibrium a false crossover may be obtained. Consider the system

$$A \rightleftharpoons B \overset{x\ y}{\rightleftharpoons} C \rightleftharpoons D \overset{y\ x}{\rightarrow} E$$

where the reaction $B + x \rightleftharpoons C + y$ is very close to equilibrium and the reaction $D + y \rightarrow E + x$ is far from equilibrium and rate determining. If the flux through the system increases the concentration of $D + y$ will decrease and x will increase, consequently the concentration of B will decrease and C will increase to maintain equilibrium and produce a false crossover between B and C.

3. Chance (1965) noted that if the flux through the respiratory chain changes from one steady-state to another, some intermediates will increase $(+)$ and others decrease $(-)$. For the case where the flux *increases*, a $(- +)$ pair in the sequence is defined as a crossover point and a $(+ -)$ pair is a negative or reverse crossover. A crossover is taken to indicate a site of interaction but Chance argues that a reverse crossover does not indicate a site of interaction and can occur anywhere between two points. The limitations must be restated. The intermediates in the sequence are conserved. The change in flux is from one steady-state to another. The conservation limitation has been widely recognized since pointed out by Scrutton and Utter (1968), but the limitations concerning steady-states has not been fully recognized. It might be useful to consider a specific case. Figure 2 shows a crossover plot for aged carrot discs and gives the percentage change in concentration of glycolytic intermediates after 6 minutes of anaerobiosis. The first crossover is consistent with an increased rate at the phosphofructokinase step. Some workers (e.g. Chapman and Graham, 1974) would argue that following Chance (1965) the reverse crossover between glyceraldehyde-3-phosphate (GAP) and 3-phospho-glycerate (3PGA) does not indicate a control point. However, this reverse crossover implies that 3PGA is being removed more rapidly than it is being formed from GAP. Thus 6 minutes after the transfer from air to nitrogen the flux through the glycolytic sequence is being restricted by a reaction between GAP and 3PGA.

III. EXPERIMENTAL DIFFICULTIES

A. COLLECTION OF DATA

The universality of metabolic pathways contrasts sharply with the diversification of metabolic controls. Closely related organisms may control the same metabolic pathway by quite different control mechanisms. This greatly increases the amount of work necessary to understand control mechanisms. However, the detailed understanding of the control of one enzyme, in one metabolic pathway in one organism presents a formidable task. Anyone who has worked with allosteric enzymes is aware that the kinetics can be drastically changed by any one of several variables. A small change in pH can change Michaelis–Menten kinetics to sigmoid kinetics. The analysis of an enzyme reaction involving two substrates and limiting the study to initial rates with zero products, requires at least 6 initial rate curves having 6 points per curve (Raval and Wolfe, 1962). In general, the number of assays increases as the nth where n is the number of substrates and effectors associated with the enzyme (Savageau, 1972). Thus to establish the complete law for a regulatory enzyme with 8 reactants and effectors, requires at least 1,679,616 assays!

B. ENZYME ASSAYS

To obtain a rough idea of which enzymes may be rate-limiting in a metabolic sequence, the activities of the individual enzymes are frequently measured. The precautions necessary to eliminate artefacts during the isolation have been carefully considered by Ap Rees (1974). The necessity to relate the maximum velocities observed to the velocities at *in vivo* concentrations of substrates is generally recognized, but few investigators have measured rates in the presence of *in vivo* concentrations of products and effectors.

A further source of error lies in the assays themselves. For example aldolase is assayed by coupling to the oxidation of NADH with α-glycerophosphate dehydrogenase:

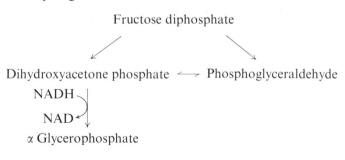

We can calculate the amount of the coupling enzyme which must be added, but it is usual to add "excess" of the coupling enzyme. In general this method gives valid results, but the coupling enzyme is usually stored in $(NH_4)_2SO_4$ and SO_4^- is an activator of some enzymes e.g. phosphoenolpyruvate carboxylase and phosphofructokinase, whilst NH_4^+ activates phosphofructokinase. A special effect may also occur in the assay of phosphofructokinase which involves the coupling enzymes aldolase, triose phosphate isomerase and α-glycerophosphate dehydrogenase.

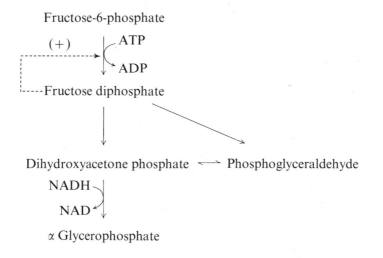

In animals, fructose diphosphate is an activator of phosphofructokinase, thus the activity of the enzyme depends on the steady-state concentration of fructose diphosphate and if "excess" aldolase is added the phosphofructo-kinase will be inhibited (Fig. 3). It is possible that the very low levels of phosphofructokinase reported in plants may be due to the use of "excess" aldolase in the assay.

The assay of pyruvate kinase involves coupling with lactate dehydrogenase.

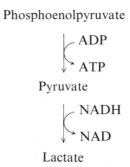

If proper precautions are not taken this assay can give erroneous results due to the presence of other enzymes which react with phosphoenolpyruvate and also to the high activity of malate dehydrogenase in extracts, which rapidly removes oxaloacetate.

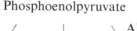

Phosphoenolpyruvate

CO_2 / | \ADP + CO_2 →ATP
Pi

Oxaloacetate   Pyruvate   Oxaloacetate

NADH → NAD   NADH → NAD   NADH → NAD

Malate   Lactate   Malate

The high activities of pyruvate kinase which are sometimes reported may be due to one or more of these side reactions.

Lastly, I mention an artefact which is produced by the stray light during spectrophotometry. The proposal that the ratio pyridoxal phosphate/pyridoxamine phosphate coordinates nitrogen and carbohydrate metabolism, suggests that the enzymes involved in the production of carbohydrate skeletons necessary for amino acid biosynthesis, should be inhibited by pyridoxal phosphate and not by pyridoxamine phosphate. Slaughter (1975), however, has reported that 3-phosphoglycerate dehydrogenase, which catalyses the first reaction in the biosynthesis of serine, is inhibited by pyridoxamine phosphate and this argues against the physiological significance of the ratio pyridoxal phosphate/pyridoxamine phosphate.

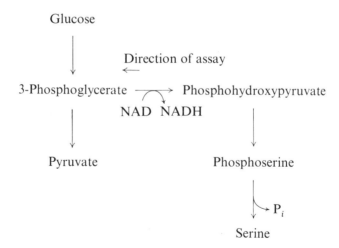

Glucose

Direction of assay ←

3-Phosphoglycerate → Phosphohydroxypyruvate
NAD NADH

Pyruvate   Phosphoserine

Pi

Serine

When studying aldolase, Mrs Sarawek observed an apparent inhibition of aldolase by pyridoxamine phosphate. Further investigation, however, has shown that an artefact is produced because pyridoxamine phosphate absorbs

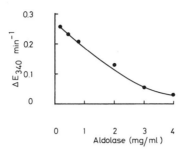

FIG. 3. Effect of aldolase on the activity of phosophofructokinase.

at 340 mμ so that the slit of the spectrophotometer has to be opened wide, thereby increasing the stray light and decreasing the "apparent" extinction coefficient of NADH producing an apparent inhibition of aldolase. It seems likely from the conditions used by Slaughter (1975) that his observation that pyridoxamine phosphate inhibits 3-phosphoglycerate dehydrogenase is an artefact. With these various limitations in mind the reported activities of a number of glycolytic enzymes are shown in Table III.

C. THE CONCENTRATIONS OF METABOLITES

The extensive compartmentation of plant tissue makes it difficult to assess the relationship between metabolic content and concentration (Ap Rees, 1974). The existence of separate pools of organic acids has been clearly demonstrated (MacLennan *et al.*, 1963) but it is generally assumed (albeit without evidence) that the glycolytic intermediates are restricted to the cytosol. Dennis and Green (1975), however, have presented evidence that proplastids from castor bean endosperm contain a number of glycolytic enzymes. This finding casts much doubt on the interpretation of data involving changes in metabolic levels.

IV. THE CONTROL OF GLYCOLYSIS

Turner and Turner (1975) have recently thoroughly reviewed the regulation of carbohydrate metabolism enabling me to concentrate on a few points.

A. ALDOLASE

In general, enzymes that catalyse equilibrium reactions possess high catalytic activities. Aldolase, which is generally assumed not to be a regulatory enzyme, is often of low activity compared with other glycolytic enzymes (Table II). This suggests that there may be problems associated with the assay of aldolase so that the *in vitro* assay underestimates the *in vivo* activity.

TABLE III

The relative activities of glycolytic enzymes. All activities have been calculated relative to the activity of aldolase

Enzyme	Potatoes[a]	Wheat seedlings[b]	Pine seedlings[c]	Red beet[d]	Red beet tumours[e]	Sycamore
Aldolase	1	1	1	1	1	1
Hexokinase	2·5	—	—	—	—	—
Hexose-p-isomerase	380	—	—	1·6	1·2	1·2
p-fructokinase	—	—	—	0·003	0·002	0·26
Triose-p-isomerase	880	—	130	36	41	—
Glyceraldehyde-p-dehydrogenase	—	60	0·7	2·2	2·4	—
p-glycerate kinase	—	240	—	5·0	4·2	—
p-glyceratemutase	56	—	—	5·0	4·6	—
Enolase	560	—	—	2·6	3·4	—
Pyruvate kinase	32	6·4	—	100	180	—

[a] Kahl et al. (1969).
[b] Firenzuoli et al. (1968).
[c] Bartels (1968).
[d] Scott et al. (1964).
[e] Fowler (1971).

Alternatively the low catalytic activity of aldolase may have physiological significance.

Work in my laboratory (Sarawek and Davies, 1977) has shown that the levels of aldolase in *Lemna* are markedly affected by the content of nitrogen in the medium. Activity is rapidly lost when NO_3^- is removed but not if the NO_3^- is replaced by NH_4^+. In the absence of a nitrogen source, *Lemna* produces a proteinaceous inactivator, which appears to have considerable specificity towards aldolase. The inactivator can be readily separated from aldolase by molecular sieving by ion exchange chromatography on DEAE-cellulose (Fig. 4). We suggest that the regulation of the aldolase by means of an inactivator (possibly a proteolytic enzyme) suggests that aldolase has a role in the control of carbohydrate metabolism. Rapoport *et al.* (1974) have also considered the possibility that aldolase in red blood cells may not be an equilibrium enzyme. I would like to suggest that the activity of aldolase has a special relationship to that of phosphofructokinase allowing the concentration of fructose diphosphate to increase when the glycolytic flux is increased and so enhance the product activation of phosphofructokinase. This possibility is supported by the data of Fig. 2 showing a dramatic rise in the concentration of fructose diphosphate when the glycolytic flux is increased by anaerobiosis. On the other hand there are no reports indicating that fructose diphosphate is an activator of phosphofructokinase from plants.

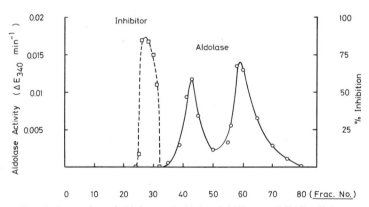

FIG. 4. Separation of aldolase and aldolase inhibitor on DEAE cellulose.

B. ADENINE NUCLEOTIDES

I have discussed elsewhere (Faiz-ur-Rahman, *et al.*, 1974) the proposition that the negative feedback system involving ATP and AMP which operates in animals is replaced in plants by a negative feedback system involving phosphoenolpyruvate. Part of the evidence for this proposition is shown in Fig. 5 where the increase in fructose diphosphate is correlated with the decrease in phosphoenolpyruvate but not with the changes in concentration

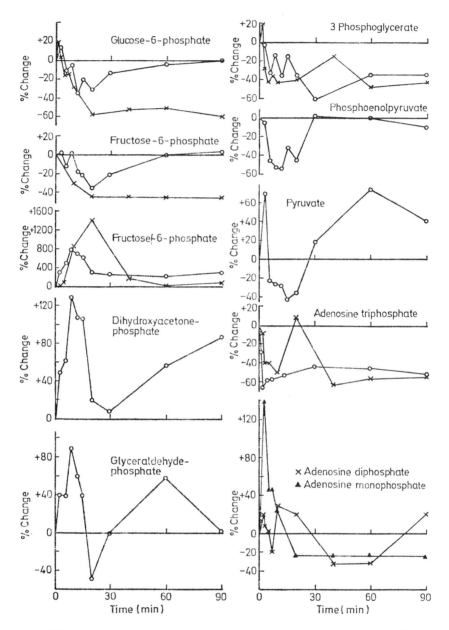

FIG. 5. Effect of anoxia on the concentration of glycolytic intermediates in aged discs of carrot. ▲—▲ ×—× values determined by 32 P method; O—O values determined by enzymic methods. (After Faiz-ur-Rahman et al., 1974.)

of ATP. The decline in the concentration of ATP is accompanied by a *decline* in the concentration of ADP and AMP. A similar decline in the level of adenine nucleotides has been noted by Kobr and Beevers (1971) when castor bean endosperm is made anaerobic (Fig. 6). It may be that what we are observing here is the purine nucleotide cycle discovered by Tornheim and Lowenstein (1972).

Adenylate deaminase

$$AMP \xrightarrow{H_2O} IMP + NH_3$$

Adenylosuccinate synthetase

$$IMP + GTP + aspartate \longrightarrow Adenylosuccinate + GDP + P_i$$

Adenylosuccinase

$$Adenylosuccinate \longrightarrow AMP + fumarate.$$

Conversion of AMP to IMP by adenylate deaminase leads to the formation of ATP from ADP by a readjustment of the adenylate kinase equilibrium.

$$2 ADP \longleftrightarrow ATP + AMP$$

The adenylate deaminase also produces NH_3 which according to Turner and Turner (1975) can relieve the inhibition of phosphofructokinase produced by phosphoenolpyruvate. There would therefore seem to be an *a priori* case for the proposition that ATP may exert control over phosphofructokinase via the purine nucleotide cycle.

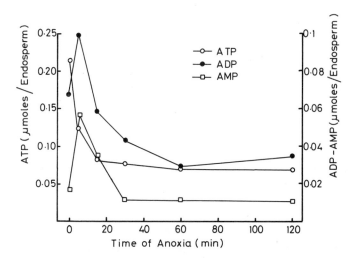

FIG. 6. Time course of adequate fluctuations in anoxic 5-day-old castor beam endosperm. (After Kobr and Beevers, 1971.)

V. The Control of pH

The fine control of cytoplasmic pH could, in principle, be achieved by a balance between carboxylation and decarboxylation

$$RH \xrightarrow{\quad CO_2 \quad} RCOOH \longrightarrow R^1COOH \xrightarrow{\quad CO_2 \quad} R^1H.$$

I have suggested that phosphoenolpyruvate carboxylase and malic enzyme can function as a pH stat because the carboxylase has an alkaline pH optimum whilst malic enzyme has an acid optimum (Davies, 1973). The sensitivity of the pH stat is determined by the slopes of the activity versus pH curves and the sensitivity is enhanced by effectors which act in opposing ways on the two enzymes. As shown in Table IV, malic enzyme and phosphoenolpyruvate carboxylase respond to effectors in a manner consistent with their proposed roles in a metabolic pH stat, although the comparison is unsatisfactory because the enzymes were obtained from different sources. More recently, Mr Bonugli working in my laboratory has shown that phosphoenolpyruvate carboxylase purified from potato tubers behaves in a manner very similar to the enzyme from etiolated maize, except that the responses to effectors were quantitatively somewhat less.

Table IV

Comparison of the responses of malic enzyme and phosphoenolpyruvate carboxylase to various effectors

	Effect on	
Effector	Potato malic enzyme (Davies and Patil, 1974)	Maize phosphoenol-pyruvate carboxylase (Wong and Davies, 1973)
Phosphate	Inhibits	Activates
Sulphate	Inhibits	Activates
Phosphoglyceric acids	Inhibit	Activate
Dicarboxylic acids	Activate	Inhibit
Aspartate	Slight inhibition	Inhibits
AMP	Inhibits	Activates

A. Interaction between glycolysis and the control of pH

Some of the interactions between glycolysis and the carboxylation—decarboxylation reactions of the pH stat are shown in Fig. 7. Phosphoenolpyruvate can be converted to pyruvate with the associated phosphorylation of ADP to give ATP. Alternatively, phosphoenolpyruvate may be converted

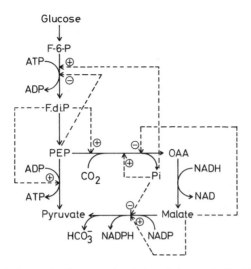

FIG. 7. Interaction between glycolysis and the metabolic pH stat.

into pyruvate by the reactions of the pH stat. The price for the control of pH is thus seen to be 8kcal per H^+.

The control of the partitioning of phosphoenolpyruvate between phosphoenolpyruvate carboxylase and pyruvate kinase is complex. Let us consider a simple case of two simultaneous reactions:

$$C \longleftarrow A \longrightarrow B$$

If the two enzymes conform to Michaelis–Menten kinetics the ratio $V_{A \to B}/V_{A \to C}$ will be constant, despite fluctuation in the concentration of A provided both enzymes have the same Km. If the two enzymes have different Kms, the ratio $V_{A \to B}/V_{A \to C}$ will be affected by changes in the concentration of A as shown in Fig. 8. If one of the reactions shows cooperativity with regard to the substrate, then the ratio $V_{A \to B}/V_{A \to C}$ will be particularly susceptible to changes in concentration of A. Superimposed on these fundamentals of simultaneous reactions are a number of feedback and feedforward controls (Fig. 7). Most of these interactions have been described elsewhere (Davies, 1973; Turner and Turner, 1975). However, a number of attempts to demonstrate that the pyruvate kinase of plants is activated by fructose diphosphate have been unsuccessful. Recently, Mr Bonugli working in my laboratory has observed an activation of pyruvate kinase by fructose diphosphate using a preparation obtained from peas and pH values below 7.

The control of glycolysis is essentially the control of flux through the system. Certain metabolites may change in concentration but these changes are incidental as the system moves from one steady-state to another. The control of pH is essentially the control of the concentration of malate. To decrease pH, the decarboxylation of malate must be blocked and the car-

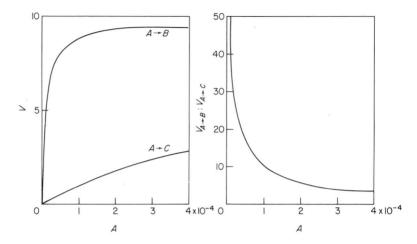

FIG. 8. Effect of substrate concentration on the velocities of simultaneous enzyme-catalysed reactions. The curves are calculated for the following conditions:

$$V_{max}A \rightarrow B = V_{max}A \rightarrow C; \quad K_mA \rightarrow B = 10^{-5}\text{ M}; \quad K_mA \rightarrow C = 10^{-3}\text{ M}.$$

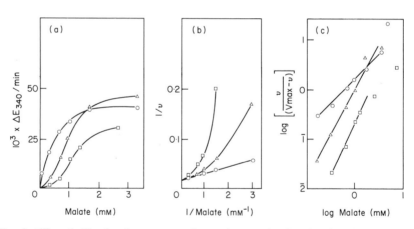

FIG. 9. Effect of pH and malate concentration on the rate of malate decarboxylation catalysed by partially purified malic enzyme. a. direct plot of rate versus malate concentration; b. double-reciprocal plot; c. Hill plot. Enzyme activity was determined as described in the text except that the malate concentration and pH were varied as indicated. Velocity is expressed as the change in E_{340}/min. 0, pH 6·9($n_H = 1$); Δ, pH 7·3($n_H = 2·0$); □, pH 7·6($n_H = 2·1$).

boxylation of phosphoenolpyruvate increased. The control of phosphoenolpyruvate carboxylase in this system is easily understood, since the reaction is the first step of the sequence and is irreversible. The control of malic enzyme as part of the proposed pH stat is more difficult to understand. For example, let us assume that for some reason the pH of the cell rises, the model requires that the production of malic acid be increased and that simultaneously the removal of malate by malic enzyme be decreased. Now malic enzyme shows sigmoid kinetics with respect to malate (Fig. 9), so that if the enzyme functions at concentrations of malate below the $S_{(0.5)}$ value, then a doubling of the malate concentration would be expected to lead to the rate of decarboxylation being *increased* approximately 4-fold, whereas the pH stat model requires that the decarboxylation be *decreased*. These and other considerations lead us to consider the possibility that malic enzyme catalyses an equilibrium reaction as in some animal tissues (Krebs, 1973).

B. UNIDIRECTIONAL INHIBITION

Newsholme and Crabtree (1973) have suggested that "fast" reactions close to equilibrium could be particularly effective in controlling flux as shown by the following example from their paper:

$$\text{System 1 "Slow"} \begin{cases} x \mathrel{\underset{1}{\overset{11}{\rightleftharpoons}}} y \quad \text{(overall flux} = 10 \text{ units)} \\[2em] x \mathrel{\underset{1}{\overset{12 \cdot 1}{\rightleftharpoons}}} y \quad \text{(overall flux} = 11 \cdot 1 \text{ units)} \end{cases}$$

$$\text{System 2 "Fast"} \begin{cases} x \mathrel{\underset{110}{\overset{100}{\rightleftharpoons}}} y \quad \text{(overall flux} = 10 \text{ units)} \\[2em] x \mathrel{\underset{110}{\overset{121}{\rightleftharpoons}}} y \quad \text{(overall flux} = 21 \text{ units)} \end{cases}$$

In system 1, a 10% increase in the rate of formation of y produces about a 10% increase in the flux whereas a 10% increase of the same reaction under the conditions of system 2 produces about a 100% change in the flux.

In general we expect inhibitors to inhibit the reaction in both directions, so that the control proposed by Newsholme and Crabtree, (1973) requires unidirectional inhibition. For thermodynamic reasons, unidirectional inhibition is impossible at equilibrium, but away from equilibrium it is theoretically possible (Frieden, 1964) and unidirectional inhibition and activation of malic enzyme has been demonstrated (Nascimento *et al.* 1975). These unusual properties can be explained on the assumption that malic enzyme has an active and a control site with structural similarities. It could be argued that the special allosteric properties of malic enzyme are necessary to permit unidirectional inhibition and activation. However, the effector is *meso*-tartrate and we have been unable to demonstrate unidirectional effect with normal metabolites. Until such effects have been demonstrated we must search for other possibilities.

C. THE EQUILIBRIUM REACTION CATALYSED BY MALIC ENZYME

Enzyme kineticists have developed rate equations for the special conditions that at time t_0 the products are zero. This condition has little significance for the biochemical physiologist and few attempts have been made to analyse the complex situation involving all the reactants, which represents reality. However, a conclusion from the work of Anderson (1974) which is important for our discussion is that the rate at which an equilibrium reaction readjusts after a small fractional perturbation is independent of which reactant is initially changed.

Consider the reaction Malate + NADP $\longleftrightarrow$ Pyruvate + NADPH + CO_2. The concentration of NADP in the cytoplasm is probably less than $\frac{1}{1000}$ of the concentration of malate. Thus the *rate* at which the system responds to a small change in the *amount* of malate will be produced by an extremely small change in the amount of NADP (less than $\frac{1}{1000}$ of the change in malate). Similarly a small change in the ratio NADP/NADPH will lead to an equal change in the ratio pyruvate/malate (assuming CO_2 is constant). However, since the concentration of malate + pyruvate is much greater than that of NADP + NADPH, there will be a large change in the *amount* of malate for a small change in the amount of NADP. Thus large and rapid changes in the amount of malate can be produced by changes in the ratio NADP/NADPH. This ratio is sensitive to changes in pH due to reactions of the general type:

$$AH_2 + NADP^+ \longleftrightarrow A + NADPH + H^+$$

Thus a high pH will favour the oxidation of AH_2. At first glance we might think that a high pH would favour the oxidative decarboxylation of malate and thus act in a direction opposite to that required by the metabolic pH stat. However, the equation for the oxidative decarboxylation of malate shows no net change in charge; the loss of the carboxyl group of malate being balanced by the loss of the charge on the quaternary nitrogen of $NADP^+$ and the equilibrium is not affected by pH. On the other hand, most of the reactions catalysed by NADP-linked dehydrogenases are affected by changes in pH—a high pH producing an increase in the ratio NADPH/NADP. This high ratio causes the equilibrium reaction catalysed by malic enzyme to move in the direction of malate accumulation—thereby producing H^+ ions and so helping to lower the pH as required by the metabolic pH stat. It should be noted that the malic equilibrium not only acts as a pH buffering system but additionally acts to maintain the ratio NADP/NADPH constant, as indicated by Krebs (1973.)

It remains to consider what significance is to be attributed to the allosteric properties of malic enzyme. I have no explanation for negative effectors such as phosphate but the action of positive factors and the cooperativity towards malate are readily understood in terms of the system adjusting to an increase in pH. According to the principle of microreversibility, if the rate of malate decarboxylation is stimulated, the rate of pyruvate carboxylation will also be stimulated. Thus the sigmoid kinetics observed with respect to malate enables the system to adjust rapidly to the production of malate. However, sigmoid kinetics will tend to reduce the speed with which the system adjusts to the removal of malate. It is difficult to assess the balance of advantage between these two situations but it should be noted that the malic enzyme of grasses lacks allosteric properties (Davies *et al.*, 1974).

In conclusion, I admit to having returned to speculation. My defence is that the difficulties in comprehending the highly non-linear mechanisms of biochemical control can produce despair rather than enthusiasm for research. Perhaps one should not be too analytical; a certain naivety may be useful.

REFERENCES

Anderson, J. H. (1974). *J. ther. Biol.* **47**, 153.

Ap Rees, T. (1974). *In* "MTP International Review of Science" (Kornberg and Phillips eds), Vol. 7 p. 84, *Biochemistry Series*, **1**, Butterworth, London.

Bartels, H. (1968). *Planta* **55**, 573.

Bücher, Th. and Rüssmann, W. (1964). *Angew. Chem. Int.* **3**, 426.

Chance, B. ed. (1965). *In* "Control of Energy Metabolism" (Chance, B., Estabrook, R. W. and Williamson, J. R., eds), p. 9. Academic Press, New York and London.

Chance, B., Holmes, W., Higgins, J. and Connelly, C. M. (1958). *Nature* **182**, 1190.

Chapman, E. A. and Graham, D. (1974). *Pl. Physiol.* **53**, 886.

Davies, D. D. (1961). "Intermediary Metabolism in Plants" Cambridge University Press, Cambridge.

Davies, D. D. (1973). *Symp. Soc. exp. Biol.* **27**, 513.

Davies, D. D. and Patil, K. D. (1974). *Biochem. J.* **137**, 45.

Davies, D. D., Nascimento, K. H. and Patil, K. D. (1974). *Phytochemistry* **13**, 2417.

Dennis, D. T. and Green, T. R. (1975). *Biochem. biophys. Res. Commun.* **64**, 970.

Faiz-ur-Rahman, A. T. M., Trewavas, A. J. and Davies, D. D. (1974). *Planta* **118**, 195.

Firenzuoli, A. M., Vanni, P., Ramponi, G. and Baccari, V. (1968). *Pl. Physiol.* **43**, 260.

Fowler, M. W. (1971). *J. exp. Bot.* **22**, 715.

Frieden, C. (1964). *J. biol. Chem.* **239**, 3522.

Heinrich, R. and Rapoport, T. A. (1974). *Europ. J. Biochem.* **42**, 89.

Higgins, J. (1965). *In* "Control of Energy Metabolism" (Chance, B., Estabrook, R. W. and Williamson, J. R., eds), p. 13. Academic Press, New York and London.

Hohorst, H. J., Krentz, F. H. and Bücher, Th. (1959). *Z. Biochem.* **332**, 18.

Kacser, H. and Burns, J. A. (1973). *Symp. Soc. exp. Biol.* **27**, 65.

Kahl, G., Lange, H. and Rosenstock, G. (1969). *Z. Naturf.* **24b**, 1544.

Kobr, M. J. and Beevers, H. (1971). *Pl. Physiol.* **47**, 48.

Kohen, E., Michaelis, M., Kohen, C. and Thorell, B. (1973). *Expl. Cell. Res.* **77**, 195.

Krebs, H. A. (1954). *Bull. Johns Hopkins Hosp.* **95**, 45.

Krebs, H. A. (1957). *Endeavour* **16**, 125.

Krebs, H. A. (1973). *Symp. Soc. exp. Biol.* **27**, 299.

MacLennan, D. H., Beevers, H. and Harley, J. L. (1963). *Biochem. J.* **89**, 316.

Nascimento, K. H. D., Davies, D. D. and Patil, K. D. (1975). *Biochem. J.* **149**, 349.

Newsholme, E. A. and Crabtree, B. (1973). *Symp. Soc. exp. Biol.* **27**, 429.

Rapoport, T. A., Heinrich, R., Jacobasch, G. and Rapoport, S. (1974). *Europ. J. Biochem.* **42**, 107.

Raval, D. N. and Wolfe, R. G. (1962). *Biochemistry* **1**, 263.

Rolleston, F. S. (1972). *In* "Current Topics in Cellular Regulation" (Horecker, B. L. and Stadtman, E. R., eds), Vol. 5, p. 47. Academic Press, London and New York.

Sarawek, S. and Davies, D. D. (1977). (In preparation.)

Savageau, M. A. (1969). *J. ther. Biol.* **25**, 365.

Savageau, M. A. (1972). *In* "Current Topics in Cellular Regulation" (Horecker, B. L. and Stadtman, E. R., eds), Vol. 7, p. 63. Academic Press, London and New York.

Scott, K. J., Craigie, J. S. and Smillie, R. M. (1964). *Pl. Physiol.* **39**, 323.

Scrutton, M. C. and Utter, M. F. (1968). *A. Rev. Biochem.* **37**, 249.

Slaughter, C. J. (1975). *Trans. Biochem. Soc.* **3**.

Tornheim, K. and Lowenstein, J. M. (1975). *J. biol. Chem.* **250**, 6304.

Turner, J. F. and Turner, D. H. (1975). *A. Rev. Pl. Physiol.* **26**, 159.

Wong, K. and Davies, D. D. (1973). *Biochem. J.* **131**, 451.

Wu, R. and Racker, E. (1959). *J. Biol. Chem.* **234**, 162.

CHAPTER 4

Regulation of Enzyme Levels During the Cell Cycle

M. M. YEOMAN, P. A. AITCHISON AND A. J. MACLEOD

Department of Botany, University of Edinburgh, Edinburgh, Scotland

I. Introduction	63
II. Synchrony in Plants and Plant Parts	64
A. Naturally Synchronous Systems	64
B. Synchronized Systems	65
C. Significance of Synchrony in the Intact Plant	68
III. Patterns of Enzyme Activity	69
A. In Developing Pollen	69
B. In Cell Suspension Cultures.	70
C. In the Artichoke Explant System	71
IV. Factors Underlying Periodic Changes in Enzyme Levels . . .	74
V. Significance of Periodic Changes in Enzyme Levels	76
A. Linear Reading and Gene Dosage	76
B. Programmed Development	77
C. General Effects of Perturbation	78
Conclusion	79
Acknowledgements	80
References	80

I. INTRODUCTION

The procession of structural and metabolic events which takes place from one mitosis to the next and results in the formation of two cells from one, constitutes the cell cycle. In the meristem, the dividing cells double in mass during each cycle and subsequently two daughter cells are formed which are equal in size to the parent at the beginning of the previous cycle. However, such a pattern is not common to all dividing cells outside the apical and lateral meristems of the intact plant (Brown and Dyer, 1972).

The constituent cells of meristems are asynchronous and at any one time many different stages of the cell cycle are represented within the dividing population. This means that one, or a group of growing meristems when sampled will present an average value for the particular parameter being measured. Although the average cell cycle time within the dividing population and the timing of particular syntheses can be determined with some accuracy, using labelling techniques with precursors of DNA, RNA and protein, it is not yet possible to examine changes in individual species of RNA or protein. This is indeed a major disadvantage because our present approach to obtaining an understanding of how cell division is regulated depends on the acquisition of a greater knowledge of how the synthesis and degradation of macromolecules is controlled. Therefore, within the limitations of the techniques presently available to the investigator, the only practical approach to this problem lies in the use of synchronous systems. Accordingly this paper is concerned with these systems and the results that have been obtained using the tool of synchrony.

II. Synchrony in Plants and Plant Parts

A. NATURALLY SYNCHRONOUS SYSTEMS

The exploitation of natural synchrony and the creation of artificially synchronized systems has been a useful approach in the study of cell division. Synchrony in eukaryote systems is always imperfect. Perhaps the closest approach to absolute simultaneity is to be seen within a natural process in plants, the development of the microspore (pollen) from the microspore mother cell (pollen mother cell) in angiosperms. Here a meiotic division is followed by a mitotic division and the product is four haploid microspores (pollen). The synchrony observed in this pair of divisions is extremely high, especially in certain genera of the *Liliaceae*, and has been exploited by Stern

TABLE I

A summary of naturally occurring synchronous systems in multicellular green plants

Development	Plant	Reference
[a]Female gametophyte in pteridophytes	*Selaginella, Isoetes*	Lyon (1901)
[a]Female gametophyte in gymnosperms	*Dioon. Zamia, Ginkgo*	Chamberlain (1935)
[a]Megagametophyte or embryo-sac in flowering plants	*Acalypha*	Maheshwari (1950)
[a]Endosperm in flowering plants	*Frittilaria, Haemanthus*	Maheshwari (1950)
Antheridia in bryophytes	*Spherocarpos*	Nevins (1932)
Sporangia in bryophytes, pteridophytes and gymnosperms	*Mnium, Botrychium, Isoetes*	Cardiff (1905); Wilson (1909); Chamberlain (1935); Eames (1936)
Microspores (pollen) in flowering plants	*Trillium, Lilium*	Huskins and Smith (1935); Sparrow et al., (1952); Erickson (1964)

[a] Syncitial (Free nuclear).

and Hotta in a series of papers (see Stern, 1966). Other systems in which synchrony occurs naturally have been reported (Table I) to display very high degrees of synchrony but so far these have not been used to study the molecular behaviour of dividing cells.

B. SYNCHRONIZED SYSTEMS

Some attempts have been made to enforce synchrony in cell populations of higher plants (Table II). Here the induction of synchrony in apical root meristems has been accomplished relatively easily (Clowes, 1965; Jakob and Trosko, 1965; Mattingly, 1966; Kovacs and Van't Hof, 1970) but the synchronized systems obtained are not very suitable for studies on the biochemical characteristics of dividing cells. In contrast to the studies with meristems, tissue and cell cultures have been manipulated to produce synchronous systems which have proved amenable to the investigation of macromolecular changes which take place as a cell traverses the cell cycle (Mitchell, 1967, 1968, 1969; Fraser and Loening, 1974). All of the tissue and cell culture systems used in this work have received a perturbation during culture which has either synchronized them (King et al., 1974) or induced a population already in a similar state to divide (Yeoman and Evans, 1967). Various perturbations have been employed (Table II) including the addition of cytostatic agents (Clowes, 1965; Jakob and Trosko, 1965; Eriksson, 1966; Mattingly, 1966; Kovacs and Van't Hof, 1970), raising or lowering the incubation temperature (Okamura et al., 1973), removal of an essential nutrient(s) which is then replaced at a later time (Roberts and Northcote, 1970; Jouanneau, 1971; Wilson et al., 1971), addition of a growth substance to freshly excised explants (Yeoman et al., 1966) or separated protoplasts (Nagata and Takebe, 1970) and deprivation of oxygen followed by a return to aerobic conditions (Constabel et al., 1974).

Inhibitors of DNA synthesis such as fluorodeoxyuridine (FUdR), 5-aminouracil (5-AU), hydroxyurea (HU) and high concentrations of thymidine (TdR) have been used to synchronize cell suspensions of *Haplopappus gracilis* (Eriksson, 1966). The treatment consists of exposing cells in the exponential phase of growth to fairly high concentrations of DNA synthesis inhibitors. Cells in S-phase cease synthesizing DNA soon after the treatment while cells in G_1, G_2 or mitosis proceed through the cell cycle until they reach the beginning of S-phase where they accumulate. Removal of the inhibitor, and in the case of FUdR addition of TdR releases the cells which proceed through the next division in a highly synchronized state (mitotic indices of 35% were observed). It was also observed by Eriksson (1966) that FUdR, HU and 5-AU used at the concentration necessary to induce synchrony all promote chromosome breakage. These effects of inducers must inevitably lead to changes in the cell population which will distort the cell cycle. Therefore such methods are considered to be unsuitable for synchronizing

TABLE II

A summary of synchronizing procedures used with higher plant cells

Plant	Culture technique	Synchronizing procedure	Reference
Vicia faba	Root meristem attached to cotyledons	Inhibition/Release	Jakob and Trosko (1965) Mattingly (1966)
Zea mays	Root meristem attached to cotyledons	Inhibition/Release	Clowes (1965)
Pisum sativum	Cultured root	Inhibition/Release	Kovacs and Van't Hof (1970)
Haplopappus gracilis	Cell Suspension	Inhibition/Release	Eriksson (1966)
Hslianthus tuberosus	Explant	None (Division induced)	Yeoman et al. (1966)
Nicotiana tabacum	Leaf protoplasts	None (Division induced)	Nagata and Takebe (1970)
Acer pseudoplatanus	Cell suspension	Kinetin exclusion/Addition	Roberts and Northcote (1970)
Acer pseudoplatanus	Cell suspension	Starvation/Growth	Wilson et al. (1971)
Nicotiana tabacum	Cell suspension	Kinetin exclusion/Addition	Jouanneau (1971)
Daucus carota	Cell suspension	Starvation/Cold	Okamura et al. (1973)
Glycine max	Cell suspension	Inhibition/Release	Constabel et al. (1974)

cell populations which are to be subsequently used for studies on the cell cycle.

Roberts and Northcote (1970), Peaud-Lenöel and Jouanneau (1971) and Jouanneau (1971) have used a "growth substance" exclusion method for synchronizing cell suspension cultures of *Acer pseudoplatanus* and *Nicotiana tabacum* respectively. These cell cultures have an obligate requirement for cytokinin to sustain cell division. Without it the culture stops dividing, whereas addition of kinetin recommences division. During the period in which the cultures are deprived of kinetin a high proportion of the cells are apparently arrested at a similar point in the cell cycle, and the addition of kinetin allows the retarded cells to recommence division together in a wave. Mitotic indices of *c.* 15% have been reported by Roberts and Northcote (1970).

King *et al.* (1973) have used a rather different technique to synchronize cultures of *Acer pseudoplatanus*. They established cultures at a low initial density (*c.* 2×10^4 cells ml^{-1}) from cells in a late stationary phase culture. The induced synchrony in cell division persisted through five to six generations, the increase in cell number at each step normally falling within the range 70–95% (King and Street, 1973). A correlation may be observed between the increase in cell number and the rise in mitotic index although unexpectedly low values for mitotic index (*c.* 7·5%) are encountered and these are seen against a relatively high background of mitotic activity during interphase (Mitotic index $= c.$ 3·0%). A possible explanation of this phenomenon is that the degree of synchrony of cytokinesis in these cultures may be much higher than that of the preceding events in the cell cycle. It is possible that an "entrainment" phenomenon is operating in which cells may move at different speeds through interphase and mitosis but stop and accumulate immediately before cytokinesis. The release of the complete dividing population leads to an extremely sudden increase in cell number.

Constabel *et al.* (1974) have described an effective method for inducing appreciable synchrony (mitotic indices reaching 25%) in cell suspension cultures of soybean (*Glycine max*) by subjecting the culture to repeated short treatments (eight times a day) with gaseous nitrogen for four days. This method has obvious potential but has so far not been exploited for studies on the cell cycle.

The synchrony observed in cultured explants removed from the Jerusalem artichoke tuber has also received attention especially in this laboratory (Yeoman and Evans, 1967). This system is different from those described above in that it is a multicellular tissue culture, each explant contains a large population of dividing cells, and it is division which is induced and not synchrony. The culture of the tissue fragments in a simple medium of mineral salts, sucrose and 2,4-D is sufficient to promote a wave of synchronous divisions. Synchrony is lost progressively with time, and it disappears completely by the fifth division. Wound effects, which are most vigorous during the first division, complicate the interpretation of some of the data. These complications give the system a disadvantage compared with cell suspension cultures. On the other hand, this explant system has the advantage that the structural and metabolic events which precede division are more synchronous than cytokinesis and entrainment effects are not detectable in explants which are excised in low intensity green light and cultured in complete darkness (Yeoman and Davidson, 1971; Davidson and Yeoman, 1974). An additional important feature is that explants are only in culture for very short periods so that there is insufficient time for nuclear aberrations to occur.

Summary of advantages of synchronous systems for the investigation of the control of enzyme synthesis

(a) Large numbers of cells pass through the same metabolic sequences at the same time (amplification). This permits sophisticated structural or chemical analyses which would be impossible with a single cell or a few cells.
(b) Representative samples can be withdrawn without a disturbance of the cyclic chain of events under study.

All of the systems described here have particular advantages and disadvantages for examining cell division. However, only three of them have been used to examine the control of synthetic events which take place during the preparation for division.

C. SIGNIFICANCE OF SYNCHRONY IN THE INTACT PLANT

Synchrony is not a common feature of intact organisms and is only observed at particular times during the development of higher plants (Table I). It is a common feature of this situation that the period of synchronous division is short and with only a few exceptions, which involve the development of antherozoids and spores, takes place in a common cytoplasm. Even in spore and antherozoid development access between nuclei via substantial cytoplasmic connections is exceptionally good. Therefore, a possible common feature of all these developmental situations is that free exchange of molecular information may take place, and it is this which is responsible for the promotion of synchrony. It is perhaps significant to remember that the high degree of nuclear division synchrony in developing endosperm is lost with the formation of intervening walls. In such "short lived" phases it seems likely that the normal regulatory processes are overridden by the act of synchronization and that the cessation of synchronous division is accompanied by a return to a state in which the basic regulatory processes are re-established. Similarly, the artificial induction of synchrony in a nonsynchronous system produces a like situation.

III. PATTERNS OF ENZYME ACTIVITY

A. IN DEVELOPING POLLEN

Compared with other groups, there are relatively few reports of enzyme patterns during the cell cycle in higher plants. To a large extent this reflects the difficulty of obtaining synchronous populations of plant cells, whether natural or induced. The earliest reports of variation in enzyme levels in plants which were demonstrably associated with particular stages in the cell cycle, i.e. variations which reproducibly occurred at the same relative time between successive cytokineses, exploited the natural synchrony occurring in developing microspores in *Lilium* and *Trillium* (Stern, 1966). In this material, thymidine kinase (TdRK) and thymidine monosphosphate kinase (dTMPK) appear transiently at a characteristic stage during microspore development in a manner that suggests that the changes are part of a strictly controlled process of differentiation (Hotta and Stern, 1963). However, this system has several unusual features. Firstly, the cell cycles involved are the two divisions of meiosis which result in the formation of a pollen tetrad, which might be regarded as atypical, and is certainly not part of a repeated sequence. Secondly, the time between cell divisions is very long, 22 days between the first and second divisions. Thirdly, despite this relative quiescence the enzymes show relatively transient peaks of activity. TdRK rises from virtually undetectable levels to a peak and back to zero in an interval representing only 5% of the cycle period. The sharpness of the peak implies an extremely high degree of synchrony in the population which also seems to be true from other data. However, as a general observation, the decay rate of specific proteins in dividing cultures of different organisms is roughly proportional to the duration of the cell cycle, which makes the rapid disappearance, if not the appearance, of activity surprising, if enzyme synthesis and degradation are solely responsible. However, the possibility that the (relatively) sharp rises and decreases are due to activation and inactivation, the latter being particularly likely, has never been completely ruled out despite attempts to do so with inhibitors (Hotta and Stern, 1963). It might be valid to regard this system as an example of tissue differentiation rather than displaying patterns characteristic of a normal cell cycle. Certainly, the stimulus for increased TdRK activity seems to be provided by exogenous factors; an increased supply of thymidine released by the surrounding anther tissue. Other enzymes show marked variations in activity in *Lilium* anthers. These changes, however, occur again in the non-synchronous tissue surrounding the microspores (Hotta and Stern, 1961; Stern, 1961). On the other hand, the enzymes concerned do show periodic (though relatively more gradual) changes in mitotic cycles in other organisms (e.g. Brent *et al.*, 1965). The system has the advantage that the long cycle time should facilitate analysis of particular portions of the cycle, but suffers from the difficulty of obtaining sufficient material for analysis.

B. IN CELL SUSPENSION CULTURES

A potentially more useful system is that developed by Street and co-workers (King *et al.*, 1973). As described earlier in this article several successive synchronous mitotic divisions may be induced in suspension cultures of *Acer pseudoplatanus* by diluting growth-restricted (stationary phase) cultures into new medium. Under these conditions TdRK showed a single peak in activity which spanned the S-phase. Activity increased from about the same time in two cultures relative to the start of S-phase, 6–10 h before it, but at very different times relative to mitosis, depending on overall cycle length (King *et al.*, 1974). In a culture with cycle length of 67 h, TdRK activity increased in late G_1, whereas in a culture with a much shorter G_1 and total cycle length of 48 h, TdRK activity increased before cytokinesis in the pre-

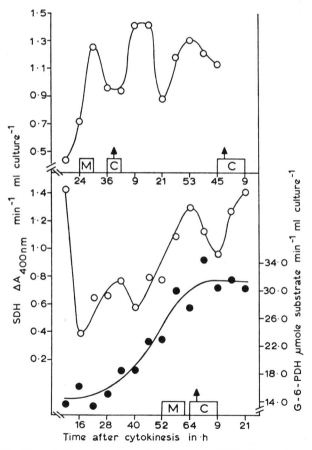

FIG. 1A and B. The activities of succinate dehydrogenase (SDH ○) and glucose-6-phosphate dehydrogenase (G6PDH ●) during two cell cycles of synchronized *Acer* cells. Upper: from culture 77. Lower: from culture 73 (King *et al.*, 1974).

ceding cycle. The close relationship between TdRK activity and DNA replication is an almost universal feature of enzyme patterns during cell cycles
(Harland, *et al.*, 1973), though the two cultures studied by King *et al.* (1974)
do seem to show differences in the times of *maximum* TdRK activity, occurring
at mitosis in one, and in mid S-phase in the other. However, the data are
presented as specific activities not absolute amounts, and the differences
could be accounted for by different relative overall rates of protein synthesis
in the two cultures. In the same material aspartate transcarbamylase (ATC)
shows a peak of specific activity which is almost exactly out of phase with
the peak of TdRK activity. Succinate dehydrogenase (SDH) displayed a
rather variable pattern of absolute activity, seeming to show two peaks in
each cycle, while Glucose-6-phosphate dehydrogenase (G6PDH) showed
an approximate doubling step in the second half of the cycle (Fig. 1) (From
King *et al.*, 1974).

C. IN THE ARTICHOKE EXPLANT SYSTEM

Periodic changes in enzyme levels have also been noted during the cell
cycle in Jerusalem artichoke tuber tissue, in which a few synchronous
divisions may be induced by growing explants in an auxin-containing
medium (Yeoman and Evans, 1967). Again TdRK, dTMPK and DNA
polymerase show cell-cycle dependent increases in activity (Harland *et al.*,
1973), which are closely correlated with the period of DNA synthesis (Fig. 2).
RNAase activity showed a more complicated pattern (Fig. 3), with a small
rise during G_1; S and G_2 followed by an extremely sharp substantial increase

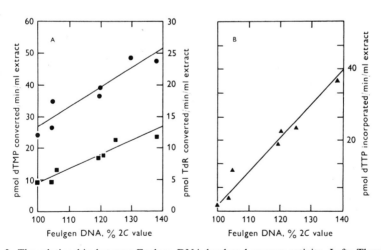

FIG. 2. The relationship between Feulgen DNA level and enzyme activity. Left: Thymidine
kinase (TdRK, ●, correlation coefficient [C.C.] = 0·923) and Thymidylate kinase (dTMPK,
■, C.C. = 0·933) Right: DNA polymerase (DNA Polym., ▲, C.C. = 0·988). Calculated
regression lines drawn through the values (Harland *et al.*, 1973).

(*c.* six-fold) at about the time of mitosis. The peak is followed by a sharp decrease after cytokinesis followed by another sudden increase coincident with the second wave of division. The initial slow increase is probably not cell-cycle dependent, as it occurred in the absence of 2,4-D (Fig. 3), under which conditions no cell division occurred (Macleod, 1976). On the other hand ATPGK, and G6PDH (Fig. 3) reach a maximum value at the end of G_1 and then decrease reaching a minimum value at about the time of mitosis. In this material the patterns are stepped/peaked (Aitchison and Yeoman, 1973; Yeoman and Aitchison, 1973). In the cases of TdRK dTMPK, G6PDH and ATPGK there is reasonable evidence that the changes are not

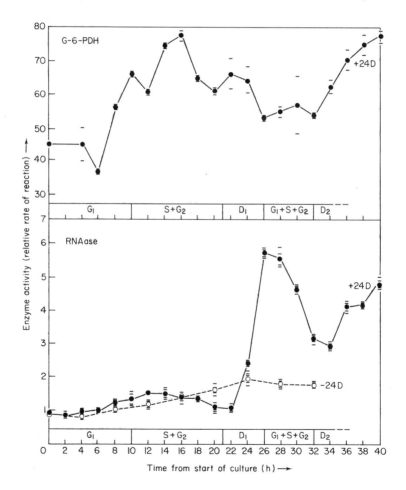

FIG. 3. Changes in the activities of G6PDH and RNAase in explants of Jerusalem arti-choke cultured with (●) and without (○) 2,4-D. Explants cultured in the absence of 2,4-D do not exhibit any significant change in the level of G6PDH (Yeoman and Aitchison, 1973). The phases of the cell cycle relate only to the series with added 2,4-D.

TABLE III

Patterns of enzyme activities in synchronous populations of higher plant cells

	Enzyme	Source	Pattern	Timing	Reference
1	TdRK	*Lilium* meiocytes	Single peak	G_1	Hotta and Stern (1963)
2	dTMPK	Natural synchrony	Single Peak	G_1	
3	TdRK	*Acer pseudoplatanus*	Single Peak	S	King *et al.* (1974)
4	ATC	Cell suspension culture	Single peak	G_2	
5	SDH	Synchrony induced	Double Peak	G_1/S G_2/M	
6	G6PDH		Step	S	
7	TdRK	*Helianthus tuberosus*	Step/Peak	S	Harland *et al.* (1973)
8	dTMPK		Step/Peak	S	
9	DNA Polym	Explant system	Step	S	
10	ATP-GK	Division induced	Step/Peak	G_1	Aitchison and Yeoman (1973, 74, 76). Yeoman and Aitchison (1973); Macleod (1976)
11	G6PDH		Step/Peak	G_1	
12	RNAase		Single peak	M	

TdRK (thymidine kinase), dTMPK (thymidylate kinase), ATC (aspartate transcarbamylase), SDH (succinate dehydrogenase), G6PDH (Glucose-6-phosphate dehydrogenase) ATP-GK (ATP-glucokinase), DNA polym. (DNA polymerase), RNAase (ribonuclease).

due to changes in the presence of activators or inhibitors (Harland, *et al.*, 1973; Aitchison and Yeoman, 1974) and that, in the case of G6PDH at least, the increase can be wholly accounted for by *de novo* synthesis (Aitchison *et al.*, 1976). These patterns are summarized in Table III. The only other report, to our knowledge, of periodic changes in particular proteins during the cell cycle in plants is of increases in histones in *Haplopappus* (Yonezawa and Tanaka, 1973) which occur during the period of DNA replication, which is a feature shared with numerous other organisms.

IV. FACTORS UNDERLYING PERIODIC CHANGES IN ENZYME LEVELS

From the restricted data available what can be deduced about the control of enzyme levels during the cell cycle in plants? Firstly, as is the case in other eukaryotes, where DNA synthesis occupies a restricted portion of the total cycle time, no *general* correlation between increases in enzyme levels and increase in DNA is observed. Some enzymes increase wholly (e.g. ATPGK and G6PDH in artichoke explants) or partially (TdRK in sycamore) during G_1, and one shows two periods of increase at the G_1/S and G_2/M interfaces (SDH in sycamore). In no case so far reported has a continuous pattern of increase, with or without a detectable rate change, been reported. These observations rule out a facile explanation of levels of activity reflecting continuous synthesis with a rate change occurring at the time corresponding to the point of replication of the structural genes, i.e. a gene-dosage effect, as has been proposed for bacteria and yeast (Masters and Pardee, 1965; Mitchison and Creanor, 1969; Mitchison, 1971). The control of enzyme levels seems an altogether more complicated process in higher eukaryotes. Even where there is an (approximate) temporal correlation between periods of enzyme increase and DNA replication (dTMPK, DNA polymerase and TdRK in artichoke, TdRK and G6PDH in sycamore), a more complicated control process is almost certainly involved.

Until recently it was an almost universal approach in cell cycle studies to explain an increase in enzyme levels mainly in terms of onset or change in rate of synthesis. While it may indeed turn out that periodic changes in rates of enzyme synthesis are a major underlying cause of these patterns, a convincing demonstration of the involvement of *de novo* synthesis has been presented only for TdRK in *Physarum polycephalum* (Oleinick, 1972) and for G6PDH in artichoke explants (Aitchison *et al.*, 1976). However, many of the patterns of enzyme changes during the cell cycle in plants (examples 1,2,3,4,5,7,8,10,11,12 in Table III) suggest that degradation (or possibly inactivation) also plays an important role in determining those patterns. All these enzymes show activity maxima, followed by a decrease in levels, i.e. they are "peak enzymes" (Mitchison, 1971), suggesting that they are unstable *in vivo*. The overall levels would be determined by a balance between synthesis and degradation and changes in this balance at different times during the cell cycle. To take a specific example, G6PDH activity in freshly isolated artichoke explants remains constant or slightly decreases for several hours and then shows a sudden increase. The early decrease not observed in all experiments) was ascribed to a loss of cells from the periphery of explants which has been invoked to explain initial decreases in other parameters (Yeoman *et al.*, 1968) but could also be due to net degradation of enzyme protein. The increase in activity was shown to be dependent on continuing RNA synthesis (Aitchison and Yeoman, 1974) and the entire increase could be accounted for by *de novo* synthesis, as was shown by the appearance of enzyme with increased buoyant density when cultures were grown in the presence of ^2H_2O (Aitchison *et al.*, 1976). However, this does not mean to say that the sudden increase is caused by a cessation of degradation or a commencement of synthesis alone. Indeed, density labelling experiments have shown that the pre-existing G6PDH is degraded over the period when net levels are increasing (Aitchison *et al.*, 1976). The disappearance of this light enzyme species during culture is shown in Fig. 4. Other reports from eukaryotes have suggested that actual enzyme levels may be determined by factors affecting stability against a background of continuous synthesis as in the case of ATCase in *Chlorella* (Schmidt, 1974). In the same organism degradation of ribulose diphosphate carboxylase seemed to

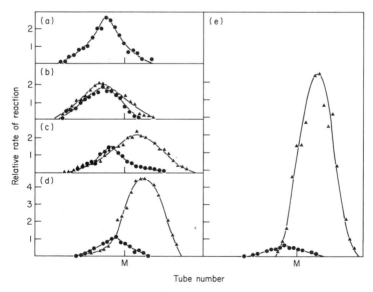

FIG. 4. Density gradient profiles of G6PDH from explants of Jerusalem Artichoke cultured in ^2H_2O. In the absence of 2,4-D (●) activity gradually decreases during culture (a, Oh; b, 18h; c, 48h; d, 72h; e, 96h). In the presence of 2,4-D, (▲) after an initial lag period, there is an appearance of enzyme with an increased buoyant density and a rapid loss of pre-existing low-density enzyme.

continue during periods when the level of activity remains constant, suggesting that constancy was only achieved by a precise balance of synthesis and degradation (Sitz *et al.*, 1973). In mammalian cells relatively rapid rates of turnover of 16 enzymes, all of them with half-lives (shown by cyclo-heximide blocking) less than the duration of the cell cycle have been reported (Bosmann, 1972). Moreover, the rates of turnover, and initial "stability times", during which no degradation was observed, could not be simply correlated with the actual patterns of cyclic changes in activities. Enzyme degradation may be a universal feature of higher eukaryote cells which has been comparatively overlooked in many discussions of the control of enzyme levels during the cell cycle. This is probably due to a preponderant interest in enzyme levels in prokaryotes, or simple eukaryotes, which are relatively easy to grow in rapidly dividing synchronous cultures, but in which, during logarithmic growth, there is relatively little protein turnover. In such cultures, only when growth is restricted does degradation, which is often very rapid, become significant.

As a converse situation it is possible that enzyme synthesis continues, possibly at a reduced rate, during periods of net loss of activity. This has been shown to occur during loss of nitrate reductase activity under non-inducing conditions in asynchronous tobacco cultures (Zielke and Filner, 1971) and it would not be surprising if such a situation were common.

It would be better to consider periodic changes in enzyme levels during the cell cycle as reflecting relatively sudden changes in the balance of synthesis and degradation (Yeoman and Aitchison, 1976).

V. Significance of Periodic Changes in Enzyme Levels

A. LINEAR READING AND GENE DOSAGE

It is clear that the majority (admittedly of relatively few cases) of plant enzymes that have been investigated display periodic changes in activity during the cell cycle. As outlined above, this implies that relatively large changes in the balance of synthesis and degradation occur at times during the cell cycle. The question should be asked, however, as to what is the significance of these changes. In bacteria and some lower eukaryotes (e.g. *Schizosaccharomyces pombe*), periodic increases in activity (or of potential for inducible enzymes) of many enzymes can be explained as either reflecting continuous availability of the genome, with an increase occurring when the structural gene is replicated (gene dosage), or reflecting an ordered sequence of transcription of structural genes related to their arrangement on the chromosome (linear reading). As mentioned above, gene dosage can be ruled out as a generally important regulatory mechanism in higher eukaryotes. The possibility of linear transcription is more difficult to test because of the lack of information on the structural arrangement of codons on eukaryote

DNA and the lack of suitable mutants to test the theory. A recent study using human lymphoblasts showed that TdRK, the structural gene which is located close to the centromere of chromosome no. 17, increases earlier in the cycle than 6-phosphogluconate dehydrogenase, the structural gene which is at some distance from the centromere of chromosome no. 1. There is no *a priori* reason for supposing that transcription should proceed distally from centromeres, nor that it should start at the same time and progress at equal rates on different chromosomes. However, increases in the products of genes all located on the same chromosome, the long arm of the human X, showed a very tentative correspondence to the believed sequence of these genes from centromere outwards (Zielke *et al.*, 1976). However, the four enzymes involved, phosphoglycerate kinase, α-galactosidase, hypoxanthine-guanine phosphoribosyltransferase and G6PDH, could only be resolved into two groups according to the timing of increase, because of the difficulty in establishing the precise timing of increases. Indeed this difficulty exists in most cell cycle studies with higher eukaryotes, due to imperfections in synchrony and in some cases lack of homogeneity in the material. In some cases, e.g. SDH and TdRK in sycamore, the timing of increase in activity (relative to cytokinesis or the start of the S-phase) seems variable between different batches of cultures (King *et al.*, 1974). It might be pertinent to ask what significance, if any, should be attached to the timing of changes in the synthesis/degradation balance, if gene dosage and linear transcription are excluded as causative factors.

B. PROGRAMMED DEVELOPMENT

One suggestion is that changes in levels of particular enzymes are part of a programmed developmental sequence which must occur to allow progression through the cycle, ultimately to cell division (Mitchison 1974; Brown, 1976), i.e. that each change determines the next stage in cell development. In some instances this may be partly valid. The frequent increases in enzymes of DNA metabolism at the start of S-phase, or slightly before, may conceivably be necessary to allow the progression of DNA replication. In *Lilium* microspores, TdRK appears shortly (24 h) before DNA synthesis commences, and only at that time (Hotta and Stern, 1963). There is evidence that, in its turn, TdRK is induced by thymidine released at that time by breakdown of polynucleotides in the surrounding tissue, and that the inducibility is itself transient. However, in other cases the increases in DNA-related enzymes seems to accompany or slightly follow the commencement of DNA replication (again with the caveat that the precise timing of such events is often difficult to determine) and it has been argued that such increases are unlikely to be significant controlling factors, at least within the cycle in which they occur (Aitchison and Yeoman, 1974). TdRK in *Lilium* microspores is also unusual in that it increases from an undetectable level, and returns to such a level in a brief interval. In most cases in plants, and probably all cases irrespective of source in which patterns have been followed through more than one cycle (the first cycle sometimes being atypical) a detectable level of activity persists throughout the cycle, and the increase whether at the plateau of a step, or at the top of a peak, amounts to a 0·8–3-fold increase from minimum levels (an exact doubling would be a feature only of a cyclical system with 100% cell division and no turnover which are not typical features of eukaryote systems). It is much easier to envisage a markedly peaked transient appearance of enzymes, such as is the case of TdRK in *Lilium* microspores, as having a temporal regulatory role, than the more normal approximate doubling. It is difficult to ascribe a regulatory role to a doubling of level of many of the enzymes of general metabolism which have been shown to increase during cell cycles. Certainly (enzymes of DNA metabolism apart) none of the work on periodic changes in enzymes in this context has given any indications of what such regulatory roles might be. Furthermore, while periodic changes do occur, the evidence for a precise temporal relationship between changes in the synthesis/degradation balance and particular stages in the cell cycle is by no means as clear from the examples presented so far from plants as it is from say bacteria. There is a need for a system with rather better synchrony than is available at present to establish this relationship. In cases in plants where a regulatory role can be ascribed to a change in enzyme levels in a (non-synchronous) tissue, the magnitude of the change is often very great. We may be justified, therefore, in asking whether the cell cycle associated periodic changes in enzyme levels that have been observed are not of any regulatory significance, but are merely a consequence of the division orientated growth of the cell populations used.

C. GENERAL EFFECTS OF PERTURBATION

Two factors merit consideration in this context. Firstly, in order to be able to monitor changes in enzyme levels at the cellular level, synchronous systems must be used to provide the amplification necessary to bring enzymes levels up to within a measurable range. But (non-natural) synchronous systems have the disadvantage that they are usually perturbed. The perturbing influence is exogenous and may act by overriding the normal cell regulatory processes, in such a way, as to ensure a temporal uniformity of response from the cell population. Neither the *Acer* suspension system, which is synchronized after a dilution shock, or the artichoke explant system, in which division is induced in a quiescent tissue by exogenous supply of auxin (in a presumably auxin-deficient tissue, dependent on auxin for division), is free from this criticism. Even the isolation of synchronous cultures by selection procedures may be suspect, as stepped changes in certain enzymes have been found in *Schizosaccharomyces pombe* cells centrifuged on a sucrose density gradient, as for selection of a synchronous population, but in which the gradients were shaken up before removal of cells (Mitchison, 1977). Such changes could have been ascribed to periodic cell-cycle dependent changes in a supposedly synchronous population, had the gradients not been mixed.

The second consideration is that cell division itself is a major perturbing influence on normal regulatory processes within a cell, and it is possible that many of the periodic changes detected in a dividing population of cells occur only in response to this perturbation. If we consider the cases for periodicity reflecting linear transcription or programmed cell development as unproven, there seems no reason why a continuous availability of the genome for transcription, with actual rates of transcription of particular genes determined by some feedback regulatory mechanism, not directly related to the cell cycle, which presumably operates in non-dividing cells, should not operate, at least for the majority of enzymes of general metabolism. In such a situation the focus of attention should not be on what determines a periodic change in the synthesis/degradation balance in favour of ac-cumulation ("switch on") but on what prevents a continuous accumulation through the cycle. We suggest that an interference with the normal regulatory system, through changes in precursors, effector supplies, compartmentation or other means, caused by changes associated with impending cell division, might be responsible. As regulatory mechanisms could vary for different enzymes, the effect could be felt at different times in the cycle, giving the appearance of a controlled periodic effect, which in reality is the opposite, an interference with regulatory systems. An adjustment in enzyme levels to allow for the effects of such an interference would be made by the normal feedback mechanisms as soon as the restraint imposed by "cell division factors" is removed, either in the next cycle, or possibly in the same cycle, for enzymes showing two periods of accumulation in each cycle.

It is interesting that it has been noted that synchronous cultures of yeast (Folkes *et al.*, 1975) and *Chlorella* (Molloy and Schmidt, 1970) can be manipulated to produce either periodic or continuous accumulation of enzyme activity during the cell cycle, suggesting that even where they may occur, periodic events are not an obligate feature of the cell cycle.

VI. Conclusion

One important question remains unanswered, are the patterns observed using synchronous system a reflection of the method chosen to achieve synchronization or are they characteristic of dividing cells in intact plants? We would suggest that this question will only be answered when techniques are developed to enable measurements of enzyme levels to be made in indi-vidual cells of a meristem, assuming of course that the position of the cell in the cell cycle can be defined at the time the measurement is made. Here, the technique of immunofluorescence would appear to be most promising and this will be included in our future approach.

ACKNOWLEDGEMENTS

The authors wish to thank the Science Research Council for a grant in support of this research. We also extend our thanks to Mr W. Foster, Mrs E. Mills and Mrs E. Raeburn for their skilled technical and secretarial assistance.

REFERENCES

Aitchison, P. A. and Yeoman, M. M. (1973). *J. exp. Bot.* **24**, 1069.
Aitchison, P. A. and Yeoman, M. M. (1974). *In* "Cell Cycle Controls" (Padilla, G. M., Zimmerman, A. M. and Cameron, I. L., eds), pp. 251–263. Academic Press, New York and London.
Aitchison, P. A., Aitchison, J. M. and Yeoman, M. M. (1976). *Biochim. biophys. Acta* **451**, 393.
Bosmann, H. B. (1972). *J. Cell Sci.* **10**, 152.
Brent, T. P., Butler, J. A. V. and Crathorn, A. R. (1965). *Nature, Lond.* **207**, 176.
Brown, R. (1976). *In* "Cell Division in Higher Plants" (Yeoman, M. M., ed.), pp. 3–46. Academic Press, London and New York.
Brown, R. and Dyer, A. F. (1972). *In* "Plant Physiology" (Steward, F. C., ed.), Vol. 6C, pp. 49–90. Academic Press, New York and London.
Cardiff, I. D. (1905). *Bot. Gaz.* **39**, 340.
Chamberlain, C. J. (1935). *In* "Gymnosperms, Structure and Evolution" University of Chicago Press, Chicago.
Clowes, F. A. L. (1965). *J. exp. Bot.* **16**, 581.
Constabel, F., Kurz, W. G. W., Chatson, B. and Gamborg, O. L. (1974). *Expl. Cell Res.* **85**, 105.
Davidson, A. W. and Yeoman, M. M. (1974). *Ann. Bot.* **38**, 545.
Eames, A. J. (1936). *In* "Morphology of Vascular Plants, Lower Groups" McGraw-Hill, New York.
Erickson, R. O. (1964). *In* "Synchrony in Cell Division and Growth" (Zeuthen, E., ed.), pp. 11–37. Wiley Interscience, New York.
Eriksson, T. (1966). *Physiologia Pl.* **19**, 900.
Folkes, B. F., Bishop, R., Box, V., Hinde, R. W., Sims, A. P. and Walls, D. (1975). 3rd European Workshop on the Cell Cycle (Abstract).
Fraser, R. S. S. and Loening, U. E. (1974). *J. exp. Bot.* **25**, 847.
Harland, J., Jackson, J. F. and Yeoman, M. M. (1973). *J. Cell Sci.* **13**, 121.
Hotta, Y. and Stern, H. (1961). *J. biophys. biochem. Cytol.* **9**, 279.
Hotta, Y. and Stern, H. (1963). *Proc. natn. Acad. Sci. U.S.A.* **49**, 648.
Hotta, Y. and Stern, H. (1965). *J. Cell Biol.* **25**, 99.
Huskins, C. L. and Smith, S. G. (1935). *Ann. Bot.* **49**, 119.
Jakob, K. M. and Trosko, J. E. (1965). *Expl. Cell Res.* **40**, 56.
Jouanneau, J. P. (1971). *Expl. Cell Res.* **67**, 329.
King, P. J. and Street, H. E. (1973). *In* "Plant Tissue and Cell Culture" (Street, H. E., ed.), pp. 269–337. Blackwell Scientific Publications, Oxford.
King, P. J., Mansfield, K. J. and Street, H. E. (1973). *Can. J. Bot.* **51**, 1807.
King, P. J., Cox, B. J., Fowler, M. W. and Street, H. E. (1974). *Planta* **117**, 109.
Kovacs, C. J. and Van't Hof, J. (1970). *J. Cell Biol.* **47**, 536.
Lyon, F. M. (1901). *Bot. Gaz.* **32**, 124–141; 170–194.
Macleod, A. J. (1976). Unpublished observations.
Maheshwari, P. (1950). *In* "An Introduction to the Embryology of Angiosperms" McGraw-Hill, New York.

Masters, M. and Pardee, A. B. (1965). *Proc. natn. Acad. Sci. U.S.A.* **54**, 64.
Mattingly, E. (1966). *In* "Cell Synchrony: Studies in Biosynthetic Regulation" (Cameron, I. L. and Padilla, G. M., eds), pp. 256–268. Academic Press, New York and London.
Mitchell, J. P. (1967). *Ann. Bot.* **31**, 427.
Mitchell, J. P. (1968). *Ann. Bot.* **32**, 315.
Mitchell, J. P. (1969). *Ann. Bot.* **33**, 25.
Mitchison, J. M. (1971). "The Biology of the Cell Cycle". Cambridge University Press, Cambridge.
Mitchison, J. M. (1974). *In* "Cell Cycle Controls" (Padilla, G. M., Zimmerman, A. M. and Cameron, I. L., eds), pp. 125–142. Academic Press, New York and London.
Mitchison, J. M. (1977). *In* "Cell Differentiation in Micro-organisms, Higher Plants and Animals" Proceedings of Deutsche Akademie der Naturforschung Leopoldina. Gustav Fischer, Jena.
Mitchison, J. M. and Creanor, J. (1969). *J. Cell Sci.* **5**, 373.
Molloy, G. R. and Schmidt, R. R. (1970). *Biochem. biophys. Res. Commun.* **40**, 1125.
Nagata, T. and Takebe, I. (1970). *Planta* **92**, 301.
Nevins, B. I. (1932). *Cellule Rec. cytol. histol.* **41**, 293.
Okamura, S., Miyasaka, K. and Nishi, A. (1973). *Expl. Cell Res.* **78**, 467.
Oleinick, N. L. (1972). *Radiat. Res.* **51**, 638.
Peaud-Lenöel, C. and Jouanneau, J. P. (1971). *In* "Les Cultures de Tissus de Plantes" Vol. 193, pp. 95–102. Coll. int. CNRS, Paris.
Roberts, K. and Northcote, D. H. (1970). *J. Cell Sci.* **6**, 299.
Schmidt, R. R. (1974). *In* "Cell Cycle Controls" (Padilla, G. M., Zimmerman, A. M. and Cameron, I. L., eds), pp. 201–233. Academic Press, New York and London.
Sitz, T. O., Molloy, G. R. and Schmidt, R. R. (1973). *Biochim. biophys. Acta* **319**, 103.
Sparrow, A. H., Moses, M. J. and Steele, R. (1952). *Br. J. Radiobiol.* **25**, 182.
Stern, H. (1961). *J. biophys. Biochem. Cytol.* **9**, 271.
Stern, H. (1966). *A. Rev. Pl. Physiol.* **17**, 345.
Wilson, M. (1909). *Ann. Bot.* **23**, 141.
Wilson, S. B., King, P. J. and Street, H. E. (1971). *J. exp. Bot.* **22**, 177.
Yeoman, M. M. (1970). *Int. Rev. Cytol.* **29**, 383.
Yeoman, M. M. and Aitchison, P. A. (1973). *In* "Cell Cycle in Development and Differentiation" Br. Soc. devl Biol. Symp. 1 (Balls, M. and Billett, F. S., eds), pp. 185–201. Cambridge University Press, Cambridge.
Yeoman, M. M. and Aitchison, P. A. (1976). *In* "Cell Division in Higher Plants" (Yeoman, M. M., ed), pp. 111–133. Academic Press, London and New York.
Yeoman, M. M. and Davidson, A. W. (1971). *Ann. Bot.* **35**, 1081.
Yeoman, M. M. and Evans, P. K. (1967). *Ann. Bot.* **31**, 323.
Yeoman, M. M., Evans, P. K. and Naik, G. G. (1966). *Nature, Lond.* **209**, 1115.
Yeoman, M. M., Naik, G. G. and Robertson, A. I. (1968). *Ann. Bot.* **32**, 301.
Yonezawa, Y. and Tanaka, R. (1973). *Bot. Mag., Tokyo* **86**, 63.
Zielke, H. R. and Filner, P. (1971). *J. biol. Chem.* **246**, 1772.
Zielke, H. R., Hong, S-C. and Littlefield, J. W. (1976). *Expl Cell Res* **97**, 426.

CHAPTER 5

Hormonal Control of Enzyme Activity in Higher Plants*

J. E. VARNER AND D. T-H. HO

*Biology Department, Washington University,
St Louis, Missouri, U.S.A.*

I. Introduction 83
II. The Cereal Grain Aleurone Layer Response to Gibberellins, Abscisic
Acid and Ethylene 84
III. The Soybean Hypocotyl Response to Auxin and Cytokinins . . 88
IV. The Pea Seedling Response to Auxin 88
V. Response of *Rhoeo* Leaf Sections to Auxin and Abscisic Acid . . 89
VI. The Oat Internode Response to Gibberellins 89
VII. Cytokinin-dependent Increases in Nitrate Reductase in *Agrostemma*
githago Embryos 90
References 91

I. INTRODUCTION

There are now hundreds of known instances in which a change in tissue concentration of a hormone and/or the exogenous addition of a hormone specifically changes the level of activity of one or more enzymes. In no case is it known precisely how the tissue controls the level of enzyme activity in response to the changed hormone concentration. Nonetheless it is instructive to review the best known of these responses as we try to develop concepts that will allow us to understand them.

We shall review the cereal grain aleurone layer responses to gibberellins, abscisic acid and ethylene; the soybean hypocotyl responses to auxin and cytokinins; the pea seedling response to auxin and ethylene; the *Rhoeo* leaf section responses to auxins and abscisic acid; the oat internode responses to gibberellins, and the *Agrostemma* response to cytokinins.

* This work was supported in part by a grant from the National Science Foundation (GB-39944).

II. The Cereal Grain Aleurone Layer Response to Gibberellins, Abscisic Acid and Ethylene

Because it is an easily isolated tissue in which all cells respond to added gibberellins, abscisic acid and ethylene, the aleurone layer is frequently used to study hormone action. Apparently the response of the isolated tissue to added gibberellins is identical to or at least similar to its response *in situ* in the seedling to gibberellins synthesized by the embryo.

The barley aleurone layer synthesizes and secretes, in response to added gibberellins, α-amylase, protease (see Yomo and Varner, 1971, for a review), ribonuclease, β-glucanase (Bennett and Chrispeels, 1972), α-glucosidase, and limit dextrinase (Hardie, 1975). Part of the secreted ribonuclease and β-glucanase is present in the aleurone layers before the addition of gibberellins (Bennett and Chrispeels, 1972), as is part of the α-glucosidase (Clutterbuck and Briggs, 1973).

This response is specifically evoked by gibberellins. The aleurone cells respond to GA_1, GA_3, GA_4, GA_7, but not to GA_8 (Clutterbuck and Briggs, 1973). Gibberellic acid, GA_3, has been used in most investigations. The response to gibberellins is prevented by the presence of abscisic acid and permitted if ethylene is present along with abscisic acid (Jacobsen, 1973). In contrast to many reports that α-amylase is produced in response to a variety of compounds, Clutterbuck and Briggs (1973) found no α-amylase produced by aleurone tissue in response to kinetin, benzylaminopurine, hydroxylamine, glutamine, ornithine, ent-kaurene, ent-kaurenol, ent-kaurenoic acid, phorone, isophorone, phenobarbitone, ATP, ADP, or cyclic-3',5'-AMP. High concentrations (80 mg/l) of helminthosporol and of helminthosporic acid evoke the production of only a fraction of the α-amylase that is produced in response to gibberellins. Nonetheless it appears that studies with helminthosporin analogues will be useful in determining the minimum structure required to evoke a response to gibberellin sensitive cells. In particular, it has been shown that the carbonyl group must be present in a certain spatial relationship but the hydroxymethyl group is not required (Coombe *et al.*, 1974).

Acid phosphatase activity in the aleurone layers increases in the absence of added gibberellins and increases somewhat more with added GA_3 (Ashford and Jacobsen, 1974). There is some acid phosphatase localized in the walls of the aleurone cells of the dry grain and further secretion of acid phosphatase into the walls occurs during imbibition and incubation in the absence of gibberellins. Release of this acid phosphatase is dependent upon the cells' production of cell wall-degrading enzymes in response to added gibberellins (Briggs, 1963; Ashford and Jacobsen, 1974). The release from the cell wall of other enzymes is likewise dependent upon cell wall degradation (Ashford and Jacobsen, 1974). Because the aleurone cell walls consist almost entirely of arabinoxylan (McNeil *et al.*, 1975) it is likely that gibberellin-dependent increases in pentosanases (Taiz and Honigman, 1976) are largely responsible for cytolysis. In addition, at least for α-amylase, a certain ionic concentration is required for rapid release (Varner and Mense, 1972). Because the only source of these ions is the aleurone cell itself the cell has an opportunity to control the rate of release of already secreted α-amylase by controlling the secretion of its ions. This secretion of ions is also determined by the concentration of added gibberellins (Jones, 1973)—perhaps through the effect of gibberellins on intracellular phosphatase level and rate of phytate hydrolysis (Clutterbuck and Briggs, 1974).

The possibility that gibberellins, abscisic acid and ethylene might change the intracellular localization of ions in aleurone tissue (Pomerantz, 1972) has received little attention. Clearly, such localization would be of great interest because of the many possible effects on metabolism. Perhaps the gibberellin-dependent sensitivity to poisoning by o-phenanthroline and the apparent requirement for exogenous calcium (Goodwin and Carr, 1972) are related to changes in the intracellular localization of iron and calcium ions. Such localization might in turn be related to the changes in phytase activity.

The increases in several enzyme activities that are due to enzyme synthesis may result from 1) the translation of existing mRNAs or 2) the translation of newly transcribed mRNAs with such transcription being gibberellin-dependent. Addition of GA_3 does not cause any marked increase in protein synthesis so there is no reason to expect a dependence on rRNA, 5S RNA, and tRNA synthesis and no evidence of such dependence (Yomo and Varner, 1971; Jacobsen and Zwar, 1974a). There is a gibberellin-dependent increase in the rate of labelling of a poly A-rich mRNA-like fraction (Jacobsen and Zwar, 1974b; Ho and Varner, 1974) and there is a parallel increase in translatable mRNA for α-amylase (Higgins et al., 1976). This increase in translatable mRNA could be due to 1) decreased rate of degradation of mRNA, 2) increased rate of synthesis of mRNA or 3) some kind of processing or activation of an inactive form of mRNA (for example, addition of the blocked and methylated 5-terminal sequence required for efficient translation of some eukaryotic mRNAs (Yang et al., 1976).

In considering these possibilities for the control of amylase mRNA translation it is interesting that further amylase synthesis is prevented when abscisic acid is added 12 hours after the addition of GA_3 (at this time amylase synthesis is not susceptible to inhibition by cordycepin). Thus, abscisic acid added at this time, directly or indirectly, inhibits translation (Ho and Varner, 1976). Cordycepin added with abscisic acid, or within a few hours after abscisic acid, prevents or relieves the abscisic acid inhibition of amylase synthesis. If cordycepin is a specific inhibitor of RNA synthesis the inhibition of amylase translation by abscisic acid must involve the synthesis of a new RNA (Ho and Varner, 1976).

We now suppose that it will not take long to resolve at what point abscisic acid inhibits translation of amylase mRNA and whether amylase mRNA is synthesized or activated after the addition of GA_3. Knowing this, however, may not tell us much about the primary action of GA_3 and of abscisic acid. The earliest observed responses of aleurone tissue to GA_3 are related to phospholipid metabolism and not dependent on protein synthesis. These responses include changes within 30 minutes in the rate of labelling of soluble nucleotides, particularly CTP (Collins et al., 1972), and increases (beginning in a few minutes and prevented if abscisic acid is added at the same time as GA_3) in the activity of phosphorylcholine cytidyl transferase and phosphorylcholine glyceride transferase as measured in cell-free membrane preparations (Johnson and Kende, 1971). These increases in enzyme activity apparently do not require RNA synthesis or protein synthesis because they are not prevented by cordycepin or the presence of amino acid analogues (Ben-Tal and Varner, 1974). Therefore, the explanation for these responses to GA_3 and to GA_3 plus abscisic acid may be quite different from that for the changes in rate of translation of amylase mRNA.

These changes in the cell-free activities of enzymes of phospholipid metabolism do not have a strict parallel in observable in vivo phospholipid metabolism. Although there is a marked GA_3 dependent (and abscisic acid inhibited) increase in the incorporation of ^{14}C-choline (Evins and Varner, 1970) and $^{32}P_i$ (Koehler and Varner, 1973) into phospholipid this does not begin until about four hours after the addition of GA_3. This lack of parallel between the in vitro and in vivo observations plus the failure to observe any net increase in the quantity of phospholipid in response to added gibberellins (Firn and Kende, 1974) raises questions about—but does not eliminate—the attractive idea that the proliferation of rough endoplasmic reticulum observed by electron microscopy is the result of a gibberellin-enhanced phospholipid synthesis.

Gibberellic acid changes the permeability of model membranes (liposomes) composed of plant components, presumably as a result of direct interaction of gibberellic acid with the phospholipid of the membrane (Wood and Paleg, 1972). Although these observations are the basis of a proposal that some or all of the in vivo effects of gibberellic acid might be explained by such interactions (Wood et al., 1974) it is difficult to see how such a model could account for the great specificity of cells for the different gibberellins. Along these lines it has been reported that phospholipids act as ionophores (Tyson et al., 1975).

Addition of ethylene partly removes the inhibition by abscisic acid of the gibberellin-dependent amylase synthesis (Jacobsen, 1973) and ethylene plus GA_3 almost completely prevents the inhibition by abscisic acid. It would be of interest to see whether ethylene would prevent the inhibition by abscisic acid of $^{32}P_i$ incorporation into phospholipid and the inhibition by abscisic acid of sucrose release (Chrispeels et al., 1973). It will also be of great interest to find at what point short-chain fatty acids (particularly C_5 and C_9 fatty acids) inhibit gibberellin-dependent sugar release from barley endosperm (Buller et al., 1976) and whether ethylene affects this inhibition.

Even though it is certain that synthesis of various hydrolases by aleurone layer tissue is dependent upon and specific for certain gibberellins it is not at all certain whether gibberellins play any direct role in the secretion of these hydrolases. In barley, appearance of rough endoplasmic reticulum is dependent upon added gibberellins (Yomo and Varner, 1971) but, in wheat, formation of the endoplasmic reticulum occurs during imbibition of the half-seeds before any addition of gibberellins (Varty and Laidman, 1976). Amylase synthesis in wheat, as in barley, does not occur in the absence of added gibberellins. Immunohistochemical localization of α-amylase in barley shows that the enzyme first appears in the rough endoplasmic reticulum associated with the nuclear envelope and the perinuclear region (Jones and Chen, 1976). How amylase proceeds from the rough endoplasmic reticulum to a position beyond the plasmalemma is uncertain. Some evidence suggests that it is released from the endomembrane system into the cytoplasm and moves across the plasmalemma molecule by molecule rather than by way of any sort of packaging (Jones and Chen, 1976). Other evidence indicates that amylase moves from the endoplasmic reticulum to the outside of the plasmalemma by way of secretory organelles (Gibson and Paleg, 1975; Firn, 1975). Whatever the mechanism for the movement of amylase molecules across the plasmalemma such movement requires respiratory energy and has not been shown to be directly under the control of any hormone (Varner and Mense, 1972). Release of amylase from the wall into the medium requires a certain ionic strength and some degree of degradation of the wall. The cell wall degradation apparently occurs only in response to gibberellins (Ashford and Jacobsen, 1974) and begins, and is most extensive, around the plasmodesmata.

As the hydrolases from the aleurone layer act on the starchy endosperm, glucose and maltose accumulate to concentrations as high as 500 milliosmolar (Jones and Armstrong, 1971). The resulting osmotic stress is sufficient to inhibit protein synthesis by causing disaggregation of polysomes (Hsiao, 1970; Armstrong and Jones, 1973). Such osmotic stress is not selective for the gibberellin-dependent protein synthesis (Chrispeels, 1973).

In barley (Jacobsen et al., 1970; Bilderback, 1971; Tanaka and Akazawa, 1970; Daussant et al., 1974; Momotani and Kato, 1967) wheat (Grabar and Daussant, 1964; Daussant and Corvazier, 1970) and rice (Tanaka et al., 1970) several isozymes of α-amylase appear in response to added gibberellins. The physical basis for the different properties of these isozymes is not known (for example are they different gene products or do they reflect post translational modifications?) nor is their physiological function understood (are they produced in different relative amounts in response to different gibberellins or to different growth conditions?). Immunoelectrophoresis techniques seem to offer the simplest way to resolve many of these questions (Daussant et al., 1974).

In spite of the many published papers claiming that cyclic AMP mediates or potentiates the action of gibberellins in the barley endosperm system, we see no evidence that would modify an earlier conclusion that cyclic AMP does not mediate the action of gibberellins (Keates, 1973).

III. THE SOYBEAN HYPOCOTYL RESPONSE TO AUXIN AND CYTOKININS

When auxins, e.g. 2,4-dichlorophenoxyacetic acid (2, 4-D), are applied to etiolated soybean hypocotyls, cells of the mature regions elongate and proliferate while the normal growing site—the apical meristem—becomes quiescent. The effect of auxin on cell elongation in this tissue can be divided into two separate events (Vanderhoef and Stahl, 1975). The first response, which begins 12 minutes after auxin addition, can be mimicked by low pH treatment. The second response, which starts about 30 minutes after auxin addition, is completely inhibited by cytokinin. The first response, which is similar to the fast response of *Avena* coleoptile to auxin, probably does not depend on protein and RNA synthesis. The second response is accompanied by a large increase in RNA, especially ribosomal RNA. This large accumulation of RNA is probably the consequence of an enhanced chromatin-directed RNA synthetic activity. The activity of α-amanitin resistant nucleolar RNA polymerase, RNA polymerase I, is enhanced five to eight-fold after 2,4-D treatment (Guilfoyle *et al.*, 1975). There is no significant effect of 2,4-D on the α-amanitin sensitive RNA polymerase II. Regarding the mechanism of this increase in RNA polymerase I activity, it is still not known whether the activity increase is due to activation or to an increase in the number of enzyme molecules. The auxin enhanced RNA polymerase I activity, solubilized from lentil roots, is the result of increased specific activity of this enzyme (Teissere *et al.*, 1975). Because RNA polymerase activity is subject to regulation by several cofactors, these observations suggest that RNA polymerase I is activated in auxin treated tissue. A lack of hormone-enhanced enzyme synthesis as measured by the incorporation of radioactive or stable isotope-labelled amino acids would provide an unequivocal answer to this problem.

Although cytokinin is able to inhibit cell elongation and RNA synthesis, the possible effect of cytokinin on RNA polymerase I activity has not been checked (Vanderhoef and Key, 1968).

IV. THE PEA SEEDLING RESPONSE TO AUXIN

When an auxin is applied to the cut surface of a pea seedling decapitated immediately below the apical hook, knob-like swellings develop that are similar to those produced in intact seedlings by treatment with ethylene (Ridge and Osborne, 1969). The auxin-induced swelling is accompanied by a many-fold increase in cellulase activity (Fan and Maclachlan, 1966) that seems to be a direct effect of the auxin rather than an effect mediated by auxin-enhanced ethylene production (Ridge and Osborne, 1969).

The increased cellulase activity is due to parallel increases in two distinctly different cellulases (Byrne *et al.*, 1975). The increases are most probably due to synthesis rather than to activation. The polyA RNA isolated from the polysomes of auxin-treated pea epicotyls when translated in the wheat embryo cell-free system yielded an antigen identical to one of the cellulases (Verma *et al.*, 1975).

The mRNA for this cellulase is located on membrane-bound polysomes and is not detectable before treatment of the seedlings with auxin (Verma *et al.*, 1975). The increase in translatable mRNA for cellulase begins immediately after the auxin treatment while the increase in cellulase activity is not detected until after 24 hours. There may be therefore, a control in this system at the translation level as well as at the transcription level (Verma *et al.*, 1975).

At the moment the pea epicotyl-auxin-cellulase system and the aleurone-gibberellin-amylase system seem to be in the best position for further study of how hormones control enzyme levels.

V. RESPONSE OF *RHOEO* LEAF SECTIONS TO AUXIN AND ABSCISIC ACID

The senescence of *Rhoeo* leaf sections maintained in darkness is associated with degradation of protein and marked increases in ribonuclease and acid phosphatase (DeLeo and Sacher, 1970). This increase in hydrolases is of general interest because it is characteristic of many detached leaves and of ripening fruit and of especial interest to us because the increase is prevented by auxins (and kinetin) and promoted by abscisic acid. The increases are prevented by inhibition of RNA synthesis and of protein synthesis (DeLeo and Sacher, 1970). The increase in ribonuclease activity in *Rhoeo* leaf sections appears to be due to *de novo* synthesis (Sacher and Davies, 1974). This system should therefore be of use in determining precisely how added hormones control the synthesis of specific enzymes.

VI. THE OAT INTERNODE RESPONSE TO GIBBERELLINS

The growth response of *Avena* stem segments to added gibberellins is spectacular and due solely to cell elongation (Kaufman, 1965). It therefore is an attractive system for studying a tissue response to gibberellins.

The gibberellin enhanced growth and cell wall synthesis is observable in one to two hours as is a gibberellin-dependent increase in cell wall plasticity (Adams *et al.*, 1975). Cycloheximide added with gibberellic acid prevents the increase in plasticity and if added six hours after the gibberellic acid immediately prevents any further increase in plasticity (Adams *et al.*, 1975).

We hope that this brief account will encourage physiologists and biochemists to examine this system.

VII. Cytokinin-dependent Increases in Nitrate Reductase in *AGROSTEMMA GITHAGO* Embryos

Cytokinins, in the absence of added nitrate, increase the activity of nitrate reductase (NR) in excised embryos of *A. githago* (Borriss, 1967; Kende *et al.*, 1971). Benzyladenine, at the concentration of 10 μM, increases nitrate reductase maximally three to five-fold over the control (Kende *et al.*, 1971). The increase is detectable 30 to 60 minutes after BA treatment and is therefore one of the fastest biochemical responses to cytokinin so far observed.

NADH-cyt c reductase and FMN-nitrate reductase activities are associated with NADH-nitrate reductase and these activities increase in response to either nitrate or benzyladenine (Dilworth and Kende, 1974a). The physicochemical properties of nitrate-induced and benzyladenine-induced nitrate reductase appear to be identical as checked by sedimentation velocity through a sucrose gradient, differential precipitation with $(NH_4)_2SO_4$, gel filtration, and gel electrophoresis (Dilworth and Kende, 1974a). Employing density-labelling techniques, Kende and Shen (1972) have shown that both benzyladenine and nitrate-induced nitrate reductase have a density shift in the presence of D_2O indicating that the nitrate reductase is *de novo* synthesized. However, the possibility of the existence of a constantly turning over nitrate reductase precursor—or inactive form—has not been completely ruled out.

When excised embryos of *A. githago* are incubated with a mixture of nitrate and benzyladenine, an additive effect on nitrate reductase is observed (Kende *et al.*, 1971). Benzyladenine-mediated nitrate reductase induction is not the consequence of leakage of nitrate from a storage pool to a metabolically active pool (Kende *et al.*, 1971). The induction of nitrate reductase by benzyladenine is not, while the nitrate induction of nitrate reductase is, dependent on the level of endogenous nitrate, suggesting that benzyladenine and nitrate act by different mechanisms to induce nitrate reductase activity. This suggestion is further strengthened by the observation that the induction by benzyladenine is much more susceptible to inhibitors of protein synthesis and RNA synthesis (Dilworth and Kende, 1974a).

Although benzyladenine does not have a significant effect on the degradation of preexisting protein in *Agrostemma* embryos (Kende and Shen, 1972), it is still possible that benzyladenine preferentially slows down the degradation of nitrate reductase. A specific nitrate reductase inhibitor that is apparently a specific protease has been found in maize (Wallace, 1974).

Unlike nitrate, benzyladenine is not able to increase nitrite reductase activity in *Agrostemma* (Dilworth and Kende, 1974b).

REFERENCES

Adams, P. A., Montague, M. J., Tepfer, M., Rayle, D. L., Ikuma, H. and Kaufman, P. B. (1975). *Pl. Physiol.* **56**, 757.
Armstrong, J. E. and Jones, R. L. (1973). *J. Cell. Biol.* **59**, 444.
Ashford, A. E. and Jacobsen, J. V. (1974). *Planta* **120**, 81.
Bennett, P. A. and Chrispeels, M. J. (1972). *Pl. Physiol.* **49**, 445.
Ben-Tal, Y. and Varner, J. E. (1974). *Pl. Physiol.* **54**, 813.
Bilderback, D. E. (1971). *Pl. Physiol.* **48**, 331.
Borriss, H. (1967). *Wiss. Z. Univ. Rostock Math. naturw. Reihe.* **16**, 629.
Briggs, D. E. (1963). *J. inst. Brew.* **69**, 13.
Buller, D. C., Parker, W. and Reid, J. S. G. (1976). *Nature* **260**, 169.
Byrne, H., Christou, N. V., Verma, D. P. S. and Maclachlan, G. A. (1975). *J. biol. Chem.* **250**, 1012.
Chrispeels, M. J. (1973). *Biochem. biophys. Res. Commun.* **53**, 99.
Chrispeels, M. J., Tenner, A. J. and Johnson, K. D. (1973). *Planta* **113**, 35.
Clutterbuck, V. J. and Briggs, D. E. (1973). *Phytochemistry* **12**, 537.
Clutterbuck, V. J. and Briggs, D. E. (1974). *Phytochemistry* **13**, 45.
Collins, G. G., Jenner, C. F. and Paleg, L. G. (1972). *Pl. Physiol.* **49**, 398.
Coombe, B. G., Mander, L. N., Paleg, L. G. and Turner, J. V. (1974). *Aust. J. Pl. Physiol.* **1**, 473.
Daussant, J. and Corvazier, P. (1970). *FEBS Letts.* **7**, 191.
Daussant, J., Skakoun, A. and Niku-Paavola, M. L. (1974). *J. inst. Brew.* **80**, 55.
DeLeo, P. and Sacher, J. A. (1970). *Pl. Physiol.* **46**, 806.
Dilworth, M. F. and Kende, H. (1974a). *Pl. Physiol.* **54**, 821.
Dilworth, M. F. and Kende, H. (1974b). *Pl. Physiol.* **54**, 826.
Evins, W. H. and Varner, J. E. (1970). *Proc. natn. Acad. Sci. U.S.A.* **68**, 1631.
Fan, D. F. and Maclachlan, G. A. (1966). *Can. J. Bot.* **44**, 1025.
Firn, R. D. (1975). *Planta* **125**, 227.
Firn, R. D. and Kende, H. (1974). *Pl. Physiol.* **54**, 911.
Gibson, R. A. and Paleg, L. G. (1975). *Aust. J. Pl. Physiol.* **2**, 41.
Goodwin, P. B. and Carr, D. J. (1972). *J. exp. Bot.* **23**, 8.
Grabar, P. and Daussant, J. (1964). *Cereal Chem.* **41**, 523.
Guilfoyle, T. J., Lin, C. Y., Chen, Y. M., Nagao, R. T. and Key, J. L. (1975). *Proc. natn. Acad. Sci. U.S.A.* **72**, 69.
Hardie, D. G. (1975). *Phytochemistry* **14**, 1719.
Higgins, T. J. V., Zwar, J. A. and Jacobsen, J. V. (1976). *Nature* **260**, 166.
Ho, D. H. and Varner, J. E. (1974). *Proc. natn. Acad. Sci. U.S.A.* **71**, 4783.
Ho, D. H. and Varner, J. E. (1976). *Pl. Physiol.* **57**, 175.
Hsiao, T. C. (1970). *Pl. Physiol.* **46**, 281.
Jacobsen, J. V. (1973). *Pl. Physiol.* **51**, 198.
Jacobsen, J. V. and Zwar, J. A. (1974a). *Aust. J. Pl. Physiol.* **1**, 343.
Jacobsen, J. V. and Zwar, J. A. (1974b). *Proc. natn. Acad. Sci. U.S.A.* **71**, 3290.
Jacobsen, J. V., Scandalios, J. G. and Varner, J. E. (1970). *Pl. Physiol.* **45**, 367.
Johnson, K. D. and Kende, H. A. (1971). *Proc. natn. Acad. Sci. U.S.A.* **68**, 2674.
Jones, R. L. (1973). *Pl. Physiol.* **52**, 303.
Jones, R. L. and Armstrong, J. E. (1971). *Pl. Physiol.* **48**, 137.
Jones, R. L. and Chen, R. F. (1976). *J. Cell. Sci.* **20**, 183.
Kaufman, P. B. (1965). *Physiol. Pl.* **18**, 703.
Keates, R. A. B. (1973). *Nature* **244**, 355.
Kende, H. and Shen, T. C. (1972). *Biochim. biophys. Acta* **286**, 118.
Kende, H., Hahn, H. and Kays, E. (1971). *Pl. Physiol.* **48**, 702.

Koehler, D. E. and Varner, J. E. (1973). *Pl. Physiol.* **52**, 208.

McNeil, M., Albersheim, P., Taiz, L. and Jones, R. L. (1975). *Pl. Physiol.* **55**, 64.

Momotani, Y. and Kato, J. (1967). *Pl. Cell Physiol.* **8**, 439.

Pomerantz, Y. (1972). *Cereal Chem.* **49**, 5.

Ridge, I. and Osborne, D. J. (1969). *Nature* **223**, 318.

Sacher, J. A. and Davies, D. D. (1974). *Pl. Cell Physiol.* **15**, 157.

Taiz, L. and Honigman, W. A. (1976). *Pl. Physiol.* **58**, 380.

Tanaka, Y. and Akazawa, T. (1970). *Pl. Physiol.* **46**, 586.

Tanaka, Y., Ito, T. and Akazawa, T. (1970). *Pl. Physiol.* **46**, 650.

Teissere, M., Penon, P., Van Huystee, R. B., Azou, Y. and Ricard, J. (1975). *Biochim. biophys. Acta* **402**, 391.

Tyson, C. A., Zande, H. V. and Green, D. E. (1975). *J. biol. Chem.* **251**, 1326.

Vanderhoef, L. N. and Key, J. (1968). *Pl. Cell Physiol.* **9**, 343.

Vanderhoef, L. N. and Stahl, C. A. (1975). *Proc. natn. Acad. Sci. U.S.A.* **72**, 1822.

Varner, J. E. and Mense, R. M. (1972). *Pl. Physiol.* **49**, 187.

Varty, K. and Laidman, D. L. (1976). Private communication.

Verma, D. P. S., Maclachlan, G. A., Byrne, H. and Ewings, D. (1975). *J. biol. Chem.* **250**, 1019.

Wallace, W. (1974). *Biochim. biophys. Acta* **341**, 265.

Wood, A. and Paleg, L. G. (1972). *Aust. J. Pl. Physiol.* **1**, 167.

Wood, A., Paleg, L. G. and Spottswood, T. M. (1974). *Aust. J. Pl. Physiol.* **1**, 167.

Yang, N. S., Manning, R. F. and Gage, L. P. (1976). *Cell* **7**, 339.

Yomo, H. and Varner, J. E. (1971). *In* "Current Topics in Developmental Biology" (Moscona, A. A. and Monroy, A., eds), pp. 111–144, Academic Press, New York and London.

CHAPTER 6

The Photocontrol of Gene Expression in Higher Plants

H. SMITH, E. ELLEN BILLETT AND A. B. GILES

University of Nottingham, School of Agriculture, Sutton Bonington, England

I. Introduction .	. . . .	93
A. The Photoregulation of Plant Development .	. . . .	93
B. Sites for the Regulation of Gene Expression .	. . .	95
C. The Perception of Light	. . . .	96
II. General Aspects of the Effects of Light on Enzyme Levels	. .	97
A. The Range of Enzymes Controlled by Light .	. . .	97
B. Techniques	. . . .	100
III. The Photoregulation of Phenylpropanoid Synthesis Enzymes	.	102
A. The Photocontrol of PAL Activity	. . . . .	102
B. The Evidence for Inactive PAL .	. . . . .	106
C. A High Molecular Weight PAL Inhibitor from Gherkins	.	107
D. Summary.	. . . .	110
IV. Regulation by Specific Enzyme Inhibitors .	. . . .	111
V. Photocontrol of Other Enzymes	. . . . . .	111
A. Ascorbic Acid Oxidase in Mustard	. . . .	111
B. Nitrate Reductase in Mustard	. . . .	111
C. Phosphoenolpyruvate Carboxylase in Sugar Cane .	. .	113
D. Lipoxygenase in Mustard .	. . . . . .	113
E. Peroxidase in Squash Membrane Preparations	. .	115
F. Invertase in Radish	. . . .	115
G. Conclusion	. . . .	116
VI. Photocontrol of Transcription and Translation .	. . .	116
A. Technical Problems .	. . . . . .	116
B. Photocontrol of the Pattern of Protein Synthesis .	. .	117
C. Photocontrol of Polyribosome Levels .	. . .	118
D. Photocontrol of Messenger-RNA Levels	. . .	121
E. Does Light Regulate Transcription? .	. . .	123
VII. General Comments	. . . .	124
Acknowledgements	. . . .	124
References .	. . . .	125

I. Introduction

A. The Photoregulation of Plant Development

The development of higher plants is particularly responsive to environmental factors, the most important of these usually being light. The absorption of light by the plant evokes a precise, specific and often profound, alteration in the pattern of development.

In recent years, photomorphogenesis has attracted increasing attention, not only for its own intrinsic interest, but because of its value as a model system for investigations into the mechanisms underlying the regulation of development. Light treatments may be applied and removed without residual effect, in contrast to the application of hormones or other chemicals, and to surgical manipulations. Furthermore, in many cases only very small amounts of light are necessary to initiate a developmental change, allowing for convenient experimentation. Thus, investigations into photomorphogenesis offer an excellent opportunity of gaining insights into the wider problem of the regulation of development in general.

Photomorphogenesis can usefully be divided into three partial processes:

1. the perception of a radiant energy stimulus;
2. the transduction of this stimulus within the cells in some way to modify the internal mechanisms regulating development;
3. the ultimate morphogenic changes themselves, which may be temporally and spatially considerably removed from the initiating environmental stimulus.

This division into three partial processes is convenient, since it serves to concentrate attention on the area of greatest current ignorance. A great deal is now known concerning the perception of photomorphogenically active radiation (i.e. the first partial process), and thousands of light-induced morphogenic changes have been extensively listed and described (i.e. the final partial process). Where information and concepts are relatively lacking is in relation to the intermediate phase, the transduction of the environmental stimulus within the cells and the consequent initiation of the developmental changes. It is with this phase we shall be mainly concerned, working on the principle that light-mediated changes in enzyme levels represent the selective regulation of gene-directed activities and thus are an essential step in the control of development.

Three very important points should be stressed at the outset however. Firstly, in very few cases indeed has it been possible to relate in a causal sense, light-mediated changes in enzyme levels to observed alterations in the pattern of development. Indeed, light often causes changes in enzyme levels and changes in product levels which bear no discernible relationship to each other. Nonetheless, it is tacitly accepted that changes in enzyme levels are representative of developmental changes, even though the precise role of the particular enzymes in development is not yet understood. Secondly, insofar as can be ascertained from the published reports, all cases of the photo-morphogenic control of enzyme levels are quantitative, rather than qualitative, with light bringing about a change in the level of an enzyme activity already present in measurable amounts. This may be indicative of a "modulatory" mode of regulation, rather than an "all-or-nothing" switching mechanism. Finally, the changes in enzyme activities caused by light are all secondary processes, and are not intimately associated with the primary action or actions of the photoreceptor molecules. The secondary nature of most of the known enzyme changes can be deduced from their long lag periods, but even the most rapid enzyme changes reported appear to be too slow to be involved in the primary action of the photoreceptors, which probably takes place within seconds of the perception of the light stimulus.

With these three provisos borne in mind, however, it is still valuable to investigate the mechanisms whereby light causes changes in enzyme levels. Enzymes are directly coded for by the genome, and the regulation of development is widely recognized as involving the selective expression of the genetic information. By determining how light regulates enzyme levels, we can, theoretically at least, determine how gene expression is controlled in photomorphogenesis.

B. SITES FOR THE REGULATION OF GENE EXPRESSION

It is now generally accepted that development involves the spatial and temporal regulation of gene expression. In operational terms, this means the control of those processes which lead to the action of specific enzymes within the cell. In bacteria, the principal site for the regulation of gene expression appears to be the synthesis of messenger-RNA (i.e. transcription). In higher organisms, on the other hand, especially in non-dividing cells, it seems likely that regulation of specific protein synthesis (i.e. translation) may be very important. In addition to these two synthetic processes, enzyme levels may conceivably be regulated by variations in the rates of degradation of the messenger-RNAs or of the proteins *per se*, by the activation of stored messenger RNAs, and by the activation, inactivation and interconversion of existing enzymes.

Thus the potential sites for gene expression are numerous and varied:

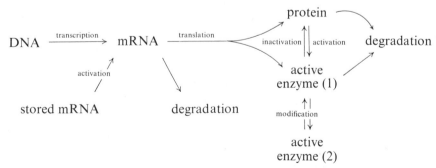

Each of the processes outlined in the above scheme is highly complex, and many different mechanisms for regulation within each level are conceivable.

One of the main problems confronting research workers in this subject is the development of techniques which enable distinctions to be made between the various potential regulatory mechanisms. All that will be attempted in this article, which is not intended to be a comprehensive review, is the description of some evidence allocating regulation of gene expression to the levels of transcription, translation and enzyme activation, inactivation and modification; precise mechanisms will not be discussed.

C. THE PERCEPTION OF LIGHT

Before embarking upon an analysis of some selected light-mediated enzyme changes, it may be useful to outline the nature of the perception of developmentally important light stimuli. There are at least two different photoreceptors responsible for the perception of light and the subsequent modulation of development. One of these, phytochrome, has been intensively studied over the past 30 years or so and a fairly comprehensive picture of its molecular properties and physiology has emerged, although its primary mode of action is still not clear. Phytochrome appears to act in two distinct ways, depending on the duration, wavelength and irradiance of the light treatment, although this may not necessarily reflect different molecular mechanisms of action. The molecule, which is a photochromic protein (molecular weight $c.$ 120 000 daltons) with a bilitriene chromophore, exists in two isomeric forms, photoconvertible according to the following scheme:

$$\text{Pr} \underset{\lambda_{max} = 730 \text{ nm}}{\overset{\lambda_{max} = 66 \text{ nm}}{\rightleftarrows}} \text{Pfr} \longrightarrow \text{Biological action}$$

Etiolated seedlings contain only Pr, and a short treatment (1–10 min) with red light ($\lambda_{max} = 660$ nm) will convert sufficient Pr to Pfr for ultimate biological action to be detected. If, however, a short period of far-red ($\lambda_{max} = 730$ nm) light is given after the red light, the Pfr is transformed to Pr and the biological action is prevented. This low-energy, red/far-red reversible phenomenon is the classical phytochrome-mediated response first shown by Borthwick et al. (1952).

Upon continuous irradiation, on the other hand, both Pr and Pfr absorb (due to their broadly overlapping absorption spectra) and a photodynamic equilibrium is achieved. At equilibrium, the proportions of Pr and Pfr are wavelength-dependent, but irradiance-independent. Hartmann (1966), in a classic series of experiments, has shown that under continuous irradiation, phytochrome is capable of regulating development by the establishment of critical photoequilibria, which for most species tested are achieved with wavelengths between 705 nm and 730 nm. Thus, continuous far-red light leads to biological action similar to that caused by brief pulses of low-energy red light.

The other photoreceptor responsible for effects on development absorbs principally in the blue, where it acts most effectively during continuous irradiation. The identity of this photoreceptor has been a topic of considerable argument, but evidence is now accumulating in favour of a flavoprotein linked to a b-type cytochrome in which electron transfer is light-dependent (see Briggs, 1976 for review). Both phytochrome and the flavoprotein are probably membrane components, or at least act upon membranes to bring about either a specific change in membrane properties, or to regulate the transmembrane transport of one or more critical metabolites.

Although the two photoreceptors responsible for the perception of light in photomorphogenesis are chemically quite different, and presumably have different primary mechanisms of action, their secondary (i.e. metabolic) and tertiary (i.e. morphogenic) effects appear to be very similar. Consequently, in this article, no distinction will be made between investigations of phytochrome and flavoprotein-photoreceptor initiated changes in enzyme levels. Many investigators have used continuous, high irradiance white light in their studies of enzyme changes. Although this is experimentally convenient, it is photomorphogenically very complex, since both phytochrome and the flavoprotein-photoreceptor will be acting and, more importantly, light absorption by the chlorophylls, carotenoids, and possibly protochlorophylls of the photosynthetic machinery will occur. It is usually assumed that light absorption via photosynthesis, although clearly an important factor in determining overall rates of growth, is not involved in the regulation of the pattern of development. Nevertheless, some interesting information has been derived from the effects of white light, and indeed, such treatments more closely approximate the conditions to which plants are exposed in nature.

II. General Aspects of the Effects of Light on Enzyme Levels

A. THE RANGE OF ENZYMES CONTROLLED BY LIGHT

In recent years a very large number of investigations into the effect of light on enzyme activities have been reported. Unfortunately, most of this work has been purely descriptive in nature and very few of these investigations have been pursued to the point where they yield useful information on the mechanisms involved. A list of enzymes whose extractable activities are known to be under photomorphogenic control is given in Table I. As can be seen, a wide range of metabolic activities, including photosynthesis, photorespiration, chlorophyll synthesis, fat degradation, starch degradation, nitrate assimilation, nucleic acid synthesis and degradation and secondary product synthesis may thus be under photocontrol exerted at the level of specific enzyme activities. Table I also gives information on the approximate lag phases which precede the first detectable change in enzyme activity, showing that most of the changes are clearly of a secondary nature. Another

TABLE I

Light-mediated changes in enzyme activity

Enzyme[a]	Plant	Effect[b]	Lag[c]	Reference
General metabolism				
NAD-kinase (in vitro)	Pisum sativum	+	0	Tezuka and Yamamoto (1969)
Lipoxygenase	Sinapis alba	−	0	Oelze-Karow et al. (1970)
Amylase	Sinapis alba	+	~6 h	Drumm et al. (1971)
Ascorbic acid oxidase	Sinapis alba	+	~3 h	Drumm et al. (1972)
Galactosyl transferase	Sinapis alba	+	~12 h	Unser and Masoner (1972)
Glyceraldehyde-3-P DH (NAD$^+$)	Phaseolus vulgaris	+	<12 h	Filner and Klein (1969)
Nitrate reductase	Sinapis alba	+	0	Johnson (1976)
PEP-carboxylase	Kalanchoe blossfeldiana	+	—	Queiroz (1969)
Malic enzyme	Kalanchoe blossfeldiana	+	—	Queiroz (1969)
Malate DH (NAD$^+$)	Kalanchoe blossfeldiana	+	—	Queiroz (1969)
Invertase	Raphanus sativus	+	~24 h	Zouaghi and Rollin (1976)
Nucleic acid and protein metabolism				
RNA polymerase (nuclear)	Pisum sativum	+	4 h	Bottomley (1970)
Ribonuclease	Lupinus alba	+	4 h	Acton (1972)
Amino acid activating enzymes	Pisum sativum	+		Henshall and Goodwin (1964)
Photosynthesis and chlorophyll synthesis				
RUdP-carboxylase	Phaseolus vulgaris	+	<24 h	Filner and Klein (1969)
Transketolase	Secale cereale	+	—	Feirabend and Pirson (1966)
Glyceraldehyde-3-P DH (NADP$^+$)	Phaseolus vulgaris	+	<12 h	Filner and Klein (1969)
Alkaline fructose diphosphatase	Pisum sativum	+	—	Graham et al. (1968)
Ribulose-5-P isomerase	Secale cereale	+	—	Feierabend and Pirson (1966)
Phosphoribulokinase	H ordeum vulgaris	+	—	Keller and Huffaker (1967)
Pyruvate, Pi dikinase	Zea mays	+	—	Graham et al. (1970)
Inorganic pyrophosphatase	Zea mays	+	~2 h	Butler and Bennett (1969)
Adenylate kinase	Zea mays	+	—	Butler and Bennett (1969)

PEP-carboxylase	*Saccharum* hybrid		+	24 h	Goatly and Smith (1974)
Succinyl-CoA synthetase	*Phaseolus vulgaris*	+	~2 h	Steer and Gibbs (1969a)
ALA-dehydratase	*Phaseolus vulgaris*	+	24 h	Steer and Gibbs (1969b)
Peroxisome and glyoxisome enzymes				
Peroxidase (*in vitro*)	*Cucurbita pepo*	−	~2 min	Penel *et al.* (1976)
Peroxidase	*Sinapis alba*	+	72 h	Schopfer and Plachy (1972)
Glycollate oxidase	*Sinapis alba*	+	6 h	van Poucke *et al.* (1969)
Glyoxylate reductase	*Sinapis alba*	+	6 h	van Poucke *et al.* (1969)
Phenylpropanoid metabolism				
Phenylalanine ammonia-lyase	*Pisum sativum*	+	1·5 h	Attridge and Smith (1967)
Phenylalanine ammonia-lyase	*Sinapis alba* (+ many others)	+	1·5 h	Durst and Mohr (1966)
Cinnamate-4-hydroxylase	*Pisum sativum*	+	4 h	Russell and Conn (1967)
p-Coumarate:CoA ligase	*Petroselinum hortense*	+	~3 h	Hahlbrock and Grisebach (1970)
Chalcone-flavanone isomerase	*Petroselinum hortense*	+	~6 h	Hahlbrock *et al.* (1971a)
UDP-apiose synthetase	*Petroselinum hortense*	+	~6 h	Hahlbrock and Wellmann (1970)
Malonyl-transferase	*Petroselinum hortense*	+	~4 h	Hahlbrock *et al.* (1976)
7-*O*-glucosyl-transferase	*Petroselinum hortense*	+	~6 h	Hahlbrock *et al.* (1976)
Apiosyl-transferase	*Petroselinum hortense*	+	~6 h	Hahlbrock *et al.* (1971a)

From Smith (1974) with additions.
[a] DH = dehydrogenase.
[b] + = stimulation, − = inhibition.
[c] Time before detectable change in activity (where determined).

generalization, which is not clear from the table, is that most light-mediated changes in enzyme activity are transient, usually with a rise in activity being followed by a subsequent decline. This is often the case even with a transition from darkness to continuous illumination.

B. TECHNIQUES

The possible processes which can contribute to a change in the extractable activity of an enzyme are enzyme synthesis, enzyme degradation, enzyme activation, enzyme inactivation and enzyme interconversion. Considerable interest is therefore concentrated on techniques which enable enzyme synthesis and turnover to be detected and measured. Aspects of this topic are dealt with in greater detail in this volume, in the articles by Daussant *et al.* (Chapter 10) and Johnson (Chapter 11), but it is of sufficient importance to warrant attention here.

There are basically four techniques which have been employed to test for an effect of light (or other developmental stimulus) on the *de novo* synthesis of an enzyme. These are: 1. the use of inhibitors of protein synthesis; 2. density labelling; 3. radioactive labelling; and 4. immunology.

The use of inhibitors has been justifiably criticized on many grounds, especially those of lack of specificity and the blanket nature of their action. It can now be fairly said that the application of inhibitors to complex multi-cellular systems cannot give reliable information on mechanism, *unless* the inhibitor *does not* inhibit the observed change in enzyme activity. If it can then be shown that the inhibitor, at the concentration used, effectively inhibits protein synthesis within that tissue, then it is valid to conclude that protein synthesis is not necessary for the change in enzyme activity. No useful conclusions can be drawn solely from experiments in which the inhibitor successfully inhibits the enzyme change.

Density labelling, a technique developed by Hu *et al.* (1962) for proteins and pioneered in plants by Filner and Varner (1967) has in recent years been considerably refined, so that it is now in principle possible to estimate rates of synthesis and turnover of specific enzymes (see Johnson, Chapter 11, for a full treatment). The major advantage of density labelling is that purification of the enzyme is not required, as it is for radiolabelling and for immunology. There are serious drawbacks, however; the incorporation of large amounts of density label, particularly ^2H_2O, has deleterious effects on the metabolism of the plants; the procedure is tedious and time consuming; and interpretations of the data can be very complex. Nevertheless the technique has been of considerable value in analysing some light-mediated changes in enzyme levels.

Radiolabelling and immunology both require the purification of the enzyme to a single homogeneous protein, the latter technique requiring the preparation of relatively large quantities of protein. Consequently, the range of enzymes which are at present amenable to these methods is rather small, and even with those which can be purified, it is a substantial endeavour. On the other hand, the interpretation of the results of these techniques is much clearer than with density labelling. Both radiolabelling (Hahlbrock and Schröder, 1975; also see below) and immunology (See Daussant *et al.*, Chapter 10) have recently been used with success in studies of the photomorphogenic regulation of enzyme levels.

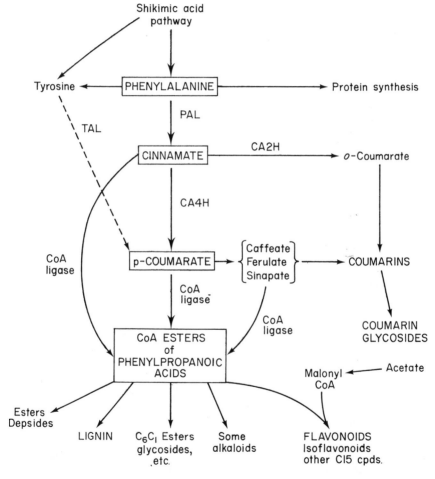

FIG. 1. The biosynthetic pathway of secondary products derived from phenylpropanoids. The intermediates in boxes represent branch points and thus potential regulatory sites. (PAL = phenylalanine ammonia-lyase; TAL = tyrosine ammonia-lyase; CA2H = cinnamic acid 2-hydroxylase; CA4H = cinnamic acid 4-hydroxylase; CoA ligase = aryl acid Coenzyme A ligases.)

III. The Photoregulation of Phenylpropanoid Synthesis Enzymes

One of the earliest biochemical observations relating to photomorphogenesis was that light treatment often caused the accumulation of large
amounts of anthocyanin (Mohr, 1957; Siegelman and Hendricks, 1957,
1958). In other plants, light was shown to increase the synthesis of related
substances, such as phenylpropanoids (Zucker, 1963; Engelsma and Meijer,
1965) and other flavonoids (Furuya et al., 1962). The pathway of phenylpropanoid biosynthesis is now well worked out and a central sequence from
phenylalanine through cinnamate and p-coumarate to Coenzyme A esters
of the phenylpropanoic acids, can be discerned (Fig. 1). These central intermediates all appear to be branch points for various other pathways of
secondary metabolism, and thus are prime candidates for regulatory steps.
The three enzymes involved in this central sequence are phenylalanine
ammonia-lyase (PAL), cinnamic acid 4-hydroxylase (CA4H), and p-
coumaroyl: coenzyme A ligase.

Table I includes those enzymes in the phenylpropanoid and flavonoid
biosynthetic pathways whose activities are known to be affected by light.
As can be seen, the three central enzymes, in all cases reported, are increased
in activity by light. Some other enzymes, e.g. in the shikimic acid pathway,
and in the malonyl: coenzyme A pathway, are not so affected.

In many, but not all cases, PAL, CA4H and the ligase respond to light
treatment in an apparently coordinated manner. This has been shown most
powerfully by Hahlbrock and co-workers using cell suspension cultures of
parsley (Hahlbrock et al., 1976). Other similar reports exist, and indeed
treatments other than light often cause coordinated changes in the three
enzymes (see Table II); the mechanisms underlying this coordination are
unknown.

A. THE PHOTOCONTROL OF PAL ACTIVITY

Of the three central enzymes, only PAL has been studied at all extensively,
and even in this case, very few investigations have proceeded beyond the
purely descriptive; nevertheless, some valuable conclusions can be made.

PAL has been found in a wide variety of plants (Camm and Towers, 1973a),
and has been extensively purified from potatoes, maize and *Rhodotorula
glutinus* (Havir and Hanson, 1975) and wheat (Nari *et al.*, 1972) where it has
a molecular weight of 320 000 daltons. It appears to have regulatory properties as evidenced by unusual kinetic behaviour (Iredale and Smith, 1974) and
end product inhibition (Attridge *et al.*, 1971). In most etiolated plants studied,
light treatment causes an increase in PAL activity which is usually transient
and occurs in three phases: a lag phase usually about 90 min; a phase of
increase in activity (3–20 h); and a phase of decrease in activity (3–20 h). The
photoreceptors involved vary with different plants.

TABLE II

Co-ordinate regulation of phenylalanine ammonia-lyase, (PAL), cinnamic acid 4-hydroxylase (CAH) and p-coumaroyl:CoA ligase

System	Stimulus	Comments	References
Parsley cell suspension cultures	Illumination	Central pathway enzymes i.e. PAL, CAH, p-coumaroyl:CoA ligase appear with similar kinetics. Branch pathway enzymes appear subsequently in a coordinated manner.	Hahlbrock et al. (1971b)
Parsley cell suspension cultures	Transfer to fresh culture medium	PAL, p-coumaroyl:CoA ligase appear with similar kinetics. CAH increases but with a different time-course.	Hahlbrock and Wellman (1973)
Soybean cell suspension cultures	Age of culture	PAL, CAH and p-coumaroyl:CoA ligase appear with similar kinetics.	Ebel et al. (1974)
Buckwheat seedlings	Excision	PAL, CAH increase in activity (one time-point).	Amrhein and Zenk (1968)
Buckwheat seedlings	Illumination	PAL, CAH appear with similar kinetics.	Amrhein and Zenk (1970)
Potato tubers	Excision and illumination	PAL, CAH increase on excision but only the appearance of PAL is stimulated by illumination	Camm and Towers (1973b)
Pea seedlings (Plumular hook)	Ethylene	PAL, CAH increase, but exhibit different lag periods and dose-response with respect to ethylene.	Hyodo and Yang (1971)
Sweet potato	Excision	PAL, CAH increase and decrease coordinately.	Tanaka et al. (1974)
Swede root	Excision	PAL, CAH increase coordinately (ligase also increases but time course not given).	Rhodes et al. (1976)

Based on Lamb, 1976, with additions.

Figure 2 shows the typical time course of PAL activity increase in etiolated gherkin hypocotyls treated with continuous blue light, in pea terminal buds treated with brief red light, and in mustard cotyledons treated with continuous far-red light. Figure 3 shows the time-courses of PAL, CA4H and ligase activity increases in parsley suspension cultures irradiated with white light; the coordination is obvious. (In this case the active wavelengths are 320–350 nm in the ultraviolet.)

Several attempts have been made to determine whether or not the increase in PAL is due to increased rates of enzyme synthesis. Many workers have reported inhibition of light-mediated increases in PAL activity by protein

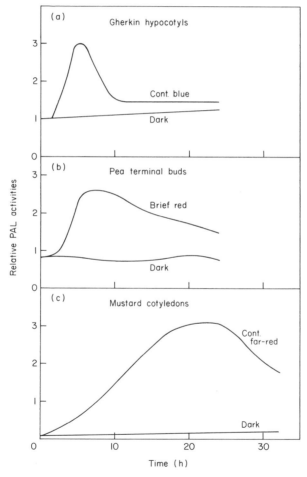

FIG. 2. Time courses of PAL changes in response to light-treatment of etiolated tissues. (a) Effect of blue light on gherkin seedlings (Attridge, unpublished results); (b) effect of brief red light on pea seedlings (from Smith and Attridge, 1970); (c) effect on continuous far-red light on mustard seedlings (from Mohr, 1970).

synthesis inhibitors applied during the lag phase (see Schopfer, 1972). Although these results are consistent with the involvement of enzyme synthesis, they do not prove the case. Zucker (1969) reported the incorporation of radiolabel into PAL in *Xanthium* leaf disks, both in the dark and the light. He also found (Zucker, 1970) that light appeared to inhibit the loss of label from PAL, suggesting that light reduced PAL turnover, rather than increased PAL synthesis. More recently, Hahlbrock and Schröder (1975) have demonstrated light-mediated increased incorporation of ^{35}S-methionine

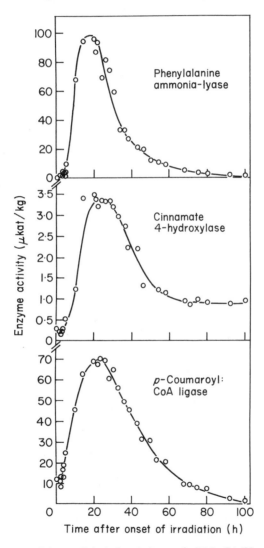

Fig. 3. Time-courses of the u.v.-light-induced changes in PAL, CA4H and *p*-coumaroyl CoA ligase in cultured parsley cells (from Hahlbrock *et al.*, 1976).

into PAL purified from parsley cells to a single homogeneous protein on SDS-acrylamide gels. This evidence strongly supports the view that light stimulates PAL synthesis at least in some cases, although the possibility still exists that the ^{35}S-methionine was not distributed randomly through the protein and had merely been added to the end of a protein whose synthesis had not been completed. This can only be tested by fingerprint analysis of the purified labelled protein.

Density labelling evidence in gherkin and mustard, on the other hand, indicates that light treatment does not operate through increased rates of PAL synthesis (Attridge and Smith, 1974; Attridge et al., 1974). This work is covered in more detail by Johnson in Chapter 11 but in essence, light treatment led to reduced incorporation of ^{2}H from ^{2}H$_2$O into PAL in mustard and gherkin. These results are not easy to interpret but, taken with the knowledge that accumulation of the end product, cinnamate, represses PAL synthesis in gherkin hypocotyl segments (Johnson et al., 1975), we have made the following suggestion. Light treatment activates pre-existing inactive PAL molecules which were synthesized before ^{2}H$_2$O was present in the cells; the increased PAL leads to temporary accumulation of end products which repress PAL synthesis; the combination of reduced PAL synthesis and the presence of active PAL molecules which are not density labelled, yields a population of PAL molecules with a lower mean buoyant density than that present in stimulated dark controls. Thix is admittedly a complicated hypothesis and has been criticized by Acton and Schopfer (1975); however, these authors failed to take account of possible light-mediated changes in the synthesis of amino acids in their experiments, and thus their data do not support the involvement of increased rates of PAL synthesis as they claim (see Chapter 11 for detailed discussion).

B. THE EVIDENCE FOR INACTIVE PAL

Engelsma (1967), working with gherkin, and Zucker (1968) working with potato, originally showed that inhibitors of protein synthesis, if applied at the time at which the light-mediated increase in PAL activity had reached its maximum, prevented the subsequent fall in activity. Zucker (1968) interpreted this as evidence for the formation of a lyase-inactivating system which was presumably proteinaceous in nature. Subsequently, Attridge and Smith (1973) showed that treatment of dark-grown gherkin seedlings with 100–500 μg/ml cycloheximide in an aerosol spray led to a massive increase in extractable PAL which began within 40 minutes of the application of the inhibitor. The response to cycloheximide treatment was up to four times greater than the maximum achieved with blue light. Significant increases in enzyme activity occurred with relatively low concentrations of cycloheximide (5–10 μg/ml) and puromycin also led to increases in activity. These data suggested that dark-grown gherkin hypocotyls contained PAL molecules which were held in an inactive state by a factor or process which depended on continual protein synthesis for its action. Judging by the rapidity of the response to cycloheximide, the inactivating factor has a rapid turnover with a half-life of around 30–90 minutes.

In this Symposium, Daussant *et al.* (Chapter 10) have reported the work of Faye who has obtained immunological evidence for inactive PAL in radish cotyledons. The data suggest that the populations of inactive PAL decrease and those of active PAL increase upon light treatment.

C. A HIGH MOLECULAR WEIGHT PAL INHIBITOR FROM GHERKINS

Englesma and van Bruggen (1971) made a brief, passing, reference to a non-dialysable substance which had been leached from excised gherkin hypocotyls and which prevented the light-mediated increase in PAL when re-applied to hypocotyls. Creasy (1976) has shown the presence of a heat-labile substance in sunflower leaf extracts which irreversibly inactivates PAL.

In our laboratory, we have obtained from gherkin hypocotyls a non-dialysable, heat-labile, substance which reversibly inhibits PAL. In the first report (French and Smith, 1975) the inhibitor was stated to be heat stable; it now seems likely that the inhibitor preparations used initially were contaminated with low molecular weight phenylpropanoids which naturally were not degraded by high temperatures. The properties of the inhibitor as recently determined (Billett and Smith, unpublished work) indicate that it could have a physiological role in regulating PAL activity.

1. *Properties of the Inhibitor*

In the work described here, endogenous PAL inhibitors were extracted from dark-grown hypocotyls in 0·2 M borate buffer pH 8·8 and centrifuged at 30 000 g for twenty minutes—the supernatant was used as the crude extract. After fractionation as outlined below, PAL inhibition was assayed using a PAL preparation extracted from three hour blue-light treated hypocotyls, precipitated with ammonium sulphate, and partially purified by affinity chromatography.

When the crude inhibitor preparation was fractionated on Sephadex G100 two major peaks of PAL inhibition were observed (Fig. 4a). One fraction runs with V_t and presumably contains a mixture of low molecular weight phenylpropanoids which are known to be potent inhibitors of PAL; the other was excluded from the gel. On running a dialysed 80% ammonium sulphate precipitate of the 30 000 g supernatant on Sepharose 4B, one major peak of inhibition was observed, quite clearly separated from PAL and excluded from the gel; a smaller, rather variable, peak of inhibitor migrated behind PAL (Fig. 4b). This would suggest that, if the inhibitor is a globular

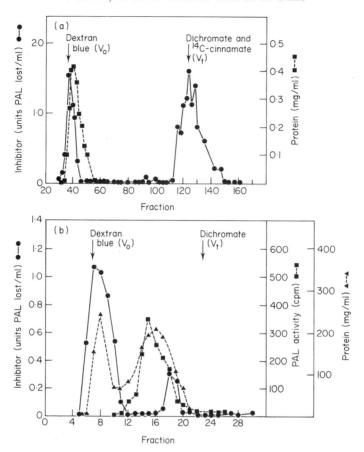

FIG. 4. Fractionation of the PAL inhibitor from gherkins. (a) Gel filtration of a concentrated 30 000 g supernatant of dark-grown hypocotyls on Sephadex G-100. (b) Gel filtration of 80% $(NH_4)_2SO_4$ precipitated material from a 30 000 g supernatant of dark-grown hypocotyls on Sepharose 4B (E. E. Billett, unpublished).

protein, it either has a molecular weight in excess of a million, or is an aggregated form of a smaller substance.

The inhibitor fraction which is excluded from Sepharose 4B is not dialysable; is destroyed by boiling or treatment at 70°C for 20 minutes; is stable at 4°C; does not bind to DEAE-cellulose equilibrated with 50 mM tris-HCl pH 8·0. The kinetic behaviour of the PAL inhibitor is extremely interesting. The inhibitor is competitive with the substrate for the active site of PAL (Fig. 5a) indicating that the inhibition is reversible. The Dixon plot (Fig. 5b) is a straight line showing that the inhibitor-PAL association is freely dissociable (Ewing and McAdoo, 1971). Similarly, enzyme concentration, at a constant level, has no effect on percentage inhibition (Fig. 5c).

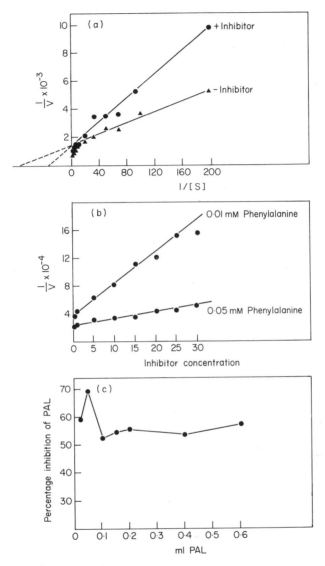

FIG. 5. Kinetic studies of inhibitor–PAL association. Inhibitor used is that excluded from Sepharose 4B in Fig. 4. (a) Lineweaver-Burke plot: when inhibitor is present V_{max} is not affected, but K_m is increased, therefore inhibition is competitive. (b) Dixon plot: the plot is a straight line, showing that the inhibitor-PAL complex is freely-dissociable. (c) Effect of enzyme concentration on inhibition at constant inhibitor concentration (E. E. Billett, unpublished).

2. Cellular Location of the Inhibitor

In the initial experiments, inhibitor preparations were always obtained from the 30 000 g supernatant. When this supernatant is centrifuged at 105 000 g for one hour, however, substantial amounts of the inhibitor are found associated with the microsomal pellet. Inhibitor can be solubilized from the microsomes by treatment with sodium cholate in the presence of KCl. This may account for the exclusion of the inhibitor from Sepharose; solubilized microsomal inhibitor is not excluded from the gel, although as yet reliable estimates of molecular weight have not been obtained. Solubilized microsomal inhibitor has similar kinetics to the inhibitor present in the 100 000 g supernatant, i.e. it is reversible and forms a freely dissociable complex with PAL. The reversibility of the gherkin PAL inhibitor clearly distinguishes it from the sunflower inhibitor (Creasy, 1976) which was also membrane-bound. The sunflower inhibitor was solubilized by Tween 20, but its action on PAL was progressive and non-reversible.

3. Specificity of the Inhibitor

The inhibitor inhibits not only gherkin PAL, but also PAL partially-purified from mustard cotyledons, radish cotyledons and pea apical buds. It does not inhibit mustard nitrate reductase, maize nitrate reductase, commercial α-amylase, gherkin NADH:cytochrome C reductase, gherkin NADPH:cytochrome C reductase (antimycin insensitive), commercial β-galactosidase and commercial β-glucosidase. Thus, it was originally thought that the inhibitor was specific to PAL. Recently, however, we have demonstrated strong inhibition of pea cinnamate 4-hydroxylase by the gherkin soluble and microsomal inhibitor preparations.

D. SUMMARY

The situation regarding the photocontrol of PAL is far from clear. There is evidence from radiolabelling of the photocontrol of PAL synthesis in parsley cells and of the photocontrol of PAL degradation in *Xanthium* leaves. Density labelling evidence in mustard and gherkin does not support light-mediated increases in enzyme synthesis and favours enzyme activation. The existence of inactive PAL is suggested from immunological studies in radish and from the cycloheximide-initiated increases in PAL activity in gherkin. Finally, an irreversible PAL inhibitor has been found in sunflower leaves and a reversible, competitive inhibitor, in gherkin hypocotyls. The gherkin PAL inhibitor preparations also inhibit cinnamate hydroxylase. Although final proof is not yet at hand, the body of evidence clearly indicates that at least in some plants, the photoregulation of PAL levels may be exerted via relatively specific macromolecular PAL inhibitors, and thus through activation of the enzyme.

IV. REGULATION BY SPECIFIC ENZYME INHIBITORS

A number of specific enzyme inhibitors or inactivators have been reported in recent years (see Wallace, Chapter 9). Some of these are specific proteases (e.g. the maize nitrate reductase protease; Wallace, 1973, 1974) and only a small number are reversible (e.g. the phosphodiesterase inhibitor from soybean callus; Brewin and Northcote, 1973). This is an important point, since rapid fluctuations in enzyme activity can only be regulated by macromolecular inhibitors if the process of inhibition is reversible.

The regulation of enzyme activity by specific enzyme inhibitors has theoretical advantages in differentiated multicellular organisms in which the overall rates of protein synthesis and degradation are relatively low. Schimke (1973) has cogently argued that only when an enzyme turns over rapidly can changes in the rate of enzyme synthesis and/or degradation result in rapid fluctuations in enzyme activity. Thus, if rapid fluctuations in the activities of important regulatory enzymes were solely caused by altered rates of enzyme synthesis, these enzymes would need to turn over rapidly. In the context of a mature plant cell, such rapid turnover may be wasteful and, in such cases, it would be economic for enzyme molecules to be maintained in an inactive state by a specific inhibitor which could be released, modified or degraded in response to metabolite or environmental changes. It is also conceivable that inhibitors which act on successive enzymes in a pathway might exist; thus coordinate regulation of successive enzymes might be due to "pathway-specific" enzyme inhibitors. This, of course, is pure speculation at this stage, but the properties of the gherkin PAL inhibitor are directing our attention along these lines.

V. PHOTOCONTROL OF OTHER ENZYMES

A. ASCORBIC ACID OXIDASE IN MUSTARD

The extractable activity of ascorbic acid oxidase in mustard cotyledons increases in response to continuous far-red light with a time-course very similar to that of mustard PAL (see Fig. 2). Density labelling evidence (Attridge, 1974; Acton et al., 1974) is consistent with the increase in this enzyme activity being due to increased rates of enzyme synthesis.

B. NITRATE REDUCTASE IN MUSTARD

The activity of nitrate reductase in mustard cotyledons, as measured by a semi- in vivo assay is rapidly increased by continuous far-red light, or by a short pulse of red light followed by darkness (Fig. 6) (Johnson, 1976). The rapidity of this response suggests that the synthesis of new messenger-RNA molecules is unlikely to be involved in the regulation.

Nitrate itself causes an increase in activity, but this induction displays an

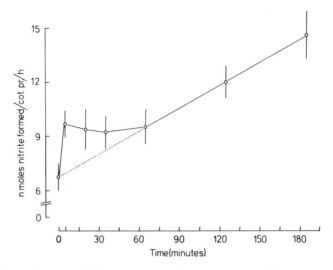

FIG. 6. Effect of 5 min red light treatment on nitrate reductase activity in etiolated mustard cotyledons (from Johnson, 1976).

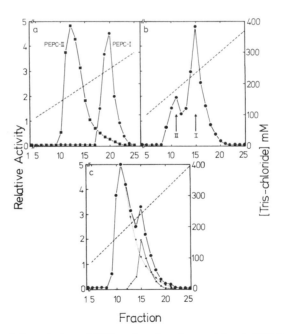

FIG. 7. DEAE-cellulose elution of PEP carboxylase from etiolated (PEPC-I) and greened (PEPC-II) sugar cane. (a) Separate elution of PEPC-I and PEPC-II from different columns using the same buffer gradient for both. (b) Elution of a mixture of PEPC-I and PEPC-II from one column. (c) Elution pattern of PEPC-II with PEPC-I added at the extraction stage, ● observed profile; dotted line: profile of a pure PEPC-II sample run separately; small peak represents the difference calculated between the other two profiles (from Goatly and Smith, 1974).

exponential time-course, whereas the phytochrome-mediated increase is linear with time. The two induction mechanisms, however, are quite different, as is shown by their differential sensitivities to cycloheximide. Cycloheximide at 100 μg/ml is capable of completely preventing the increase in nitrate reductase activity brought about by nitrate, but has only a marginal inhibitory effect on the phytochrome-mediated increase. Although great care must always be exercised in the interpretation of inhibitor data, in this case it seems reasonably certain that the light-mediated increase in enzyme activity does not require continued protein synthesis. This is evidence, therefore, for phytochrome-mediated activation of nitrate reductase in mustard cotyledons.

C. PHOSPHOENOLPYRUVATE CARBOXYLASE IN SUGAR CANE

When sugar cane nodes are grown in total darkness, the leaves which are produced contain only one form of PEP-carboxylase as separated on DEAE-cellulose. The leaves of light-grown sugar cane plants also contain only one form of PEP-carboxylase on DE-52, but its binding characteristics are quite different (Fig. 7). Mixing experiments (also in Fig. 7) show that the two forms, designated PEPC-I (dark) and PEPC-II (light), are distinct isoenzymes. They also differ in catalytic and regulatory properties. PEPC-II is substantially activated by glucose-6-phosphate, whereas PEPC-I is totally unaffected; differences also occur in the response of the two enzymes to NaCl inhibition (Goatly and Smith, 1974). It appears as if PEPC-I from dark-grown leaves has the characteristics of the dark CO_2-fixation enzyme commonly found in C-3 plants, whereas PEPC-II found in light-grown leaves has the characteristics of the photosynthetic enzyme present in C-4 plants.

When dark-grown leaves are transferred to continuous white light, the total activity of PEPC increases over a relatively long time course, together with other parameters of photosynthesis. When enzyme preparations are fractionated at various points during this time-course, the appearance of PEPC-II is observed (Fig. 8). In an attempt to test for synthesis of PEPC-II, density labelling experiments were performed. However, under no circumstances was any ^2H from ^2H_2O incorporated into PEPC, even though acid phosphatase, an internal marker enzyme, did become labelled in the light (Goatly et al., 1975). The simplest interpretation of these results therefore, is that light causes the interconversion of PEPC-I to PEPC-II, leading to a change in properties and an increase in catalytic activity.

D. LIPOXYGENASE IN MUSTARD

In mustard cotyledons, lipoxygenase activity increases steadily during growth in the dark. If continuous far-red light is given, there is no change in the time-course of increase in enzyme activity until just over 33 hours after the start of imbibition. At this point, far-red light completely and suddenly

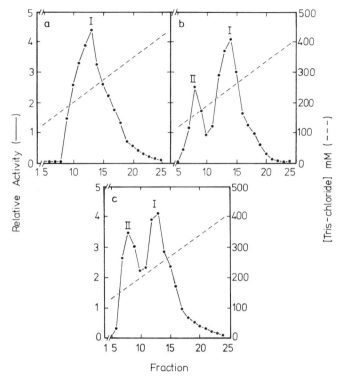

FIG. 8. DEAE-cellulose elution profiles of sugar cane PEP-carboxylase during greening. (a) O h, (b) 20 h, and (c) 40 h after transfer of leaves from darkness to continuous white light. I = PEP carboxylase I; II = PEP-carboxylase II (see Fig. 7). (From Goatly *et al.*, 1975.)

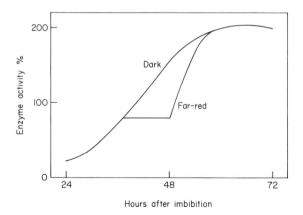

FIG. 9. The photocontrol of lipoxygenase activity in mustard seedlings. Seedlings were grown in either continuous darkness or continuous far-red light from the start of imbibition (from Oelze-Karow and Mohr, 1970).

prevents all further increase in enzyme activity for a period of c. 15 hours. At about 48 hours after imbibition, the enzyme activity begins to increase, even in far-red light, at a rate higher than that found in the dark (Fig. 9) (Oelze-Karow and Mohr, 1970).

This phenomenon is clearly a most impressive case of the photocontrol of enzyme activity. The response to far-red light is virtually instantaneous when given between 33·25 and 48 hours after the start of imbibition; given at any other time, far-red is totally ineffective. Although nothing is known of the mechanisms underlying these changes, their rapid and abrupt nature are suggestive of effects on enzyme activation or inactivation. Changes operating through enzyme synthesis would be expected to exhibit much smoother transitions.

E. PEROXIDASE IN SQUASH MEMBRANE PREPARATIONS

Recently Penel et al. (1976) have demonstrated a phytochrome-mediated in vitro modulation of peroxidase activity in a membrane preparation from zucchini squash (Cucurbita pepo) hypocotyls. The membrane fraction was prepared by treatment of a 20 000 g supernatant with 6 mM $MgCl_2$ followed by a further centrifugation at 20 000 g, from which the pellet was used. If the first supernatant was given red light together with the Mg^{++}, the resulting pellet had substantial amounts of phytochrome associated with it, presumably precipitated from the soluble phase together with the Mg^{++}-aggregated membranous material. The peroxidase activity of such phytochrome-containing pellets was subsequently found to be depressed by red light treatment for one minute. The depression in activity was only 20–30%, but was reversible by brief far-red light. Furthermore, only pellets with associated phytochrome responded in this way.

Unfortunately, the aggregated membranous material prepared by this technique, although previously used in several phytochrome-binding studies (Boisard et al., 1974; Marmé et al., 1973, 1974), has never been adequately characterized. Quail (1975a, b) has, however, shown that such in vitro binding of phytochrome is dependent on the presence of degraded ribonucleoprotein material. Thus, this in vitro photomodulation of peroxidase activity probably bears no real relationship to the primary action of phytochrome at the membrane. Nevertheless, the modulation of enzyme activity is real, although small in extent, and presumably reflects an alteration in the physical or catalytic properties of the enzyme, or perhaps a change in the conformation of the aggregated membranous material in the vicinity of the peroxidase.

F. INVERTASE IN RADISH

A very interesting case of enzyme regulation by light has recently been demonstrated for invertase (β-fructosidase) in radish (Raphanus sativus)

(Zouaghi and Rollin, 1976). Invertases are often found in multiple forms and two major classes can usually be recognised; insoluble or cell wall invertases, which normally have acidic pH optima, and cytosolic invertases, which may have acidic or alkaline pH optima. In radish roots and hypocotyls, continuous far-red light causes a rise in the cell wall-associated activity and a concomitant drop in the cytosolic activity. In the hypocotyl, but not the roots, there is also a light-mediated increase in total invertase activity. Treatment with cycloheximide (28 µg/ml) prevented the increase in total activity, but had no effect on the changes in the two fractions.

The authors conclude from this data that phytochrome regulates the transfer of cytosolic enzyme to the cell wall where it becomes sufficiently tightly bound to resist extraction by M·NaCl, or by citrate–phosphate buffers containing either Tween 20 (0·5%) or Triton X-100 (0·5%). The transfer thus appears to involve the formation of covalent bonds between the invertase molecules and the cell walls.

G. CONCLUSION

The most obvious conclusion to be derived from the above discussion is that the various light-mediated changes in enzyme levels cannot be due to regulation at a single site in the scheme shown on page 95. Thus, evidence exists for light-mediated effects on enzyme synthesis, enzyme degradation, enzyme activation, enzyme inactivation, enzyme interconversion, and even enzyme movement from one cellular location to another. Simplistic hypotheses implying that the photoreceptors operate exclusively at the level of gene transcription must, therefore, be rejected. Nevertheless, the possibility still exists that light might regulate the levels of *some* enzymes through effects on mRNA levels in addition to regulation of gene expression at later stages.

VI. PHOTOCONTROL OF TRANSCRIPTION AND TRANSLATION

The above discussion has brought together evidence that the activities of various specific enzymes may be regulated by light at a number of levels. Another approach to this question is to search directly for effects of light on the processes of transcription and translation *per se*. As yet, very little such work has been done, presumably because of the technical difficulties involved. Recently, however, a number of new and potentially very valuable techniques have been developed, and they are beginning to be applied to the question of the photocontrol of development.

A. TECHNICAL PROBLEMS

The ultimate objective of this approach is to measure the *in vivo* rates of synthesis of individual, identifiable messenger-RNA molecules, and individual, identifiable proteins, as affected by various irradiation treatments. If the synthesis of a sufficient number of mRNAs and proteins could be quantified then it would be possible to build up a picture of the overall effects of light on the regulation of gene expression at the transcriptional and translational levels. Clearly, such an approach may be applied to control by any factor and is not restricted to photocontrol. A number of major technical problems need to be overcome, however, before such critical measurements can be made.

The fractionation, purification, identification and quantification of individual mRNA molecules from higher plants is clearly the most intransigent of these problems. The discovery, a few years ago, that at least a proportion of mRNAs had tracts of poly-adenylic acid (poly-A) attached to them (see Matthews, 1973; Brawerman, 1974), provided a very useful tool for separating such mRNAs from the rest of the nucleic acids present. Poly-A-RNA may be isolated by chromatography on columns of oligo-dT-cellulose (Brawerman, 1974) and the total amount of poly-A-RNA may be estimated by hybridization with labelled poly-uridylic acid (Covey and Grierson, 1976). Even so, it is still not feasible to attempt to isolate individual mRNA molecules except in tissues which are synthesizing large quantities of a particular protein.

In recent years it has become possible to translate mRNAs *in vitro* using a plant cell-free protein synthesizing system derived from wheat germ (Roberts and Paterson, 1973). Coupled with acrylamide gel electrophoresis of the labelled products, usually under denaturing conditions (i.e. with SDS and/or urea), *in vitro* protein synthesis can be used to demonstrate differences in the populations of mRNAs being translated. Immunoprecipitation may be used to identify individual proteins as products and thus, in principle, the amount of mRNAs in the preparations coding for individual proteins can be determined. Already, this approach has been successfully used in a study of the role of gibberellins in the induction of α-amylase in barley seeds (Higgins *et al.*, 1975). Using combinations of these methods it should theoretically be possible to answer the question posed at the beginning of this Section. Such an enterprise, however, would require very large resources, particularly in connection with the purification of proteins and the raising of antibodies, and it is more realistic to expect that only rather limited objectives will be achieved. The rest of this Chapter outlines some preliminary attempts in this direction.

B. PHOTOCONTROL OF THE PATTERN OF PROTEIN SYNTHESIS

Although it has been known for many years that transfer of plants from darkness to white light increases the rate of overall protein synthesis, no attempts have as yet been made to detect a changed *pattern* of protein synthesis after light treatment. There is evidence from acrylamide gel electrophoretic studies, that the protein complement of light-treated plants is different from that of etiolated plants, reflecting particularly the presence of chloroplasts in the former and etioplasts in the latter (Cobb and Wellburn, 1973, 1974), but no studies have been made on the regulation of the synthesis of these components. In experiments not yet published, Hobson and Smith have shown, by dual label feeding, that white light treatment substantially modifies the pattern of proteins synthesized in etiolated *Phaseolus vulgaris* leaves. In these experiments, one batch of plants was fed ^{14}C-leucine into the leaves, and another ^{3}H-leucine in an identical manner. The first batch was then given white light treatment for various lengths of time; both batches were combined and the soluble proteins extracted and fractionated on DEAE-cellulose. Very large changes in the ratio of ^{14}C:^{3}H along the column were seen indicating light-mediated alterations in the rates of synthesis of various proteins. However, no such effects were seen with brief red light, or with continuous far-red light, indicating that phytochrome was not involved (Hobson and Smith, unpublished).

C. PHOTOCONTROL OF POLYRIBOSOME LEVELS

The assessment of *in vivo* rates of protein synthesis is an extremely difficult task. However, it has often been assumed that the *in vivo* rate of protein synthesis will be reflected in the overall proportion of the total ribosomes present as polysomes (e.g. Marcus and Feeley, 1965; Leaver and Key, 1967; Williams and Novelli, 1968; Mascarenhas and Bell, 1969). Pine and Klein in 1972 presented evidence that red light increased the proportion of polysomes in etiolated bean leaves, and this effect was mediated via phytochrome. Unfortunately however, their methods did not prevent ribonuclease action and thus only very small proportions of polysomes were obtained. Subsequently, better mthods of ribosome extraction from plant tissues were developed (Davies *et al.*, 1972). Using these methods, we have confirmed the results of Pine and Klein (1972) and shown that both continuous far-red and continuous white light increase the level of polysomes in bean (*P. vulgaris*) leaves (Smith, 1976).

Figure 10 shows the sucrose gradient profiles obtained with the various light treatments and Fig. 11 gives the time courses. Analysis of the density gradients show that continuous white light leads to an increase in the average size of the polysomes, and a similar trend is evident with brief red light followed by darkness, and continuous far-red. Double-labelling experiments described fully by Smith (1976) indicate more strongly that red light treatment leads to an increase in the average size of polysomes.

In order to test whether continued mRNA synthesis was necessary for this response, leaves were treated with 500 µg/ml cordycepin (3'-deoxyadenosine), three hours before light treatment. Previous experiments had shown that cordycepin applied under these conditions inhibited both total and poly-A-

RNA synthesis by 85–95% within three hours. Figure 12 shows that the red light effect on polysome levels is not inhibited at all by cordycepin, and the effect of continuous far-red light is only partially prevented. Thus, continuous messenger-RNA synthesis is not necessary for the light-mediated increase in polysome proportions.

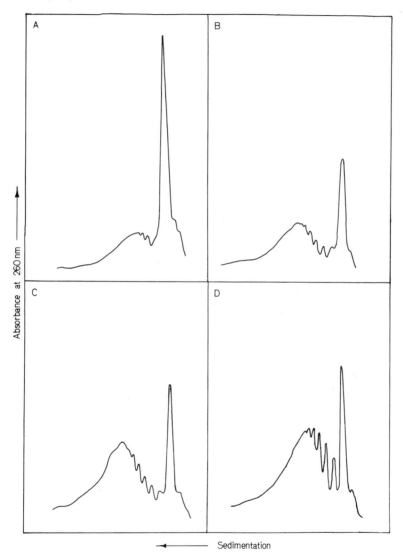

Fig. 10. The effect of light on polyribosome proportions in bean leaves. Absorbance scans of polyribosome gradients produced with extracts from dark-grown control leaves (A), dark-grown leaves treated with 10 min red light and returned to darkness for 4 h (B), dark-grown leaves irradiated with continuous far-red light for 4 h (C) and dark-grown leaves irradiated with continuous white light for 4 h (D) (from Smith, 1976).

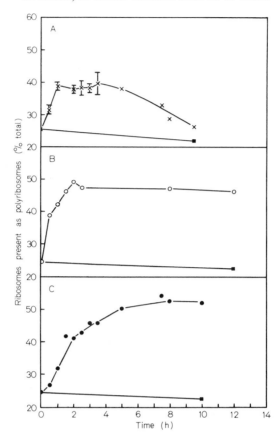

FIG. 11. Time-course of the light-mediated increases in polyribosome proportions in bean leaves. Leaves were treated either (A) with 10 min red light and returned to darkness (X) or (B) with continuous far-red light (O), or (C) with continuous white light (●). (■) The gradual drop in polyribosome proportions in the dark controls (from Smith, 1976).

Travis *et al.* (1972, 1974) have shown that both white light, and brief red light-treatment of etiolated maize leaves increase the subsequent capacity for protein synthesis *in vitro* of isolated monoribosomes. They also showed that the increased protein synthetic capacity was associated with changes in the properties of the ribosomes, indicating that light regulates the availability of initiation factors.

The data for bean leaves also seem best accounted for on the basis of phytochrome-mediated regulation of initiation. Thus, after light treatment, ribosomes tend to "pile-on" to existing small polysomes, increasing the average length of the polysomes and, presumably, increasing the rate of protein synthesis. Messenger-RNA synthesis is not necessary for these effects, at least within the first few hours of light treatment. On the other hand, there is no reason from this work to assume that light effects on mRNA synthesis

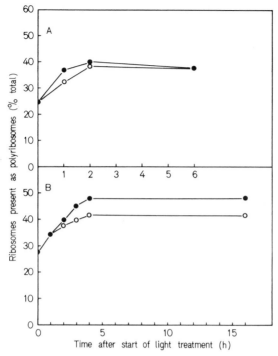

FIG. 12. Effect of cordycepin on the light-mediated increases in polyribosome proportions. Seedlings grown in darkness were either treated with cordycepin (O) or left untreated (●); 2 h later batches were either given 10 min red light and returned to darkness (A), or irradiated with continuous far-red light (B) (from Smith, 1976).

do not occur. Thus, the conclusion from this section is that light, acting through phytochrome, is able to regulate the rate of protein synthesis through the modification of ribosome properties, or through controlling the availability of initiation factors. In other words, light can regulate gene expression at the translation level. It would seem difficult, however, to account for a changed *pattern* of protein synthesis on this basis, unless some selectivity device is implicated at the initiation step of translation.

D. PHOTOCONTROL OF MESSENGER-RNA LEVELS

The central core of the Mohr (1966) hypothesis, that phytochrome regulates gene transcription, has turned out to be most difficult to test. Even now, there is no evidence, apart from circumstantial and misleading inhibitor studies, that phytochrome can act to regulate mRNA formation. It is known that light, acting through phytochrome, regulates the synthesis of ribosomal-RNA (Jaffe, 1969; Koller and Smith, 1972; Thien and Schopfer, 1975). However, the time scale of this phenomenon is of the order of hours and it cannot be considered a primary consequence of phytochrome action.

1. *Photocontrol of poly-A-RNA levels*

Recently, Grierson and Covey (1975) have measured the absolute amounts of poly-A-RNA in developing leaves of *Phaseolus aureus*. Although continuous white light leads to a significantly larger proportion of poly-A-RNA in the light-grown plants, the response is rather slow, no effect of the light being observed within six hours. Thus, there do not appear to be quantitative changes in the synthesis of total poly-A-containing mRNAs sufficiently large to be detected by the hybridization technique. However, it might be expected that any light-mediated changes in transcription would appear small against the background of mRNA synthesis necessary for the normal functioning of the cell. Thus it is necessary to search for changes in the pattern of mRNAs present.

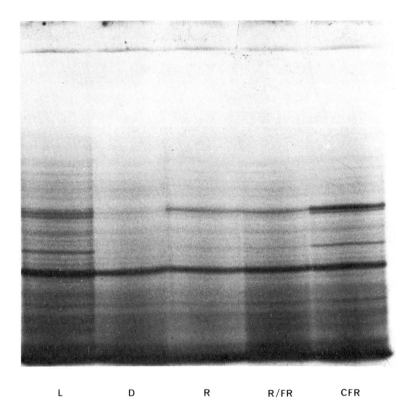

L   D   R   R/FR   CFR

Fɪɢ. 13. Electrophoretic separation on 15% acrylamide-SDS gels of proteins made *in vitro* in response to polysomal poly(A)-RNA isolated from bean leaves given the following treatment. Plants were: D, grown in continuous darkness; L, given 16 h white light; R, given 5 min red light plus 4 h darkness; R/FR, given 5 min red light, plus 10 min far red light plus 4 h darkness; FR, given 4 h continuous far-red. The ^{35}S-labelled products were detected by autoradiography (A. B. Giles, unpublished).

2. *Photocontrol of mRNA Availability*

In order to test for light-mediated changes in mRNA populations, poly-A-RNA has been extracted from bean (*P. vulgaris*) leaves and translated in a cell-free protein synthesizing system. Previously, Tobin and Klein (1975) showed that total cell poly-A-RNA, isolated from *Lemna gibba* after 24 hours white light, programmed the synthesis of a slightly different range of polypeptides *in vitro* than similar preparations from dark-grown material.

Giles, Grierson and Smith (1977) extracted poly-A-RNA from polysomes prepared from bean leaves given various irradiation treatments. After purification, the poly-A-RNA was fed to a wheat germ S-30 protein-synthesizing system with ^{35}S-methionine as the labelled substrate. Figure 13 shows the autoradiographs of the products separated by gel electrophoresis. Clearly, only a few changes are induced by light, and these are probably quantitative, rather than qualitative. Furthermore, although continuous white and far-red light have similar effects, no red/far-red reversibility was detected. Nevertheless, the changes brought about by white and far-red light demonstrate that a changed *pattern* of mRNA molecules are present in the polysomes, fitting in with the data obtained earlier for changed patterns of proteins synthesized *in vivo*.

The question thus becomes—is the changed pattern of mRNAs associated with the polysomes the result of changes in the relative rates of gene transcription, or is it due to some other post-transcriptional control? Preliminary evidence suggests that the latter alternative is more likely. When *total* bean leaf poly-A-RNA from dark-grown or white light treated seedlings was compared in the *in vitro* protein synthesizing system, no differences, qualitative or quantitative, could be discerned. Taken at face value, this result indicates that the mRNA molecules responsible for the changed pattern of protein synthesis observed after white light treatment, were already present in the dark grown cells. Light, therefore, merely mobilizes the mRNAs to the polysomes.

This conclusion is consistent with the earlier finding that light regulates the process of initiation (Smith, 1976). It should be stressed, however, that the technique of cell-free protein synthesis is open to several objections. A major problem is the quantification of the amounts of individual polypeptides synthesized. Autodiography is notoriously imprecise and even with fluorography, interpretation of the data is often dependent on subjective judgement. Other problems not yet satisfactorily solves relate to the provision of the most appropriate initiation factors and to premature chain termination.

E. DOES LIGHT REGULATE TRANSCRIPTION?

Although the technique of cell free translation of mRNA is still imperfect it nevertheless offers the only approach at present to this question, which is in essence part of the central problem of developmental biology—to what extent is development in eucaryotes determined by the regulation of gene transcription? Based on the admittedly very meagre evidence which is now at hand, transcriptional control would appear to have only a minor role, if any, in the photoregulation of plant development. Obviously this is a very premature statement but, as yet, no good evidence exists for the light-mediated regulation of the synthesis of mRNA molecules, either qualitatively, or quantitatively. The evidence is more in support of post-transcriptional effects on initiation of protein synthesis, although even this is circumstantial in that actual changes in the amounts of initiation factors have not yet been investigated. Furthermore, nothing is known as to how selectivity can be incorporated into a control mechanism operating on initiation. There is clearly room for much more intensive investigation of the photocontrol of transcription and translation, and the methods, if perhaps not the resources, are now becoming available.

VII. General Comments

The principal conclusion which can be drawn from this survey is that light does not control enzyme activity through only one mechanism. This rather obvious statement is nevertheless important, since it should dissuade the elaboration of overfacile hypotheses. At the present time, there is no direct evidence for the regulation of transcription by light, but this may be due primarily to the technical difficulties associated with such work. Improved methodology should help towards a more dependable answer in the near future. The photocontrol of translation, exerted in some way through the control of initiation, is perhaps more solidly based, but even here, the relative importance of such regulation is unknown. The best evidence exists for the regulation of enzyme activation/inactivation/interconversion by light, but in each case final proof of such control will depend upon the isolation, characterization and quantification of the activating and inactivating components.

It would perhaps be more satisfying intellectually if we could point to a single site at which light (or at least the individual photoreceptors) control gene expression. Plants have not evolved to fulfil Occam's razor, however, and it is probably more satisfying to the plant to be able to regulate enzyme levels at several different points. Advantages in flexibility and economy are clearly evident.

ACKNOWLEDGEMENTS

The authors wish to acknowledge the financial support from the Science Research Council for the previously unpublished work present in this article. They are also grateful to Dr W. Wallace, for helpful discussions, and to Miss Jane Holmes and Mr M. Foulstone for assistance in preparation of the manuscript.

REFERENCES

Acton, G. J. (1972). *Nature New Biol.* **236**, 255.
Acton, G. J. and Schopfer, P. (1975). *Biochim. biophys. Acta* **404**, 231.
Acton, G. J., Drumm, H. and Mohr, H. (1974). *Planta* **121**, 39.
Amrheim, N. and Zenk, M. H. (1968). *Naturwissenschaften* **55**, 394.
Amrheim, N. and Zenk, M. H. (1970). *Z. Pfl. physiol.* **63**, 384.
Attridge, T. H. (1974). *Biochim. biophys. Acta* **362**, 258.
Attridge, T. H. and Smith, H. (1967). *Biochim. biophys. Acta* **148**, 805.
Attridge, T. H. and Smith, H. (1973). *Phytochemistry* **12**, 1569.
Attridge, T. H. and Smith, H. (1974). *Biochim. biophys. Acta* **343**, 452.
Attridge, T. H., Stewart, G. R. and Smith, H. (1971). *FEBS Letts.* **17**, 84.
Attridge, T. H., Johnson, C. B. and Smith, H. (1974). *Biochim. biophys. Acta* **343**, 440.
Boisard, J., Marmé, D. and Briggs, W. R. (1974). *Pl. Physiol.* **54**, 272.
Borthwich, H. A., Hendricks, S. B., Parker, M. W., Toole, E. H. and Toole, V. K. (1952). *Proc. natn. Acad. Sci. U.S.A.* **38**, 662.
Bottomley, W. (1970). *Pl. Physiol.* **45**, 608.
Brawerman, G. (1974). *A. Rev. Biochem.* **43**, 621.
Brewin, N. J. and Northcote, D. H. (1973). *Biochim. biophys. Acta* **320**, 104.
Briggs, W. R. (1976). *In* "Light and Plant Development" (Smith, H. ed.), Butterworth, London (in press).
Butler, L. G. and Bennett, V. (1969). *Pl. Physiol.* **44**, 1285.
Camm, E. E. and Towers, G. H. N. (1973a). *Phytochemistry* **12**, 961.
Camm, E. E. and Towers, G. H. N. (1973b). *Can. J. Bot.* **51**, 824.
Cobb, A. H. and Wellburn, A. R. (1973). *Planta* **114**, 131.
Cobb, A. H. and Wellburn, A. R. (1974). *Planta* **121**, 273.
Covey, S. N. and Grierson, D. (1976). *Planta* **131**, 75.
Creasy, L. L. (1976). *Phytochemistry* **15**, 673.
Davies, E., Larkins, B. A. and Knight, R. H. (1972). *Pl. Physiol.* **50**, 581.
Drumm, H., Elchinger, I., Möller, J., Peter, K. and Mohr, H. (1971). *Planta* **99**, 265.
Drumm, H. Brüning, K. and Mohr, H. (1972). *Planta* **106**, 259.
Durst, F. and Mohr, H. (1966). *Naturwissenschaften* **53**, 531.
Ebel, J., Schaller-Hekeler, B., Knobloch, K., Wellmann, E., Grisebach, H. and Engelsma, G. (1967). *Naturwissenschaften* **54**, 319.
Engelsma, G. and Meijer, G. (1965). *Acta. bot. neerl.* **14**, 54.
Engelsma, G. and Van Bruggen, J. M. H. (1971). *Pl. Physiol.* **48**, 94.
Ewing, E. E. and McAdoo, M. H. (1971). *Pl. Physiol.* **48**, 366.
Feierabend, J. and Pirson, A. (1966). *Z. Pfl. physiol.* **55**, 235.
Filner, B. and Klein, A. O. (1969). *Pl. Physiol.* **43**, 1587.
Filner, P. and Varner, J. E. (1967). *Proc. natn. Acad. Sci. U.S.A.* **58**, 1520.
French, C. J. and Smith, H. (1975). *Phytochemistry* **14**, 963.
Furuya, M., Galston, A. W. and Stowe, B. B. (1962). *Nature* **193**, 456.
Giles, A. B., Grierson, D. and Smith, H. (1977). *Planta* (in press).
Goatly, M. B. and Smith, H. (1974). *Planta* **117**, 67.
Goatly, M. B., Coombs, J. and Smith, H. (1975). *Planta* **125**, 15.
Graham, D., Grieve, A. M. and Smillie, R. M. (1968). *Nature* **218**, 89.
Graham, D., Hatch, M. D., Slack, C. R. and Smillie, R. M. (1970). *Phytochemistry* **9**, 521.
Grierson, D. and Covey, S. N. (1975). *Planta* **127**, 77.
Hahlbrock, K. (1974). *Biochim. biophys. Acta* **362**, 417.
Hahlbrock, K. and Grisebach, H. (1970). *FEBS Letts.* **11**, 62.
Hahlbrock, K. and Schröder, J. (1975). *Archs Biochem. Biophys.* **166**, 47.
Hahlbrock, K. and Wellman, E. (1973). *Biochim. biophys. Acta* **304**, 702.

Hahlbrock, K., Sutter, A., Wellmann, E., Ortmann, R. and Grisebach, H. (1971a). *Phytochemistry* **10**, 109.
Hahlbrock, K., Kuhlen, E. and Lindl, T. (1971b). *Planta* **99**, 311.
Hahlbrock, K., Knobloch, K., Kreuzaler, J. R., Potts, R. M. and Wellmann, E. (1976). *Europ. J. Biochem.* **61**, 197.
Hartmann, K. M. (1966). *Photochem. Photobiol.* **5**, 349.
Havir, E. A. and Hanson, K. R. (1975). *Biochemistry* **14**, 1620.
Henshall, J. D. and Goodwin, T. W. (1964). *Phytochemistry* **3**, 677.
Higgins, T. J. V., Zwar, J. A. and Jacobsen, J. V. (1976). *Nature* **260**, 166.
Hu, A. S. L., Bock, R. M. and Halvorson, M. O. (1962). *Ann. Biochem.* **4**, 489.
Hyodo, H. and Yang, S. F. (1971). *Archs Biochem. Biophys.* **143**, 338.
Iredale, S. E. and Smith, H. (1974). *Phytochemistry* **13**, 575.
Jaffe, M. J. (1969). *Physiologia Pl.* **22**, 1033.
Johnson, C. B. (1976). *Planta* **128**, 127.
Johnson, C. B., Attridge, T. H. and Smith, H. (1975). *Biochim. biophys. Acta* **385**, 11.
Keller, C. J. and Huffaker, R. C. (1967). *Pl. Physiol.* **42**, 1277.
Koller, B. and Smith, H. (1972). *Phytochemistry* **11**, 1295.
Lamb, C. J. (1976). Ph.D. Thesis, University of Cambridge.
Leaver, C. J. and Key, J. L. (1967). *Proc. natn. Acad. Sci. U.S.A.* **57**, 1338.
Marcus, A. and Feeley, J. (1965). *J. biol. Chem.* **240**, 1675.
Marmé, D., Boisard, J. and Briggs, W. R. (1973). *Proc. natn. Acad. Sci. U.S.A.* **70**, 3861.
Marmé, D., Mackenzie, J. M., Boisard, J. and Briggs, W. R. (1974). *Pl. Physiol.* **54**, 263.
Mascarenhas, J. P. and Bell, E. (1969). *Biochim. biophys. Acta* **179**, 199.
Matthews, M. B. (1973). *In* 'Essays in Biochemistry" (Campbell, P. N. and Dickens, F., eds), Vol. 9, pp. 59–102, Academic Press, London and New York.
Mohr, H. (1957). *Planta* **49**, 389.
Mohr, H. (1966). *Photochem. Photobiol.* **5**, 469.
Mohr, H. (1970). *Naturw. Rdsch.* **23**, 187.
Nari, T., Mouttet, C. H., Pinna, M. H. and Ricard, J. (1972). *FEBS Letts.* **23**, 220.
Oelze-Karow, H. and Mohr, H. (1970). *Z. Naturf.* **25b**, 1282.
Oelze-Karow, H., Schopfer, P. and Mohr, H. (1970). *Proc. natn. Acad. Sci. U.S.A.* **65**, 51.
Penel, C., Greppin, H. and Boisard, J. (1976). *Pl. Sci. Letts.* **6**, 117.
Pine, K. and Klein, A. O. (1972). *Devl. Biol.* **28**, 280.
Quail, P. H. (1975a). *Planta* **123**, 223.
Quail, P. H. (1975b). *Planta* **123**, 235.
Queiroz, O. (1969). *Phytochemistry* **8**, 1655.
Rhodes, M. J. C., Hill, A. C. R. and Wooltorton, L. S. C. (1976). *Phytochemistry* **15**, 707.
Roberts, B. and Patterson, B. (1973). *Proc. natn. Acad. Sci. U.S.A* **70**, 2330.
Russell, D. W. and Conn, E. E. (1967). *Archs Biochem. Biophys.* **122**, 256.
Schimke, R. T. (1973). *Adv. Enzymol.* **37**, 135.
Schopfer, P. (1972). *In* "Phytochrome" (Mitrakos, K. and Shropshire Jr., W., eds), pp. 485–514, Academic Press, London and New York.
Schopfer, P. and Plachy, C. (1973). *Z. Naturf.* **28**, 296.
Siegelman, H. W. and Hendricks, S. B. (1957). *Pl. Physiol.* **32**, 393.
Siegelman, H. W. and Hendricks, S. B. (1958). *Pl. Physiol.* **33**, 185.
Smith, H. (1976). *Europ. J. Biochem.* **65**, 161.
Smith, H. and Attridge, T. H. (1970). *Phytochemistry* **9**, 487.
Steer, B. T. and Gibbs, M. (1969a). *Pl. Physiol.* **44**, 775.

Steer, B. T. and Gibbs, B. (1969b). *Pl. Physiol.* **44**, 781.

Tanaka, Y., Kojima, M. and Uritani, I. (1974). *Pl. Cell Physiol.* **15**, 843.

Tezuka, T. and Yamamoto, Y. (1969). *Bot. Mag., Tokyo* **82**, 130.

Thien, W. and Schopfer, P. (1975). *Pl. Physiol.* **56**, 660.

Tobin, E. M. and Klein, A. O. (1975). *Pl. Physiol.* **56**, 88.

Travis, R. L., Lin, C. Y. and Key, J. L. (1972). *Biochim. biophys. Acta* **277**, 606.

Travis, R. L., Key, J. L. and Ross, C. W. (1974). *Pl. Physiol.* **53**, 28.

Unser, G. and Masoner, M. (1972). *Naturwissenschaften* **59**, 39.

van Poucke, M., Barthe, F. and Mohr, H. (1970). *Naturwissenschaften* **56**, 132.

Wallace, W. (1973). *Pl. Physiol.* **52**, 197.

Wallace, W. (1974). *Biochim. biophys. Acta* **341**, 265.

Williams, G. R. and Novelli, G. D. (1968). *Biochim. biophys. Acta* **155**, 183.

Zouaghi, M. and Rollin, P. (1976). *Phytochemistry* **15**, 897.

Zucker, M. (1963). *Pl. Physiol.* **38**, 575.

Zucker, M. (1968). *Pl. Physiol.* **43**, 365.

Zucker, M. (1969). *Pl. Physiol.* **44**, 912.

Zucker, M. (1970). *Biochim. biophys. Acta.* **298**, 331.

CHAPTER 7

Isozymes: Genetic and Biochemical Regulation of Alcohol Dehydrogenase*

J. G. SCANDALIOS

Department of Genetics, North Carolina State University, U.S.A.

I. Introduction 129
II. Terminology 130
   A. Isozymes 130
III. Alcohol Dehydrogenase as a Model System 131
   A. The Enzyme 131
   B. Temporal Control of ADH Expression 133
   C. Spatial Distribution of ADH 134
   D. Genetic Control of ADH 138
   E. Purification of Maize ADH 141
   F. Properties of ADH Isozymes 143
   G. Regulation of ADH Activity 146
References 153

I. INTRODUCTION

Studies on the genetic and biochemical control of multiple molecular forms of enzymes (isozymes) are important for an understanding of their genetic origin, their role(s) in the physiology of differentiated cells, and their evolutionary significance.

The important characteristics of an enzyme are its properties as a catalyst, the regulation of its synthesis and destruction, its intracellular locations, and its tissue distribution. Consequently, isozymes can provide valuable information on the control of gene expression during the development of differentiated cells and on the nature of metabolic control in these cells. Knowledge of the genetic control, structure and kinetic properties of isozymes will aid us in our efforts to answer some of the basic questions per-

* Research has been supported, in part, by U.S. A.E.C. Contracts AT (11-1) 1338 and AT (38-1)-770, by NIH Grant GM-22733, and by the North Carolina Agricultural Experiment Station.
   Paper No. 4954 of the Journal Series of the North Carolina Agricultural Experiment Station, Raleigh, North Carolina.

taining to the effects of genes on enzyme structure and function, gene regulation and cell differentiation, and the nature and necessity of isozymes *per se*.

In this discussion, I will primarily focus upon our own investigations on the genetic, biochemical and developmental control of the enzyme alcohol dehydrogenase of *Zea mays* L. which has been the subject of intense investigations in my laboratory for several years.

II. TERMINOLOGY

A. ISOZYMES

1. Definition

The term *isozyme* was introduced by Markert and Møller (1959), to describe multiple enzyme forms with similar or identical substrate specificity occurring within the same organism. More recently, Markert (1968) proposed that the term isozyme might be modified by such adjectives as allelic, nonallelic, multimeric (hetero- or homo-), conformational, conjugated, etc. to reflect more precisely our current knowledge of a particular isozyme system. The subject of definition and classification has been discussed in detail elsewhere (Shaw, 1969; Scandalios, 1974; Eppenberger, 1975). Here, I use the term isozyme in its broad operational sense as proposed by Markert (1968).

2. Occurrence

The occurrence of isozymes is common and widespread in the biological world. Multiplicitly of enzymes seems to be the rule rather than the exception as previously thought. Among plant species examined, the number of enzymes exhibiting isozymic forms is very large and a substantial amount of the polymorphism observed is genetic (Scandalios, 1974).

3. Molecular bases of isozyme formation

Two general categories can be distinguished by which multiple forms of proteins (enzymatic or non-enzymatic) can be generated:

(a) by mechanisms operating at the level of the genome which are then transcribed onto mRNAs and then code for different polypeptides. Random assembly of subunit polypeptides following translation results in the formation of functional multimers (homo- or hetero-);

(b) by epigenetic mechanisms operating at the translational or post-translational levels to modify polypeptides to varying degrees (e.g. covalent modification, conformational changes, adenylation, deamidation, phosphorylation, and selective degradation).

The two mechanisms are not mutually exclusive, and in fact both are operative in most cases studied.

III. Alcohol Dehydrogenase as a Model System

A. THE ENZYME

1. Characteristics

Alcohol dehydrogenase (alcohol: NAD oxidoreductase, E.C.1.1.1.1; abbreviated ADH) is distributed widely among animals, plants, and micro-organisms (Sund and Theorell, 1963). Irrespective of source, the enzyme catalyses the following reaction:

$$RCH_2OH + NAD \rightleftharpoons RCHO + NADH_2$$

The reaction is not highly specific with respect to substrate. The enzyme can react with a large number of substrates including normal and branched-chain aliphatic and aromatic alcohols (primary and secondary), carbonyl compounds, and with NAD, NADP and their analogues. The most thoroughly investigated ADH is that from horse liver by Theorell and his co-workers. ADH is a metalloenzyme with four atoms of Zn^{++} per molecule; two bound at the active sites believed to take part in the H^+ transfer between the coenzyme and substrate, and two tightly bound to maintain the ternary structure of the molecule (Theorell, 1967).

In higher plants, alcohol dehydrogenase is probably physiologically necessary for anaerobic glycolysis which may be of importance in the metabolism of the resting seed and during early germination, as well as in flooded roots. However, after seed germination the increase in aerobic respiration may obviate the need for ADH.

2. ADH Heterogeneity

Alcohol dehydrogenase from various sources is known to exist in multiple molecular forms (isozymes). A number of mechanisms have been found to account for the formation of ADH isozymes. Horse liver ADH displays the highest degree of complexity, and two different mechanisms have been proposed as being responsible for the apparent heterogeneity. (i) Three dimeric isozymes result from the random association of two different subunits (Pietruszko and Theorell, 1969) which have been shown to differ in primary structure (Jörnvall, 1970) and are consequently products of different structural genes. (ii) Additional forms arise *in vitro* as a consequence of varied subunit conformations which lead to interconversions between several of the enzyme forms (Lutstorf and von Wartburg, 1969). Seven ADH isozymes have been ascertained in human liver extracts; six have been isolated and purified and shown to be dimeric products of three subunits

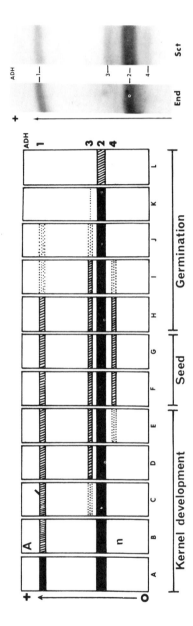

FIG. 1. A composite schematic zymogram of the developmental pattern of the alcohol dehydrogenase isozymes in maize scutella. A = endosperm 16–30 days after pollination (for comparison); B = scutellum 16–33 days after pollination; C = scutellum 16–33 days after pollination; D = scutellum 45–49 days after pollination; E = scutellum 50–56 days after pollination; F = scutellum from dry resting seed; G = scutellum from dry resting seed; H,I,J,K,L are respectively, scutella from 1,2,3,4, and 10 days after germination (sporophytic development). 0 = point of sample insertion; migration is anodal. On the right is representative zymogram; End = endosperm 18 days after pollination; Sct = scutellum from 18 hr imbibed seed.

(Schenker *et al.*, 1971). In *Drosophila melanogaster* the number of multiple forms of ADH have been reported to increase on incubation with NAD due to differential binding of the coenzyme to the basic isozymes (Ursprung and Carlin, 1968). Genetically specified isozymes have also been reported in yeast (Lutstorf and Megnet, 1968), wheat (Hart, 1971), maize (Scandalios, 1966; 1967; Schwartz and Endo, 1966), and *Peromyscus* (Felder, 1975).

B. TEMPORAL CONTROL OF ADH EXPRESSION

In the milky endosperm and scutellum of inbred strains of maize (16–18 days after pollination) there are two electrophoretically distinct forms of alcohol dehydrogenase, ADH-1 and ADH-2; these are the predominant forms during kernel development. However, as kernel development progresses, two additional ADH isozymes become apparent and prominent in the scutellum of the mature caryopsis and during the first days of germination, (Fig. 1). A time-course of ADH activity in the scutellum (Fig. 2) indicates that the maximum is reached after 24 h soaking of the seed; subsequently, the activity undergoes a rapid decline, and by the tenth day after imbibition, only one-eighth of the peak activity remains. The specific activity in the

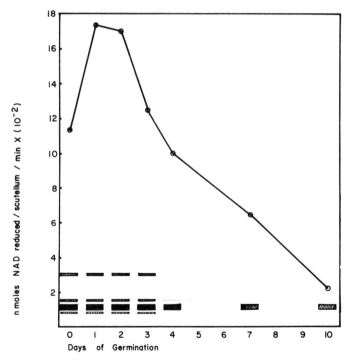

FIG. 2. Time course of ADH activity in scutella of germinating maize seeds. The corresponding zymogram pattern is shown at the bottom.

scutellum at its maximum is 3 to 4-fold higher than in the milky endosperm of the same inbred strain. The insert in Fig. 2 shows the fate of the different ADH isozymes during the specified developmental period. It should be pointed out here that the exact timing of appearance and disappearance of the isozymes varies slightly, depending on the total genetic background of the inbred line used.

C. SPATIAL DISTRIBUTION OF ADH

The isozymes ADH-1 and ADH-2 are the most common forms of alcohol dehydrogenase in all tissues examined at various developmental stages, but ADH-1 is not found in endosperm of mature kernels or during germination. Additionally, as indicated in the previous section, all four isozymes (ADH-1, ADH-2, ADH-3, and ADH-4) are present in the scutellum of imbibed seeds, but ADH-1, ADH-3 and ADH-4 are not found in that tissue after the third day of germination (see Fig. 2). Similarly, the milky endosperm of developing kernels has both ADH-1 and ADH-2 isozymes, but ADH-1 clearly disappears from this tissue as the kernel matures. This change is not due merely to dilution, but to actual disappearance of ADH-1, since the relative staining intensity (the amount of formazan dye precipitated at the site of a particular isozyme on the gel) of ADH-2 which is present is as high, if not higher, as in the milky endosperm stage; also, incubation of the zymogram in the enzyme reaction mixture beyond the optimum time does not result in any trace of ADH-1 (Fig. 3a). The pattern shown in Fig. 3b is characteristic of the leaves, stem, and roots of 7-day-old seedlings, although the staining intensity varies slightly from tissue to tissue. Both ADH-1 and

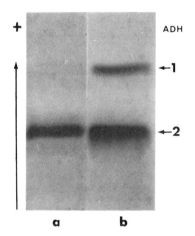

FIG. 3. Zymogram of ADH isozyme pattern in extracts from 7-day-old seedlings. a = endosperm; note absence of ADH-1; b = shoot extracts, pattern is the same for leaves, stem, and roots at this stage of development, although the staining intensity varies. Migration is anodal.

ADH-2 are present in adult tissues of the same inbred line (Fig. 4). However, the staining intensity is more equally distributed between the two isozymes as compared to the seedling tissues (Fig. 3) where ADH-2 shows greater staining intensity than ADH-1. The flag leaf of adult plants (35–50 days after planting) is the only case where both ADH-1 and ADH-2 are present

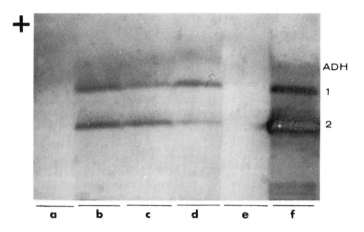

FIG. 4. Zymogram of ADH isozyme pattern in some tissues of the adult maize plant (2–4 mo.). a = leaves, 1–5 from bottom; b = leaves, 6–9 from bottom; c = husks; d = flag leaf; e = root; f = whole kernel extract (reference sample). Although no isozymes are seen in a and e, there is some ADH activity in the extracts. Note that ADH-1 and ADH-2 are more equally active as compared to earlier developmental stages (Fig. 3). Migration is anodal.

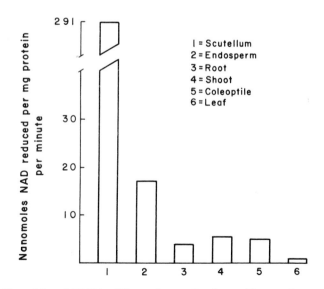

FIG. 5. Specific activity of ADH in different tissues of maize seedlings at the seventh day of germination.

but with ADH-1 exhibiting more activity (Fig. 4d). In older leaves and roots (Fig. 4a; 4e) no ADH activity is detected zymographically, but its presence, though minimal, can be detected spectrophotometrically. The total specific activity of ADH in different tissues from young seedlings and from adult plants is shown in Fig. 5 and Table I. In all cases ADH activity is significantly lower when compared to that of the milky endosperm or scutella of mature kernels.

As the different tissues age and assume a more aerobic metabolism their component ADH activity declines significantly (Fig. 6); this decline is correlated with the gradual disappearance of the corresponding tissue specific isozymes (Scandalios and Felder, 1971).

The most dramatic qualitative change observed in this system to date is the appearance of a unique ADH isozyme, ADH-5 (previously called ADH-T), in the maternally derived pericarp of developing kernels. The existence of ADH-5 is transient, being found only between days 19 and 40 of kernel development (Fig. 7).

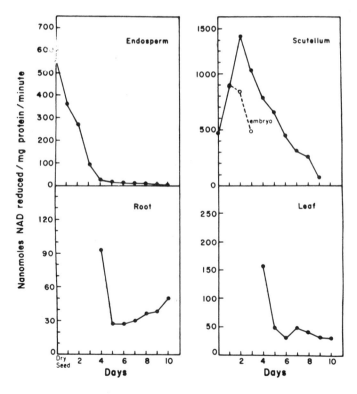

FIG. 6. Time course of ADH activity in those tissues available during the first 10 days of sporophytic development.

TABLE I

Tissue specificity of alcohol dehydrogenase in maize

Tissue	Specific activity[a]
Root	4·6
Sheath	4·2
Silk	1·6
Husk	3·2
Florets	2·7
Internodes (1–3 from top)	3·8
Internodes (4–11 from top)	4·3
Internodes (12–16 from top)	1·8
Leaves (3–5 from top)	6·7
Leaves (6–9 from top)	5·5

(Scandalios and Felder, 1971)

[a] Specific activity is expressed as nanamoles NAD reduced per mg protein per min. Tissues are from 40-day-old plant.

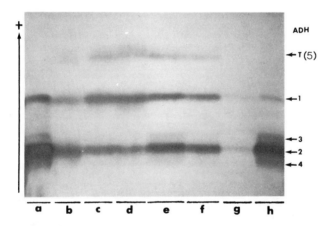

FIG. 7. Zymogram of the ADH isozyme complement of the pericarp at different stages of kernel development. Note the presence of isozyme ADH-5 characteristic of this tissue at the specific developmental stages. a and h are scutellar samples (markers). b = 19-day pericarp; c = 25-day pericarp; d = 30-day pericarp; e = 33-day pericarp; f = 38-day pericarp; g = 42-day pericarp. Migration is anodal.

Thus, in maize we have demonstrated the existence of five distinct isozymes of alcohol dehydrogenase which are under spatial and temporal control in their expression. The predominant isozymes are ADH-1 and ADH-2, the latter being the most active form. Cell fractionation experiments show that none of the ADH forms are associated with any subcellular organelles in any tissue or developmental stage, suggesting that maize ADH is a cytosolic enzyme.

D. GENETIC CONTROL OF ADH

1. Genetics

Of the five forms of alcohol dehydrogenase in maize only the two most common isozymes, ADH-1 and ADH-2, have been studied in depth from a genetic viewpoint. Both of these isozymes are polymorphic. There are two electrophoretic variants for each of the two isozymes, a relatively fast and a relatively slow one (Scandalios, 1967). Usually, an inbred maize strain contains either the fast variants of ADH-1 and of ADH-2 or else the slow variants of either enzyme. Heterozygotes between any two such variants result in kernels which express the parental types codominantly, but in addition a hybrid enzyme is generated at the ADH-2 position; no hybrid form is generated by ADH-1 in heterozygotes (Fig. 8). After appropriate genetic crosses were made, this phenotypic situation has been explained (Scandalios, 1967) by assuming two loci, *adh-1* and *adh-2*, each existing as two alleles, with *adh-1^F* coding for the fast variant and *adh-1^S* the slow variant of *adh-1*, and *adh-2^F* and *adh-2^S* specifying the fast and slow variants, respectively, of ADH-2. To account for the usual correlation in the relative electrophoretic mobilities of the ADH-1 and ADH-2 variants, a close linkage between *adh-1* and *adh-2* has been assumed, *adh-1^F* normally being linked with *adh-2^F*, and *adh-1^S* with *adh-2^S*.

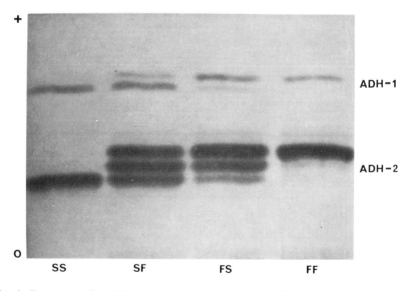

FIG. 8. Zymogram of ADH isozymes from single kernel milky endosperm (3n) extracts from homozygous S/S and F/F parents and their reciprocal heterozygotes (S/F and F/S). Note the apparent gene-dosage effects in the reciprocal F_1s. The dosage holds true for both ADH-1 and ADH-2. Migration is anodal.

Genetic evidence (Scandalios, 1969) for the two-loci hypothesis initially came from the recovery of three apparent recombinant phenotypes (ADH-1F/ADH-2S; ADH-1S/ADH-2F; and ADH-1FS/ADH-2F) resulting from self pollinating F_1 plants (*adh-1^F adh-2^F/adh-1^S adh-2^S*). The frequency with which these aberrant phenotypes were recovered is indeed low (1·491 × 10^{-6}) suggesting very close linkage. Such close linkage is not with precedent (Bahn, 1967). The important thing here is that the association between the electrophoretically fast and slow variants at the *adh-1* and *adh-2* loci is *not absolute*, as claimed by others (Schwartz, 1969), and that the recovery of recombinants, though rare, is compatible with the two gene hypothesis. It is also conceivable that the frequency of recombinational events is actually greater than observed and that recombinants are most frequently lethal in the inbred strains usually employed in such genetic studies. Thus, the rare recombinants observed really reflect the "frequency of recovery" of the aberrant types and likely do not reflect the true frequency of the "recombinational events."

Recently, we began screening open pollinated, randomly mating populations of maize in hopes of recovering possible aberrant ADH phenotypes. From American populations we have recovered an additional recombinant type (Fig. 9). Most interestingly, on screening exotic Mexican varieties we have discovered a situation where a variety is homozygous at the *adh-2* locus, but the *adh-1* gene is segregating, presumably due to the maintenance of the recombinant ADH-1S/ADH-2F in that variety, since the most frequent phenotypes are ADH-1F/ADH-2F and ADH-1FS/ADH-2F, and the least frequent is ADH-1S/ADH-2F (Fig. 10). These results lend further support to our hypothesis that ADH-1 and ADH-2 are controlled by two distinct, but closely linked loci. We are presently utilizing this material for more detailed genetic analysis. Additionally, we hope to resolve this problem further in the near future by peptide mapping of the purified variant enzymes.

Genetic analysis has not yet been done with respect to ADH-3, ADH-4, or ADH-5. That these may be epigenetic products, though unlikely based on preliminary experiments, has not been unequivocably eliminated at this point. Additionally, there are a few instances where mutations having affected the electrophoretic mobility of ADH-2 did not simultaneously affect ADH-3 (Fig. 11), suggesting that it is independent of ADH-2.

2. Gene Dosage and Subunit Structure

When milky endosperm from F_1 kernels is subjected to electrophoresis the resulting ADH zymograms reveal one of two kinds of heterozygote patterns depending on the direction of the cross (see Fig. 8). This is due to the triploid nature of the endosperm where the maternal genomic contribution is twice that of the pollen parent. Thus, the cross F♀ × S♂ results in a hybrid pattern with an activity concentration in the fast bands; the reciprocal

140   J. G. SCANDALIOS

cross S♀ × F♂ shows an activity concentration in the slow bands. Assuming ADH to be a dimeric enzyme from the genetic data, the expected binomial distribution of band activity in heterozygotes is $(2/3♀ + 1/3♂)^2$ for the 3n endosperm, and $(1/2♀ + 1/2♂)^2$ for the 2n scutellum. In fact, this is the situation observed. Of course, the formation of a hybrid ADH-2 isozyme is strong evidence for its dimeric structure. The fact that ADH-1 allelic interactions do not generate a hybrid molecule left the nature of its quaternary

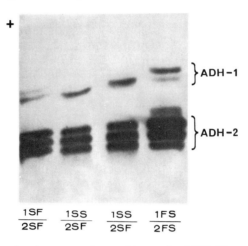

FIG. 9. Zymogram showing apparent recombinants for ADH. The two centre samples are from kernels that are apparently heterozygous for the *adh-2* gene, but homozygous for the *adh-1^S* gene. Phenotypes are indicated at bottom. Migration is anodal.

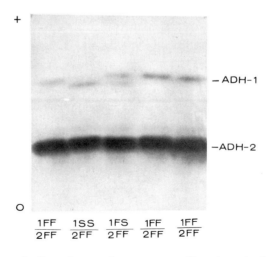

FIG. 10. Zymogram of milky endosperm from an open pollinated, randomly mating "exotic" variety of Central American maize. Each sample is from a single kernel. Note that they are homozygous for ADH-2F, but that the *adh-1* gene is segregating to give the phenotypes observed See text for discussion. Migration is anodal.

structure open (i.e. it could be either a monmer or a dimer). However, more recent data on relative molecular weights of ADH-1 and ADH-2 support the second assumption, that ADH-1 is also a dimeric enzyme. Although we reported (Scandalios, 1965) the first successful *in vitro* hybridization of a plant enzyme, our efforts with ADH have not been as rewarding, at least to our satisfaction. However, there has been a recent report (Fischer and Schwartz, 1974) that ADH-2 has been hybridized *in vitro*.

Thus, all evidence suggests that maize ADH probably exists functionally as a dimeric molecule.

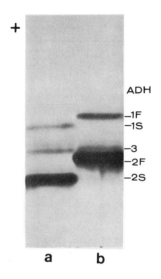

FIG. 11. Zymogram showing that an apparent mutation affecting the electrophoretic mobility of ADH-2 does not affect ADH-3, suggesting the probable genetic independence of the two isozymes. a. and b. indicate extracts from individual kernels of each genotype. F and S indicate the allelic variants at each locus. Migration is anodal.

E. PURIFICATION OF MAIZE ADH

Of the five ADH isozymes, ADH-2 is most amenable to purification because it is the most prevalent isozyme in the maize seed in terms of activity. Furthermore, it is well defined genetically and developmentally. Consequently it seemed logical to us to establish initial purification procedures utilizing ADH-2 in the hope that these would prove suitable in our efforts eventually to purify to homogeneity, and to characterize, each of the isozymes in terms of their physicochemical properties and possible physiological roles.

By use of fairly conventional methods, we have succeeded in purifying the allelic isozymes of ADH-2 (Felder and Scandalios, 1973). Table II summarizes the results of the purification of the allelic isozymes ADH-2F and ADH-2S.

It is interesting to note that the specific activity of ADH-2S is only about one-sixth as great as ADH-2F even though the slow variant was purified to a greater degree (654– and 216-fold, respectively). This fact confirms our previous report (Felder and Scandalios, 1971) that ADH-2S has a lower specific activity (i.e. 101·5 vs. 16·5); it has also been confirmed by other investigators (Marshall *et al.*, 1973). Thus, the lowered activity of ADH-2S compared to its allelic counterpart (ADH-2F) appears to be a reflection of the inherent structural differences between the isozymes. This situation is

TABLE II

Results of purification of maize alcohol dehydrogenase

ADH-2F					
Fraction	Protein[a]	Units[b]	Specific activity[c]	Recovery (%)	Purification
Crude supernatant	13,364	13,000	0·47	—	—
Protamine sulphate	9,747	12,737	0·63	100	1·4
$(NH_4)_2SO_4$	2,264	13,475	2·88	100	6·12
Sephadex G-100	558	8,200	7·12	63	15·1
DEAE-cellulose	19·8	4,160	101·5	32	216
ADH-2S					
Crude supernatant	11,088	700	0·026	—	—
Protamine sulphate	7,669	735	0·046	100	1·8
$(NH_4)_2SO_4$	598	561	0·455	80	17·6
Sephadex G-100	225	493	1·06	70	41·1
DEAE-cellulose	10	342·5	16·45	49	654

[a] Total mg of protein.
[b] Total units (one U = A_{340} nm/min.)
[c] Specific activity is μmoles of NADH formed per min per mg of protein in the standard reaction mixture.

TABLE III

Specific activity of alcohol dehydrogenase in the liquid endosperm of kernels from reciprocal F_1 crosses

Phenotype	Activity[a]
ADH-1F/ADH-2F	843·0
ADH-1FS/ADH-2FS	654·0
ADH-1SF/ADH-2SF	453·0
ADH-1S/ADH-2S	417·0

[a] Expressed as nmoles NAD reduced per min per mg of protein.

clearly reflected in the ADH activities of crude extracts of liquid endosperm from variant strains and their reciprocal hybrids (Table III).

F. PROPERTIES OF THE PURIFIED ADH ISOZYMES

1. *Stability of the Enzyme*

Purified enzyme loses activity when stored in $(NH_4)_2SO_4$ suspension at 3°C, but is completely stabilized for at least a two-month period in 50% ethylene glycol or glycerol in 0·025 M tris-HCl buffer, pH 7·0. NAD addition also has a stabilizing effect on the enzyme. However, β-mercaptoethanol causes instability of the pure enzyme (Fig. 12).

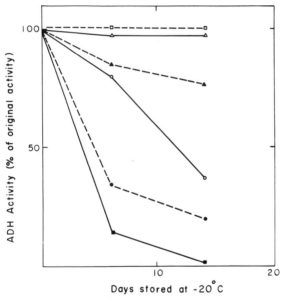

FIG. 12. Stability of purified ADH-2F. Forty units of enzyme per ml were stored in 25 mM tris-HCl, pH 7·0, containing various other compounds and tested for activity after storage for varying periods of time. (□) 50% ethylene glycol. (△) 50% glycerol. (▲) 5×10^{-5} M NAD. (○) buffer only. (●) 1 mM β-mercaptoethanol. (■) 10 mM β-mercaptoethanol.

2. *Optimum pH for Activity*

The activity of both the forward and reverse reactions was determined at various pH values, and the enzyme exhibited a rather broad pH range of activity (Felder and Scandalios, 1973). The reverse acetaldehyde reduction reaction has a slightly more acidic pH optimum (pH 8·5) than the forward reaction (pH 9·0); at pH 6·0, the reverse reaction occurs quite readily, but the enzyme is unable to oxidize ethanol at this pH.

3. Substrate Specificity

The enzyme can utilize various alcohols as substrates, but ethanol is the best substrate; methanol is not oxidized at all. For longer chain alcohols, the primary alcohol is always a better substrate than the secondary form. Formaldehyde is only reduced 4% as readily as acetaldehyde; anisaldehyde is not reduced at all.

4. Kinetic Constants

The Michaelis constants were determined for ethanol, NAD, acetaldehyde and NADH (Felder and Scandalios, 1973). The actual K_m values are summarized in Table IV. The K_m values for the various substrates are similar for both ADH-2F and ADH-2S, never showing more than a 3-fold difference.

TABLE IV

Summary of kinetic data for ADH-2F and ADH-2S

Isozyme	K_m(molarity)			
	NAD	NADH	Ethanol	Acetaldehyde
ADH-2F	7.5×10^{-5}	8.9×10^{-5}	1.35×10^{-2}	2.9×10^{-2}
ADH-2S	2.8×10^{-5}	8.5×10^{-5}	8.6×10^{-3}	5.0×10^{-2}

(Felder and Scandalios, 1973).

The K_m values obtained for the purified ADH-2 isozymes did not differ significantly from the apparent K_{ms} obtained for the same isozymes from partially purified preparations. Consequently a comparison of the above values with apparent K_{ms} for the partially purified ADH-1 isozymes may be useful. Such a comparison reveals that ADH-1S has a slightly higher K_m for ethanol than does ADH-1F, but the difference (about 2-fold) is small. The differences in K_m for ethanol are not however significant between ADH-1 and ADH-2. The apparent K_m for acetaldehyde is somewhat lower for ADH-1 than it is for ADH-2. ADH-1 also seems to favour the backward reaction (acetaldehyde → ethanol).

5. Molecular Weight Determinations

Molecular weights were estimated by sucrose density gradient centrifugation, sephadex gel filtration, and acrylamide gel electrophoresis. All methods were in close agreement, giving mol. wt values in the range of 60 000 ± 2500 for all of the ADH isozymes.

6. Amino Acid Analysis

The amino acid composition of the acid hydrolysate of ADH-2F and ADH-2S are compared in Table V. There is a similarity in the amino acid composition of the two proteins. The discrepancies observed in the relative compositions of glycine and alanine may be due to the incomplete resolution of these two amino acids on the ion exchange column. However, the close correspondence in the concentration of most amino acids suggests that ADH-2F and ADH-2S are very similar in amino acid composition.

TABLE V

Amino acid composition of the acid hydrolysates of ADH-2F and ADH-2S. Values are reported in residues/ 60,000 mol. wt.

Amino acid	ADH-2F	ADH-2S
ASP	61	61
THR	37	39
SER	43	39
GLU	62	61
PRO	37	40
GLY	55	62
ALA	93	73
VAL	46	49
CYS	—	—
MET	trace	trace
ILU	26	19
LEU	48	45
TYR	12	12
PHE	22	22
LYS	36	36
HIS	11	13
ARG	26	28

7. Thermolability of ADH Isozymes

In addition to lowered catalytic activity, ADH-1S and ADH-2S are less heat stable than the respective allelic forms, ADH-1F and ADH-2F (Fig. 13). ADH-1 is much more heat labile than is ADH-2. The heat stability values are similar for purified or partially purified ADH-2. ADH-1 was only partially purified.

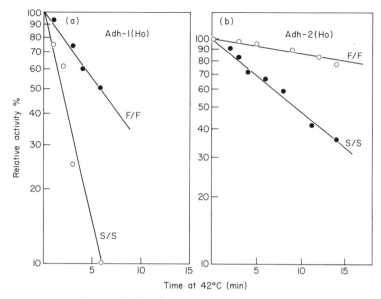

FIG. 13. Rate of heat-inactivation for the fast (F) and slow (S) variants of ADH-1 (a), and for ADH-2 (b). Small test tubes containing 0·1 ml of gel extract were incubated at 42°C for varying times, washed with reaction mixture, cooled and assayed in a recording spectrophotometer. (HO) = isozymes derived from homozygotes.

G. REGULATION OF ADH ACTIVITY

The expression of the ADH isozymes in maize is dependent on both tissue specificity and ontogenetic development. For this reason, the ADH system affords an opportunity to study the regulation of gene expression in a higher eukaryote. Possible controls for the expression of the *adh* genes in maize are likely exerted at the transcriptional, translational, and post-translational levels. We have chosen to focus our attention initially on the possible factors that may regulate the gene product (i.e. the enzyme), and subsequently work our way back to attempt to elucidate those factors which may exert their regulatory function on the translation of the specific mRNAs and on the transcription of the message from that portion of the genome which codes for the specific ADH isozymes.

1. Anaerobic Induction of ADH

Alcohol dehydrogenase of maize is known to increase under anaerobic conditions (Hageman and Flesher, 1960). The exact mechanism for the increase however, is not known, nor was it known whether the observed increase was due to total ADH or to some specific isozyme. What is known is that under anaerobic conditions (natural or induced) the Krebs cycle is blocked, leading to accumulation of pyruvate. Excess pyruvate may then be decarboxylated to form acetaldehyde which, acting as a substrate, induces ADH. If this be the case, anaerobiosis could possibly be used as a "tool" to study mechanisms involved in the regulation of the expression of the different ADH gene products in maize.

When various tissues (i.e. scutellum, root, shoot) taken from anaerobically grown seedlings were examined for ADH, significant increases (5 to 8-fold) in ADH activity were apparent (Nielsen and Scandalios, 1971); however, no new isozymes were detected. Hence, the apparent increase in ADH activity under anaerobiosis is not due to activation or "induction" of new isozymes. However, of all the ADH isozymes, the activity of ADH-1 increases most significantly (i.e. ADH-1 increases 9-fold; ADH-2 increases 3-fold).

Anaerobic induction of ADH is dependent on both stage of develpment and tissue specificity. Induction (% comparison) in the root and shoot is significantly higher (3 to 8-fold) than in the scutellum depending on the stage of development (Lai and Scandalios, unpublished data).

2. *Post-translational Regulation of ADH*

What are the factors or mechanisms regulating the pattern of ADH activity during early sporophytic development? Unlike most enzymes whose activity increases sharply after germination, alcohol dehydrogenase activity of maize scutella (see Fig. 2) and in germinating peas (Suzuki and Kyuwa, 1972) and wheat (Leblová and Stiborová, 1975) declines rapidly during this period. The slight increase observed during the first hours of germination is probably due to the anaerobic metabolism characteristic of that developmental period.

TABLE VI

Effect of cycloheximide and actinomycin D on development of maize scutellar alcohol dehydrogenase during seed germination

Treatment	H after start of water imbibition		
	24	72	168
	ADH activity (nmoles NAD reduced/scutellum/min) $\times\ 10^{-2}$		
Intact scutellum	23·4	11·5	0·8
Excised scutellum grown in nutrient[a]	23·4	13·5	2·5
Excised scutellum grown in nutrient with cycloheximide (10 µg/ml)	23·4	14·4	2·0
Excised scutellum grown in nutrient with actinomycin D (50 µg/ml)	23·4	14·0	0·75

(Ho and Scandalios, 1975).
[a] Scutellum was excised from 24-h-old seedling and cultured in Hoagland's solution.

However, the subsequent and rapid decline in ADH activity suggests at least two possibilities for the control of ADH: (1) faster degradation (or inactivation) than formation of active enzyme molecules; or (2) only degradation (or inactivation) without further formation of active enzyme molecules.

(a) *Effect of protein and RNA synthesis inhibitors.* The time course of ADH activity after germination remains unchanged in the presence of 10 µg/ml cycloheximide or 50 µg/ml actinomycin D, known protein and RNA synthesis inhibitors, respectively (Table VI). In contrast, 2 µg/ml cycloheximide and 50 µg/ml actinomycin D inhibit the increase of malate dehydrogenase activity completely in maize scutella within a short period of incubation (Yang and Scandalios, 1975) indicating that the insensitivity of ADH to these inhibitors cannot be due to general ineffectiveness of these inhibitors (e.g. because of lack of penetration). These data indicate that ADH molecules are not synthesized during seed germination (Ho and Scandalios, 1975).

(b) *Turnover studies.* To confirm the above suggestion that there is no synthesis of ADH occurring, we did an experiment employing gel electrophoresis and density labeling in conjunction with isopycnic equilibrium sedimentation (Ho and Scandalios, 1975). Results from such experiments clearly show that ADH molecules from scutella of seeds germinated in the presence of $^{15}NH_4Cl$ and D_2O for 36 h have exactly the same apparent buoyant density ($\rho = 1.297$) as the ADH molecules from scutella of seeds germinated in $^{14}NH_4Cl$ and H_2O (Fig. 14). Thus, no measurable amount of heavy isotopes was incorporated into the ADH molecules within the first 36 h of germination. Catalase in scutella of the same inbred strain and at the same developmental period showed a density shift as large as 0·019 g/ml when seeds were germinated in the presence of $K^{15}NO_3$ and D_2O under identical conditions as described above (Quail and Scandalios, 1971). Density shifts were also observed for the soluble and mitochondrial forms of malate dehydrogenase under similar conditions (Yang and Scandalios, 1975). Thus, the unchanged buoyant density of ADH molecules is not due to the failure to replace H by D, or ^{14}N by ^{15}N in the amino acid molecules during an active period of synthesis. Consequently, ADH is not synthesized *de novo* during the time examined, indicating that the *adh* genes are repressed either before or at the onset of the germination process.

(c) *Regulation by an endogenous ADH-specific inhibitor.* Having established that there is no turnover of ADH molecules during early sporophytic development; what then are the possible mechanisms controlling the decrease in ADH activity?

Reciprocal mixing experiments between scutellar extracts from early and late stages of germination resulted in a significant decrease in ADH activity

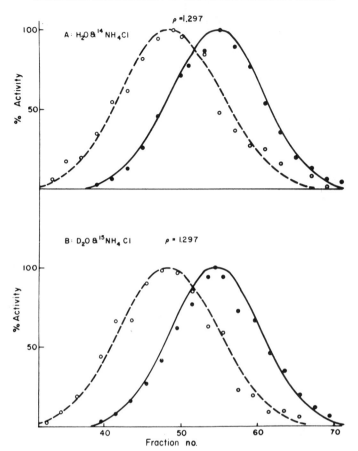

FIG. 14. Equilibrium distribution in CsCl gradients of scutellar extract ADH from seeds germinated for 36 h on either A $^{14}NH_4Cl$ in H_2O (control) or B 10 mM $^{15}NH_4Cl$ in 70% D_2O. Activity of lactage dehydrogenase (marker) (●—●); activity of alcohol dehydrogenase (○—○). Note absence of any density shift.

of earlier scutellar development extracts, suggesting the presence of an inhibitory substance for ADH (Fig. 15).

The inhibitor is specific for ADH, and pure or crude maize ADH preparations are effectively inhibited by endogenous ADH inhibitor isolated from root, shoot, or leaf (Table VII); thus, ADH inhibitor activity is present in the scutellum, root, shoot and leaf. Malate dehydrogenase, isocitrate dehydrogenase, and other enzymes present in the same tissue and developmental stages were not inhibited by this inhibitor. Interestingly, the inhibitor also inhibits ADH from other sources, though to varying degrees (e.g. yeast ADH, 68%; barley ADH, 81%; horse liver ADH, 4%; and pea ADH, 89%).

The inhibitor is thermolabile, but trypsin insensitive and is partially dialyzable. β-mercaptoethanol protects ADH activity against the inhibitor,

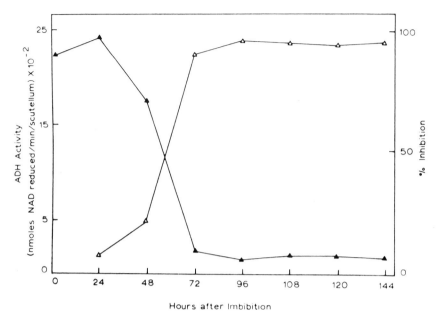

FIG. 15. Developmental time course of alcohol dehydrogenase activity and its specific inhibitor in maize scutella during germination. ADH activity (▲); inhibitor activity (△) is the percent inhibition of ADH by reciprocal mixing of scutellar extracts from 2-h imbibed seeds and later developmental stages. Mixtures were incubated at 25°C for 3 h.

TABLE VII

Tissue distribution of ADH inhibitor in maize

Treatment	Residual ADH activity %	Inhibition %
Scutellar extract	100	0
SE + root extract	45	55
SE + RE + SH[a]	91	9
SE + RE + NAD	64	36
SE + leaf extract	65	35
SE + LE + SH	88	12
SE + LE + NAD	85	15
SE + shoot extract	48	52
SE + ShE + SH	85	15
SE + ShE + NAD	90	10

(Ho and Scandalios, 1975).
[a] SE: scutellar extract; RE: root extract; SH: β-mercaptoethanol; LE: leaf extract; ShE: shoot extract.
β-Mercaptoethanol (100 mM) and NAD (350 μg/ml) were used. Scutellar extract was prepared from 3-h imbibed seeds. Root, shoot, and leaf extracts were prepared from 8-day-old seedlings.

while ethanol, acetaldehyde, or NAD^+ have no significant protective effect against the inhibitor (Table VIII). It is conceivable that the observed "inhibition" could be due to non-specific degradation by a protease. We believe this not to be the case based on the following evidence: (a) the "inhibitor" does not affect any of the other maize enzymes tested (see above); (b) the inhibition reaction is more rapid than most typical proteolytic reactions (80% of the ADH activity is eliminated in approximately 20 minutes); and (c) β-mercaptoethanol, which protects ADH from inhibition (Table VIII), has no effect on scutellar protease activity as measured by azocasein hydrolysis. This evidence does not rule out the possibility of a minor, specific protease for ADH, although such a phenomenon would be interesting in and of itself.

That the inhibitor is not a phenolic compound is supported by the fact that treatment of extracts with polyvinylpolypyrrolidone (PVP) does not eliminate or even decrease the inhibitory activity.

ADH activity in scutellar extracts from different stages of germination also decreases with time without reciprocal mixing if the extracts are incubated at 37°C. This decrease is probably also due to the presence of the ADH inhibitor, since the degree of inhibition in ADH activity is greater for the scutellar extracts from later stages of germination (Fig. 16). That the inhibitor can be synthesized in the scutellum and is not merely transported from other parts of the seedling is supported by the fact that excised scutella

TABLE VIII

Properties of ADH inhibitor in maize scutellum

Treatment	Residual ADH activity %	Inhibition %
Experiment 1		
Enzyme only	100	0
Enzyme + inhibitor	9	91
Enzyme + inhibitor + β-SH[a] (50 mM)	81	19
Enzyme + inhibitor + β-SH (10 mM)	63	37
Enzyme + inhibitor + DTT (3·5 mM)	33	67
Enzyme + inhibitor + NAD (0·14 mg/ml)	25	75
Enzyme + inhibitor + Ethyl alcohol (0·5%)	16	84
Experiment 2		
Enzyme only	100	0
Enzyme + inhibitor	14	86
Enzyme + inhibitor (boiled)	84	16

(Ho and Scandalios, 1975).

Enzyme represents scutellar extract from 2-h imbibed seed. Inhibitor represents scutellar extract from 8-day-old seedling; no detectable ADH activity was found in this preparation.

[a] β-SH: β-mercaptoethanol; DTT: dithiothreitol.

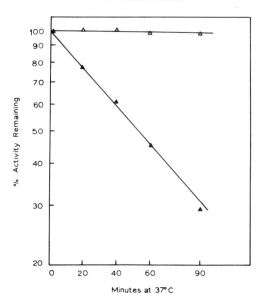

FIG. 16. Stability of alcohol dehydrogenase activity at 37°C. ADH activity from 2 h imbibed seed (△—△); ADH activity from 4-day-old seedlings (▲—▲). Enzyme was extracted from scutella.

cultured on nutrient media had even higher inhibitor activity than a comparable extract from scutella from intact seedlings (99·5% and 86% inhibition, respectively).

The inhibitor results clearly show that there is a concomitant decline of ADH activity as the inhibitor level increases after germination. This fact accompanied with the results from the metabolic inhibitor and density labelling experiments, lead us to conclude that the control of the decline in ADH activity after germination does not result from reduced levels of new synthesis of ADH molecules, but from the build up of a specific inhibitor to the enzyme. Preliminary results indicate that the level of inhibitor does not decrease as ADH activity increases under anaerobic conditions. Consequently, the inhibitor probably is not directly responsible for the effect of anaerobiosis on ADH; *de novo* synthesis of the enzyme is a more likely cause and is presently being investigated. The inhibitor may play a more subtle regulatory role by modulating the levels of ADH activity in specific tissues at given periods of development. This may obviate the necessity of more complex regulatory mechanisms such as "turning on and off of the *adh* genes" being exerted constantly.

We are presently engaged in purifying the inhibitor by affinity chromatography on immobilized ADH in an effort to further and more precisely characterize it. Additionally, we have been screening inbred maize strains for quantitative variants of the inhibitor for purposes of examining its genetic basis, and its genetic relationship to the *adh* structural genes. Both aspects appear promising to date.

REFERENCES

Bahn, E. (1967). *Hereditas* **58**, 1.
Eppenberger, H. M. (1975). *In* "Biochemistry of Animal Development" Vol. 3, pp. 217–255. Academic Press, New York and London.
Felder, M. R. (1975). *In* "Isozymes: Developmental Biology" (Markert, C. L., ed.), pp. 455–471. Academic Press, New York and London.
Felder, M. R. and Scandalios, J. G. (1971). *Molec. Gen. Genet.* **3**, 317.
Felder, M. R. and Scandalios, J. G. (1973). *Biochim. biophys. Acta* **318**, 149.
Fischer, M. and Schwartz, D. (1974). *Biochim. biophys. Acta* **364**, 200.
Hageman, R. H. and Flesher, D. (1960). *Archs Biochem. Biophys.* **87**, 203.
Hart, G. (1971). *Molec. Gen. Genet.* **3**, 61.
Ho, D. T. H. and Scandalios, J. G. (1975). *Pl. Physiol.* **56**, 56.
Jörnvall, H. (1970). *Nature* **225**, 1133.
Leblová, S. and Stiborová, M. (1975). *Biol. Pl.* **17**, 268.
Lutstorf, U. and Megnet, R. (1968). *Archs Biochem. Biophys.* **126**, 933.
Lutstorf, U. and von Wartburg, J. P. (1969). *FEBS Letts.* **5**, 202.
Markert, C. L. (1968). *Ann. N. Y. Acad. Sci.* **151**, 14.
Markert, C. L. and Møller, F. (1959). *Proc. natn. Acad. Sci. U.S.A.* **45**, 753.
Marshall, D. R., Broue, P. and Pryor, A. J. (1973). *Nature* **244**, 16.
Nielsen, G. and Scandalios, J. G. (1971). *Pl. Res.* **71**, 97.
Pietruszko, R. and Theorell, H. (1969). *Archs Biochem. Biophys.* **131**, 288.
Quail, P. H. and Scandalios, J. G. (1971). *Proc. natn. Acad. Sci. U.S.A.* **68**, 1402.
Scandalios, J. G. (1965). *Proc. natn. Acad. Sci. U.S.A.* **53**, 1035.
Scandalios, J. G. (1966). *Genetics* **54**, 359.
Scandalios, J. G. (1967). *Biochem. Genet.* **1**, 1.
Scandalios, J. G. (1969). *Science* **166**, 623.
Scandalios, J. G. (1974). *Ann. Rev. Pl. Physiol.* **25**, 225.
Scandalios, J. G. and Felder, M. R. (1971). *Devl. Biol.* **25**, 641.
Schenker, T. M., Teeple, L. J. and von Wartburg, J. P. (1971). *Europ. J. Biochem.* **24**, 271.
Schwartz, D. (1969). *Science* **164**, 585.
Schwartz, D. and Endo, T. (1966). *Genetics* **53**, 709.
Shaw, C. R. (1969). *Int. Rev. Cytol.* **25**, 297.
Sund, H. and Theorell, H. (1963). *In* "The Enzymes" (Boyer, P. D., Lardy, H. and Myrbäck, K., eds), Vol. 7, pp. 25–85. Academic Press, New York and London.
Suzuki, Y. and Kyuwa, K. (1972). *Physiologia. Pl.* **27**, 121.
Theorell, H. (1967). *In* "The Harvey Lectures" Ser. 61, p. 17. Academic Press, New York and London.
Ursprung, H. and Carlin, L. (1968). *Ann. N. Y. Acad. Sci.* **151**, 456.
Yang, N. S. and Scandalios, J. G. (1975). *Archs Biochem. Biophys.* **171**, 575.

CHAPTER 8

Conformation Changes and Modulation of Enzyme Catalysis

J. RICARD, J. NARI, J. BUC AND J.-C. MEUNIER

Laboratoire de Physiologie Cellulaire Végétale associé au C.N.R.S., Université d'Aix-Marseille II, Marseille, France

I. Regulation of Monomeric Enzymes 156
II. Unexpected Effects of Subunit Interactions. 166
III. Regulation of Enzyme Activity Through Protein-protein Interactions . 170
IV. Conclusions 174
References 174

One of the commonly accepted ideas of modern molecular biology is that regulatory enzymes are made up of several subunits. The modulation of their activity would then be exclusively mediated through subunit interactions and conformation changes. Two basic concepts have been extensively used in order to understand the molecular bases of enzyme regulation: the concept of a pre-equilibrium, in the absence of any ligand, between conformational states with symmetry conservation, and the concept of induced-fit (Koshland *et al.*, 1966). The first concept implies that the active site already exists on the free enzyme in the absence of any binding process. The second concept postulates that the active site is created upon the collision of a protein with a ligand and thus, that the conformation change is needed in order to get a productive enzyme-ligand complex.

These two basic ideas are not mutually exclusive. They can be considered as limiting cases of a more general scheme shown in Fig. 1. When the first concept is applied to a polymeric enzyme made up of identical subunits one gets the classical allosteric model of Monod *et al.* (1965). When it is the second concept which is applied to the polymeric enzyme, one gets the so-called sequential induced-fit models of Koshland *et al.* (1966). Any of these classical models allows one to express how subunit interactions and conformation changes control the fractional saturation of an enzyme by its substrate, but none of them tell us anything about the way they control the rate of catalysis, which is nevertheless the point of main biological importance.

156   J. RICARD ET AL.

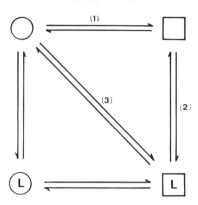

Fig. 1. Pre-equilibrium and induced-fit of a protein molecule. The reaction path 1, 2 corresponds to the shift of a pre-equilibrium between enzyme forms. The path 3 defines a typical induced-fit process.

Although there is no doubt that the regulation of many enzymes is effected by subunit interactions, it has also been recently shown that a conformation change occurring far from pseudo-equilibrium conditions can generate a pseudo-allosteric behaviour even for a one-sited enzyme (Ainslie *et al.*, 1972; Shill *et al.*, 1975; Ricard *et al.*, 1974b; Meunier *et al.*, 1974).

Lastly, much attention has been paid in recent years to the way the conformational constraints between different enzymes, within a multi-enzyme complex, control the activity of these enzymes.

The aim of this paper is to discuss in a non-mathematical way these rather new aspects of enzyme regulation, namely, regulation of monomeric enzymes, unexpected effects of subunit interactions, and regulation mediated through different protein interactions.

I. Regulation of Monomeric Enzymes

From a kinetic point of view, the regulatory behaviour of an enzyme appears as a departure from the usual Michaelis–Menten behaviour. The most frequent departures are an upward ("positive cooperativity") or a downward ("negative cooperativity") curvature of the reciprocal plots. Although the random binding of substrates on a rigid, or nearly rigid, enzyme can, in theory, generate these types of deviations, it seems unlikely that this effect can be experimentally found because of the frequent occurrence of asymptotic or degeneracy conditions (Dalziel, 1958; Cleland, 1970; Petersson, 1969; Petersson *et al.*, 1972). It is also possible, however, to explain the existence of a deviation from Michaelian behaviour of a one-sited enzyme by postulating the existence of conformational transitions between enzyme forms (Rabin, 1967; Witzel, 1967; Ainslie *et al.*, 1972; Shill and Neet, 1975; Rubsamen *et al.*, 1974; Ricard *et al.* 1974b; Meunier *et al.*, 1974). An

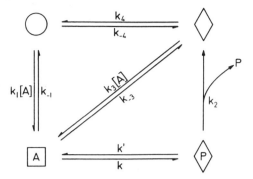

FIG. 2. Enzyme memory for a one-substrate, one-product monomeric enzyme.

interesting type of conformational transition which has been proposed recently (Ricard *et al.*, 1974b) to explain a kinetic cooperativity of a monomeric enzyme implies the existence of a molecular memory of the enzyme. This concept will now be discussed at length. The reaction model of a monomeric, one–substrate, one-product enzyme exhibiting memory phenomena is shown in Fig. 2. It is assumed that the enzyme "recalls" for a while the conformation stabilized by the product before relapsing to the initial conformation. If these two conformations (the circle and the rhombus in Fig. 2) are able to react with the substrate at different rates, the mnemonical enzyme, that is the enzyme exhibiting the memory phenomenon, will be characterized by reciprocal plots having either an upward or a downward curvature (Ricard *et al.*, 1974b).

This concept of enzyme memory can be obviously extended to the case of an enzyme conditioning the conversion of several substrates into several products through a compulsory reaction sequence. To do so, one has to define first this type of conformational transition which we call the mnemonical transition. Three requirements have to be fulfilled to define such a transition:

1. the free enzyme exists under two different conformation states in dynamic equilibrium (the circle and the rhombus of Fig. 3);
2. the collision of a ligand with any of these enzyme forms induces a new conformation (the square of Fig. 3) which is required for productive ligand binding;
3. another ligand M can be bound competitively, at the same site on one form only (the rhombus of Fig. 3) without any further conformation change.

The mnemonical transition is shown in Fig. 3. It obviously combines induced-fit phenomena (steps 2 and 3) with the existence of a pre-equilibrium shifted by the binding of a ligand. Now, if this mnemonical transition is inserted into an ordered reaction sequence, in such a way that L is the first

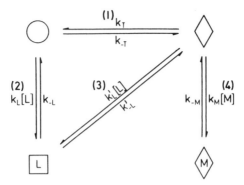

FIG. 3. The mnemonical transition.

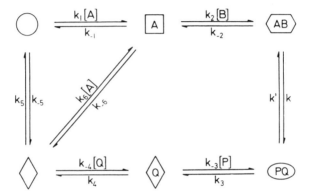

FIG. 4. Enzyme memory for a two-substrate, two-product monomeric enzyme following an ordered reaction mechanism.

substrate and M the last product, one gets a generalization of the concept of mnemonical model. For instance, if the mnemonical transition is a component of a compulsory two-substrate, two-product mechanism, one has the situation depicted in Fig. 4.

For a polymeric enzyme, the non-Michaelian behaviour is usually considered as a consequence of quaternary constraints and site–site interactions. For a one-sited mnemonical enzyme, departure from linear reciprocal plots appears as a consequence of a cooperation between two different conformation states of the same protein in the overall reaction process. It is obvious that, as for cooperativity between sites, the cooperation between conformation states of the enzyme can be either positive or negative. It has been shown analytically (Ricard et al., 1974b) and this point will not be discussed further here, that whatever the number of the substrates and that of the products, the extent of cooperation, Γ, between the two free enzyme forms can be expressed as

$$\Gamma = \frac{2k_T}{k_L k_L'^2} (k_L' - k_L) \tag{1}$$

The significance of the three rate constants k_L, k_L', k_T is to be found in the general scheme of Fig. 3. From eqn (1) it is obvious that the concavity of the reciprocal plots, that is the extent of cooperation between enzyme forms (the circle and the rhombus) is independent of the equilibrium constant between these enzyme forms, but is the consequence of enzyme memory. As a general rule, the substrates, and all but the last product, do not affect the cooperation. In the presence of the last product M (Fig. 3) and regardless the number of the substrates and products, the extent of cooperation becomes

$$\Gamma = \frac{2k_T}{k_L k_L'^2} \{k_L' - k_L(1 + K_M[M])\} \tag{2}$$

where K_M is the dissociation constant of the enzyme-M complex. Γ is thus a linear function of the concentration of M.

Thus, in the absence of any added product, the cooperation between the two enzyme forms will be positive if $k_L < k_L'$, negative if $k_L > k_L'$ and null if $k_L = k_L'$. This is exemplified by computer simulation in Fig. 5A. The last product to be released from the enzyme surface acts as a cooperation effector (eqn 2). This property can be used to strengthen a negative cooperation or to reverse a positive cooperation. These effects are shown by computer simulation in Fig. 5B.

It is obviously of cardinal importance to determine whether any real enzyme exhibits this type of conformational transition. As a matter of fact, it seems to be so. Four isohexokinases have been isolated from wheat germ (Meunier et al., 1971; Higgins and Esterby, 1974) and obtained in a homogeneous state. Two of them (LI and LII) have a mol. wt of 50 000, the two others (HI and HII) a mol. wt of 110 000. Both LI and LII are made up of only one polypeptide chain, while HI and HII are made up of two, apparently identical, chains.

It can be shown, by equilibrium dialysis experiments, that LI binds [^{14}C] glucose, but not [^{14}C] Mg ATP^{2-}. This result suggests that the binding of the substrates is ordered or at least effectively ordered*. Both the Klotz and Scatchard plots show the existence of only one binding site for glucose per enzyme molecule. In addition, these plots are linear (Fig. 6). However, if the Lineweaver–Burk plots are linear with respect to the reciprocal of MgATP^{2-}, they exhibit a downward curvature with respect to the reciprocal of glucose (Fig. 7). The effect of product glucose-6-phosphate on the concavity of the reciprocal plots is shown in Fig. 8A. The downward curvature

---

* A random model is defined as effectively ordered if a reaction path carries most of the reaction flux.

160

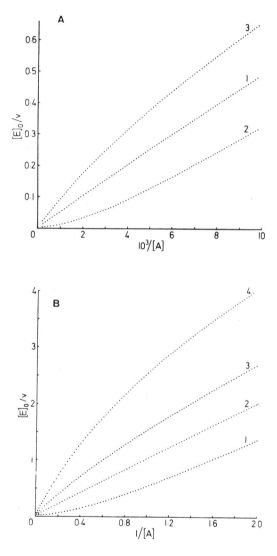

FIG. 5. Cooperation of a two-substrate, two-product mnemonical enzyme (computer outputs). A. Different shapes of the reciprocal plots: 1. Michaelian behaviour ($k_L = k'_L$), 2. positive cooperation ($k'_L > k_L$), 3. negative cooperation ($k_L > k'_L$) B. Reversal of cooperation by the last product: 1. no product, 2, 3, 4 increasing concentrations of the product. The numerical values of the rate constants can be found in Ricard *et al.* (1974).

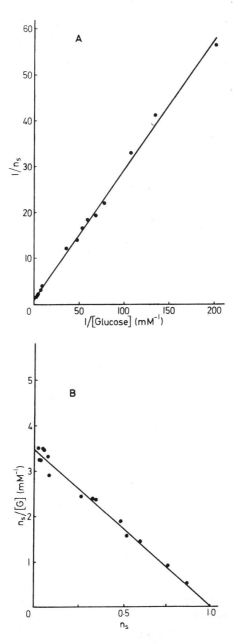

FIG. 6. Binding of [^{14}C] glucose on the wheat germ hexokinase LI. A. Klotz plot B. Scatchard plot. The experimental conditions can be found in Meunier *et al.* (1974).

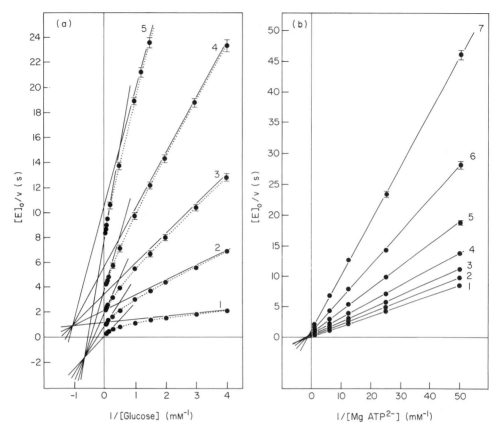

FIG. 7. Reciprocal plots for wheat germ hexokinase LI. A. Plots with respect to reciprocal of glucose. Dotted lines are obtained by non-linear least-squares fitting. Straight lines are asymptotes and tangents (obtained after fitting). B. Plots with respect to reciprocal of $MgATP^{2-}$. Plots are obtained by linear least-squares fitting. The experimental conditions can be found in Meunier *et al.* (1974).

of the reciprocal plots is obviously increased when the concentration of glucose-6-phosphate is increased. This can be even more clearly seen by estimating the Γ values of the plots and plotting these values (expressed in s M^2) against glucose-6-phosphate concentration (Fig. 8B). As predicted by equation 2 the extent of cooperation Γ increases linearly with glucose-6-phosphate concentration. The whole set of above results, along with many others (Meunier *et al.*, 1974) which have not been discussed here, allows one to conclude that wheat germ hexokinase LI is a mnemonical enzyme. The simplest scheme consistent with this conclusion is shown in Fig. 9. Since conformational equilibria involved in the mnemonical transition are considered to be an explanation of the departure from Michaelian behaviour, shifting or altering these equilibria by slight concentrations of denaturing

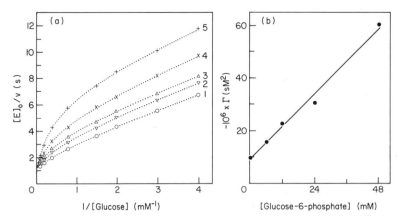

FIG. 8. Effect of glucose-6-phosphate on the cooperation of wheat germ hexokinase LI. A. Effect of glucose-6-phosphate on the curvature of the plots. 1. No glucose-6-phosphate; 2, 3, 4, 5 increasing concentrations of glucose-6-phosphate. B. Variation of the extent of cooperation Γ with glucose-6-phosphate concentration. The experimental conditions can be found in Meunier *et al.* (1974).

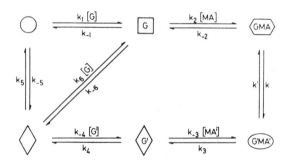

FIG. 9. The reaction model of wheat germ hexokinase LI. G = glucose, G′ = glucose-6-phosphate, MA = $MgATP^{2-}$, MA′ = $MgADP^{-}$.

agents would modify or even suppress the concavity of the primary plots. This is what is obtained when urea or sodium dodecylsulphate (SDS) are used (Fig. 10). The linearization of the Lineweaver–Burk plots ("desensitization") by urea or SDS is clearly in agreement with the existence of the conformational transition required for enzyme memory.

Mnemonical models such as those of Figs 2 and 4 can predict, during the transient phase, either bursts or lags. A burst *can* be obtained if $k_L > k'_L$ and a lag is *of necessity* obtained if $k_L < k'_L$. The amplitude of the burst and the length of the negative induction time are strongly dependent on the glucose concentration. This can be shown analytically but can be easily understood intuitively. For low substrate concentrations (if $k_L > k'_L$) the dominant reaction pathway will be that of Fig. 11a and the reaction will

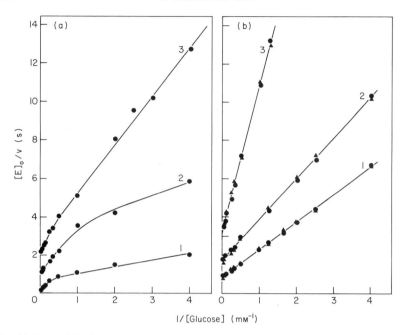

164   J. RICARD ET AL.

FIG. 10. Desensitization of wheat germ hexokinase by urea and SDS. A. Cooperation for various concentrations of MgATP^{2-}. B. Suppression of cooperation by urea (▲) or SDS (●). The experimental conditions can be found in Meunier *et al.* (1974).

exhibit no burst. For very high substrate concentrations the main pathway will now be that of Fig. 11c and the concentration of one of the free enzyme conformation (the circle conformation) will soon become negligible. Again, the reaction will show no burst. Lastly, for a certain range of substrate concentration the free enzyme will occur under the two conformational states (circle and rhombus) and the burst will correspond to the progressive shift, conditioned by the catalytic process itself, from the circle to the rhombus conformation (Fig. 11b). This situation is exactly what is found with wheat germ hexokinase LI (Fig. 12).

A very important question for the biologist is to know whether enzyme memory can represent a possible device for the regulation of metabolic processes. Since the mnemonical transition can give rise either to a positive or a negative cooperation between conformation states, the "advantages" of enzyme memory are similar to those exhibited by allosteric enzymes (sharp response of the enzyme for a narrow range of substrate concentrations) or by enzymes exhibiting negative cooperativity (comparatively high reaction rate for very low substrate concentrations). However, the mnemonical enzymes have a unique property: that of having their "regulatory behaviour" modulated by the last product of the reaction sequence. This property is obvious by simple inspection of Fig. 11. For low substrate concentrations,

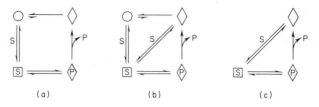

FIG. 11. Variation of the burst with substrate concentration for a mnemonical enzyme.

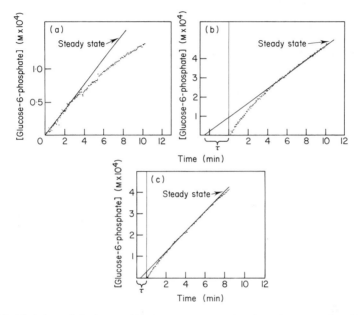

FIG. 12. Variation of the burst with glucose concentration for wheat germ hexokinase LI.
A. Glucose concentration 10^{-5} M (no burst). B. Glucose concentration 0·1 M (large burst).
C. Glucose concentration 0·6 M (small burst). Reaction is followed by monitoring decoloration
of cresol red. Measurements are effected with a stopped-flow spectrophotometer connected with
a computer.

the free enzyme occurs under two different conformations (the circle and
the rhombus), one of them only (the rhombus) being able to bind the product
(Fig. 11a). For high substrate concentrations the free enzyme does appear
under one conformation only (the rhombus) able to bind the product
(Fig. 11c). This property implies that compared with a Michaelian enzyme,
a mnemonical enzyme can be more weakly inhibited by the product at low
substrate concentration than at high substrate concentration. Since many
enzymes in the living cell catalyse the transformation of a substrate in the
presence of the corresponding product, the mnemonical transition is a device
that can represent an "advantage" evolved for the control of a metabolic
chain.

II. Unexpected Effects of Subunit Interactions

As outlined above, the classical models of Monod and Koshland tell us nothing about how the rate of catalysis is modulated by subunit interactions. We have developed in recent years a thermodynamic formalism that allows one to link the rate of catalysis with thermodynamic parameters directly related to subunit interactions. The principles of this method have been presented elsewhere (Ricard *et al.*, 1974a) and will not be discussed here. However, an interesting conclusion of this theoretical work is that subunit interaction does not affect in the same way the rate of product appearance and the fractional saturation of the enzyme by the substrate. This is clearly seen by computer simulation (Fig. 13) of rate equations and binding isotherms.

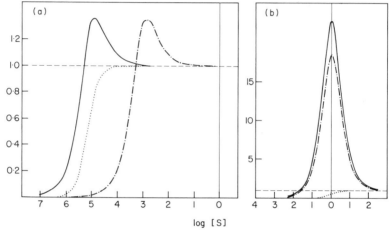

Fig. 13. Relative rates and binding isotherms for a dimeric enzyme. A. Simple sequential model of Koshland. B. Allosteric model of Monod. Dotted lines (.....) correspond to the binding isotherms. Full lines (——) correspond to the pseudo-equilibrium rate, and broken lines (----) to the steady-state rates, respectively. Curves are obtained by computer simulation. See Ricard *et al.* (1974).

We have applied this type of reasoning to the study of the regulatory properties of phenylalanine ammonia-lyase (Nari *et al.*, 1974). The enzyme has been isolated from wheat seedlings, purified and obtained in homogeneous state on analytical centrifugation and polyacrylamide gel electrophoresis. Phenylalanine ammonia-lyase of wheat seedlings has a mol. wt of 320 000 and a sedimentation constant at infinite dilution of 12 S. The protein molecule appears to be made up to two pairs of unidentical subunits. Two of them have a mol. wt of 75 000 (α-chains), the two others a mol. wt of 85 000 (β-chains). Havir and Hanson (1973, 1975) have also obtained phenylalanine ammonia-lyase in a homogeneous state from maize, potato and *Rhodotorula glutinis*. They have found that these enzymes are made up of four apparently identical polypeptide chains. They have thus suggested that in the wheat enzyme the light α chains could result from the partial proteolysis of the β-chains during enzyme purification (Havir and Hanson, 1973). This possibility cannot be dismissed at the moment. However, it is worth noting that Kalghatgi and Subba Rao (1975) have purified from *Rhizoctonia solani* a phenylalanine ammonia-lyase made up of two pairs of unidentical chains, with mol. wt of 70,000 (α) and 90,000 (β), respectively. It is thus quite possible that depending on the nature of the organism, phenylalanine ammonia-lyase can be made up of either identical or unidentical polypeptide chains.

Phenylalanine ammonia-lyase of wheat germ seedlings exhibits a negative cooperativity (Fig. 14). However, if the reaction rates are measured in the presence of both the substrate (L-phenylalanine) and a substrate analogue (D-phenylalanine or benzoic acid) this negative cooperativity can be suppressed or even reversed in the presence of the inhibitor (Fig. 15).

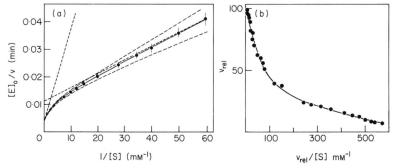

FIG. 14. Negative cooperativity of phenylalanine ammonia-lyase. A. Lineweaver-Burk plot. The full line is obtained by non-linear least-square fitting. The dotted straight lines (. . . .) are the asymptote and the tangent. The broken lines are the envelopes of expected error. B. Eadie plot. The experimental conditions can be found in Nari *et al.* (1974).

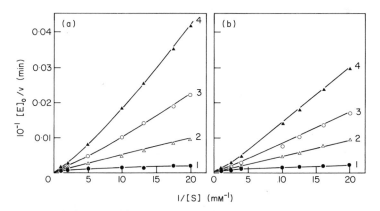

FIG. 15. Reversal of cooperativity by D-phenylalanine or benzoic acid. A. Effect of D-phenylalanine. Curve 1 no D-phenylalanine. Curves 2,3,4 increasing concentrations of D-phenylalanine. B. Effect of benzoic acid. Curve 1 no benzoic acid. Curves 2,3 increasing concentrations of benzoic acid. The experimental conditions can be found in Nari *et al.* (1974).

We have developed kinetic methods that allow one to discriminate among reaction schemes. These methods have been applied to the wheat phenylalanine ammonia-lyase. We have been able to conclude that although made up of four polypeptide chains, it behaves as a dimer. Each protomer (160 000), in the sense of Monod *et al.* (1965) would consist of two polypeptide chains; an α-chain (75 000) and a β-chain (85 000). The detailed analysis of many rate data such as those of Fig. 16 show that the enzyme can bind one molecule of D-phenylalanine only, but two molecules of benzoic acid.

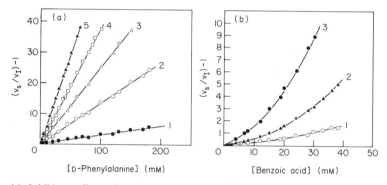

FIG. 16. Inhibitory effects of D-phenylalanine and benzoic acid. A. Effect of D-phenylalanine. The linearity of the plots (obtained for different concentrations of substrate) indicates that one molecule of D-phenylalanine is bound per dimer. B. Effect of benzoic acid. The parabolic character of the plots (obtained for different concentrations of substrate) indicates that two molecules of benzoic acid are bound per dimer. V_s is the steady-state rate in the absence of inhibitor, v_1 is the steady-state rate in the presence of both the substrate and the inhibitor. The interested reader is referred to Nari *et al.* (1974).

Moreover, the enzyme cannot form any hybrid enzyme–substrate–inhibitor complex. The available results tentatively suggest that phenylalanine ammonia-lyase follows neither the classical allosteric model of Monod *et al.* (1965), nor the simple sequential model of Koshland *et al.* (1966) but a partially concerted model of subunit interactions (Nari *et al.*, 1974). The binding of the substrate induces a conformation change of the corresponding subunit as well as the appearance of a new conformation, A', of the unliganded protomer. Since no hybrid enzyme—substrate—inhibitor complex is formed, it appears very likely that the new conformation A' is still able to bind the substrate but unable to bind D-phenylalanine or benzoic acid. Since the unliganded molecule of L-phenylalanine ammonia-lyase can bind one molecule of D-phenylalanine, but two molecules of benzoic acid, it must be concluded that the binding of D-phenylalanine on a subunit induces a conformation change of the unliganded subunit, the new conformation (A'') being unable to form a complex with any molecule of inhibitor. In the case of benzoic acid on the other hand, the new conformation of the unliganded protomer is still able to bind an inhibitor molecule. These reaction models are shown in Fig. 17. They predict the reversal of the negative cooperativity. This effect can be shown to occur both analytically and by computer simulation (Fig. 18).

It is worth noting that the partially concerted mechanism is certainly not specific of the wheat enzyme for it has been found to occur with the *Rhizoctonia* phenylalanine ammonia-lyase. However in this case Kalghatgi and Subba Rao (1975) were unable to find an inhibitor which can bind two molecules per enzyme molecule.

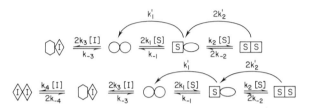

FIG. 17. Reaction scheme of phenylalanine ammonia-lyase. The substrate molecule S binds to the first subunit and induces the square conformation of the corresponding subunit and a new conformation (the ellipse conformation, A') of the unliganded subunit. That A' conformation can still bind a substrate molecule but not an inhibitor molecule. Thus, a hybrid enzyme–substrate–inhibitor complex cannot form. Upper model: one molecule of D-phenylalanine (I) can be bound per dimer. The hexagon (A'') conformation can bind neither the substrate (S) nor the inhibitor (I). Lower model: two molecules of benzoic acid (I) can be bound per dimer. The hexagon (A'') conformation can bind the inhibitor (I).

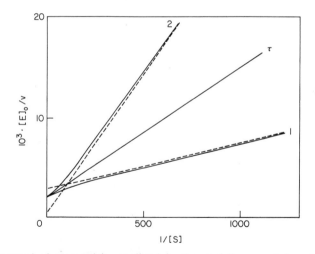

FIG. 18. Reversal of cooperativity predicted by the partially concerted model. Curve 1 is obtained for "low" inhibitor concentration (negative cooperativity), while curve 2 is obtained for "high" inhibitor concentration (positive cooperativity). It is assumed that one inhibitor molecule is bound per dimer (Upper model of Fig. 17). Whatever the inhibitor concentration, the reciprocal plots should have the same tangent, τ at the origin, and the asymptotes (dotted lines) should intersect at the same point on the tangent τ. Curves, asymptotes and tangents are obtained by computer simulation. See Nari *et al.* (1974).

The allosteric model predicts only lags during the transient phase, but the partially concerted models generate either bursts or lags (Mouttet *et al.*, 1974). A transient kinetic study of the phenylalanine ammonia-lyase reaction has shown the occurrence of a "slow" burst when enzyme and substrate are rapidly mixed in a stopped-flow apparatus (Fig. 19).

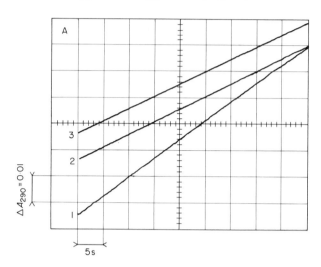

FIG. 19. Transient kinetics of phenylalanine ammonia-lyase reaction. Oscilloscope trace 1 corresponds to the transient appearance of cinnamate. Oscilloscope traces 2 and 3 correspond to the steady-state appearance of the cinnamate. Obviously, the rate of product appearance is higher under pre-steady-state conditions, than under steady-state conditions (burst). The experimental conditions can be found in Mouttet *et al.* (1974).

III. REGULATION OF ENZYME ACTIVITY THROUGH PROTEIN-PROTEIN INTERACTIONS

It has appeared increasingly obvious, in recent years, that many enzymes in the living cell occur as multi-enzyme complexes. It is then of cardinal importance to know how protein–protein interactions modulate the enzymic activity within the complex.

In chloroplasts, ferredoxin and the flavoprotein NADP reductase occur as 1:1 complex (Foust *et al.*, 1969). This complex participates, as a component of photosystem I, in the electron transfer from water to NADP. However, both the flavoprotein and the complex catalyse the electron transfer in the reverse direction from NADPH to various acceptors such as ferricyanide, dichlorophenol indophenol, NAD, cytochrome f (Forti *et al.*, 1968). While in the forward direction the electrons are transferred first to ferredoxin, then to flavoprotein and to pyridine nucleotide, in the non-physiological backward direction the electrons are not transferred to ferredoxin. The inability of NADPH to reduce ferredoxin is obviously due to the rather large span (0·15V) of oxidation-reduction potentials between pyridine nucleotide and ferredoxin. Although this non-heme iron protein does not play by itself any role in the hydride ion transfer between NADPH and the dye dichlorophenol indophenol, it considerably modifies the kinetics of this process.

The Lineweaver–Burk plots describing the hydride ion transfer by the flavoprotein alone are shown in Fig. 20. For low NADPH concentrations, the plots are parallel straight lines, and their intercepts are a linear function of the reciprocal of the dye concentration. These results, along with others, indicate that the hydride ion transfer from NADPH to dichlorophenol indophenol follows a ping-pong mechanism, that is the flavoprotein is reduced by NADPH, NADP is released, and then the hydride ion is transferred to the dye. However, for high concentrations of NADPH an inhibition by excess substrate occurs (Fig. 20). This inhibition is due to the fact that the flavoprotein accepts a second hydride ion, as already outlined (Zanetti *et al.*, 1969). This process certainly corresponds to the reduction of the disulphide bond and is not associated with any inactivation of the enzyme.

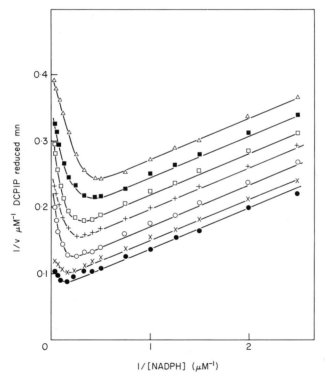

FIG. 20. Reciprocal plots for the hydride ion transfer catalysed by NADP reductase. The plots are obtained for various concentrations of the acceptor dichlorophenol indophenol (DCPIP). The steady-state rate measurements are obtained by the stopped-flow technique. The experimental conditions will be given elsewhere.

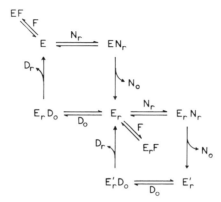

FIG. 21. Tentative scheme of the hydride transfer. E = oxidized enzyme, E_r = half-reduced enzyme (2 electrons), E_r' = fully-reduced enzyme (4 electrons), N_r = reduced NADP, N_0 = oxidized NADP, D_0 = oxidized dye, D_r = reduced dye, F = ferredoxin. The cycle involving E and E_r corresponds to the reduction of the flavin. The other cycle involving E_r and E_r' implies reduction of the disulphide bond.

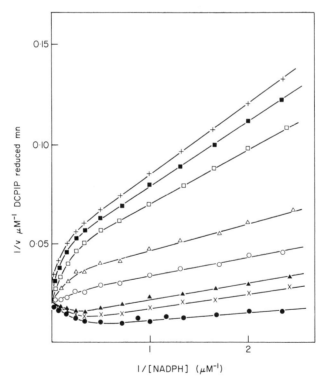

FIG. 22. Effect of ferredoxin on the hydride ion transfer. The plots are obtained for different concentrations of ferredoxin. The upper curves (with downward curvature) describe the effect of high ferredoxin concentrations and the lower curves the effect of low ferredoxin concentrations (inhibition by excess substrate). The experimental conditions will be given elsewhere.

The flavoprotein occurs under three forms: the oxidized, the half-reduced and the fully-reduced states. The reaction proceeds through two cycles as shown in Fig. 21.

However, if ferredoxin is present in the reaction medium, at low ionic strength, the hydride ion transfer is strongly decreased. Moreover, the inhibition by excess NADPH is replaced by a downward curvature of the reciprocal plots (Fig. 22).

To explain that ferredoxin dramatically changes the kinetics of hydride ion transfer, we have performed direct binding experiments of NADP and ferredoxin to the flavoprotein. It then turned out that there is a competition between these ligands. The binding of NADP brings about the dissociation of the complex, and conversely the binding of NADP does not allow that of ferredoxin. This competition (Fig. 23) can be formulated by a binding equation, the validity of which can be checked experimentally (Fig. 24).

$$\text{EF} \; \underset{k_{-2}}{\overset{k_2\,[\text{F}]}{\rightleftharpoons}} \; \text{E} \; \underset{k_{-1}}{\overset{k_1\,[\text{N}]}{\rightleftharpoons}} \; \text{EN}$$

Binding isotherm

$$\frac{\psi}{1-\psi} = K_2\,e_0 \left(\frac{f_0}{e_0} - \psi\right)\left(1 - \frac{K_1\,n_0}{K_1\,n_0 + 1}\right)$$

Fig. 23. Competition between ferredoxin and NADP. Binding isotherm for ferredoxin. Ψ = fractional saturation of flavoprotein by ferredoxin, e_0 = total concentration of flavoprotein, f_0 = total concentration of ferredoxin, n_0 = total concentration of NADP, K_1 = affinity constant of NADP (k_1/k_{-1}), K_2 = affinity constant of ferredoxin (k_2/k_{-2}).

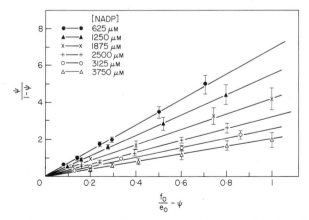

Fig. 24. Competition between ferredoxin and NADP. Binding plots. The slope of the plots decreases when concentration of NADP increases.

The surprising conclusion of this study is that the bi-enzyme complex does not participate in the hydride ion transfer. Ferredoxin simply controls the concentration of the flavoprotein which is the active species. This conclusion is summarized in the tentative scheme of Fig. 21.

IV. CONCLUSIONS

Several general ideas have emerged in the course of this article and it is now of interest to summarize them. Three levels of complexity can be distinguished in the organization of devices which control enzyme activity. The spontaneous conformation changes of some monomeric enzymes constitute a device of a first level of complexity. We have paid, in recent years, much attention to the problem of enzyme memory and shown this type of device to have similarities with allostery. However, a unique property of the cooperation between two conformation states of an enzyme is to be controlled by the concentration of the last product to be released from the active site during the reaction cycle.

A second level of complexity can be found in the quaternary constraints and subunit interactions that occur within a polymeric enzyme molecule. These interactions efficiently modulate the rate of product appearance. They can give rise not only to sigmoidal or anticooperative kinetics, but also to other phenomena such as substrate inhibition, or reversal of cooperativity by a ligand. Phenylalanine ammonia-lyase is certainly a good example of a reversion of a negative cooperativity by an analogue (D-phenylalanine or benzoic acid) of the substrate.

The association of an enzyme with another protein constitutes a device of a third degree of complexity. This association can either confer to the enzyme novel properties, or more simply, can control, through association–dissociation phenomena, the concentration of the active species. This seems to be the case for the flavoprotein-ferredoxin complex of chloroplasts, which is dissociated by NADP. There is no doubt that intracellular association of an enzyme with other proteins can represent an efficient way of modulating its activity.

REFERENCES

Ainslie, R. E., Schill, J. P. and Neet, K. E. (1972). *J. biol. Chem.* **247**, 7088.
Cleland, W. W. (1970). *In* "The Enzymes" (Boyer, P. D., ed.), Vol. 2, pp. 1–65. Academic Press, London and New York.
Dalziel, K. (1958). *Trans. Faraday Soc.* **54**, 1247.
Forti, G. and Sturani, E. (1968). *Europ. J. Biochem.* **3**, 461.
Foust, G. P., Mayhew, S. P. and Massey, V. (1969). *J. biol. Chem.* **244**, 964.
Havir, E. A. and Hanson, K. R. (1973). *Biochemistry* **12**, 1583.
Havir, E. A. and Hanson, K. R. (1975). *Biochemistry* **14**, 1620.
Higgins, T. and Esterby, J. S. (1974). *Europ. J. Biochem.* **45**, 147.

Kalghatgi, K. K. and Subba Rao, P. V. (1975). *Biochem. J.* **149**, 65.
Koshland, D. E., Nemethy, G. and Filmer, D. (1966). *Biochemistry* **5**, 365.
Meunier, J. C., Buc, J. and Ricard, J. (1971). *FEBS Letts.* **14**, 25.
Meunier, J. C., Buc, J., Navarro, A. and Ricard, J. (1974). *Europ. J. Biochem.* **49**, 209.
Monod, J., Wyman, J. and Changeux, J. P. (1965). *J. molec. Biol.* **12**, 88.
Mouttet, Ch., Fouchier, F., Nari, J. and Ricard, J. (1974), *Europ. J. Biochem.* **49**, 11.
Nari, J., Mouttet, Ch., Pinna, M. H. and Ricard, J. (1972). *FEBS Letts.* **23**, 220.
Nari, J., Mouttet, Ch., Fouchier, F. and Ricard, J. (1974). *Europ. J. Biochem.* **41**, 499.
Petersson, G. (1969). *Acta chem. scand.* **23**, 2717.
Petersson, G. and Nylen, V. (1972). *Acta chem. scand.* **26**, 420.
Rabin, B. R. (1967). *Biochem. J.* **102**, 226.
Ricard, J., Mouttet, Ch. and Nari, J. (1974a). *Europ. J. Biochem.* **41**, 479.
Ricard, J., Meunier, J. C. and Buc, J. (1974b). *Europ. J. Biochem.* **49**, 195.
Rubsamen, H., Khandker, R. and Witzel, H. (1974). *Hoppe-Seyler's Z. Physiol. Chem.* **355**, 1.
Schill, J. P. and Neet, K. E. (1975), *J. biol. Chem.* 2259.
Witzel, H. (1967). *Hoppe-Seyler's Z. Physiol. Chem.* **348**, 1249.
Zanetti, G. and Forti, G. (1969). *J. biol. Chem.* **244**, 4757.

CHAPTER 9

Proteolytic Inactivation of Enzymes

W. WALLACE

Department of Agricultural Biochemistry, Waite Agricultural Research Institute, University of Adelaide, South Australia

I. Introduction	177
II. Occurrence and Main Properties of Inactivating Enzymes . . .	178
A. Tryptophan Synthase Inactivation in Yeast	178
B. Group Specific Proteases in Animal Tissues	178
C. Nitrate Reductase Inactivating Enzyme in the Maize Root . .	180
D. Other Examples of Specific Protease Action in Higher Plants . .	183
III. Characterization of Inactivating Enzymes	183
A. Active Site Inhibitors	183
B. Estimation of Protease Activity	184
C. Esterase Activity	187
IV. Specificity of Inactivating Enzymes	187
V. Evidence for Inhibitors of Inactivating Enzymes. . . .	189
VI. Intracellular Compartmentation of the Inactivating Enzymes . .	190
VII. Control of Inactivating Enzymes during Cell Extraction . . .	192
Acknowledgement.	193
References	193

I. INTRODUCTION

The term inactivation is used in this review to describe the irreversible loss of catalytic activity that follows denaturation of an enzyme. Since 1972 considerable evidence has been reported for the occurrence of proteases which specifically inactivate one or a limited group of enzymes. The two most extensive studies are those of Holzer and his colleagues on yeast and that of Katunuma and associates in animal tissues. A brief resume of these is given as a preface for the consideration of a similar type of protease in the maize root which inactivates nitrate reductase. The main aim of the review is to describe the characteristics of these inactivating enzymes and the methodology involved in establishing their proteolytic action. No attempt is made to speculate on the function of the maize root inactivating enzyme in the cell. There is good evidence that the yeast and animal proteases do function *in vivo* (Holzer, 1975; Kominami and Katunuma, 1976) and the reader is referred to a review of Holzer *et al.* (1975) for discussion on the possible function and control of such intracellular proteases. Terms such as nitrate reductase inactivating enzyme or maize root inactivating enzyme are used for descriptive purposes only and appear more convenient than alternatives such as "a proteolytic enzyme which inactivates nitrate reductase". In the maize root, as in yeast, the proteolytic activity could be that of a more general protease which may or may not have been characterized.

II. Occurrence and Main Properties of Inactivating Enzymes

A. TRYPTOPHAN SYNTHASE INACTIVATION IN YEAST

In the early studies on an enzyme fraction in yeast which inactivated tryptophan synthase it was found to consist of two components named Inactivase I and Inactivase II (Katsunuma *et al.*, 1972). Both were heat labile, non-dialysable and precipitable by $(NH_4)_2SO_4$. In addition to the inactivation of tryptophan synthase the Inactivases also degraded haemoglobin, albumin and casein. It was shown with these proteins and with a partially purified sample of tryptophan synthase that the action of the Inactivases resulted in the release of trichloroacetic acid soluble material which absorbed at 280 nm (Schott and Holzer, 1974). Inactivase II degrades azocoll (see Section IIIB) and exhibits esterase activity against ATEE and BAEE (see Section IIIC). It is inhibited by DFP and PMSF (see Section IIIA).

It has subsequently been demonstrated by Saheki and Holzer (1974) that Inactivase I and II were equivalent to an earlier described yeast protease A (EC 3.4.23.8) and yeast protease B (EC 3.4.22.9). They reported that yeast protease B had maximum activity against tryptophan synthase at pH 5·0 while the optimum pH with azocoll was 7·0. The influence of pH on the inactivation of tryptophan synthase by protease A was not reported but its optimum activity on acid denatured haemoglobin was at pH 3·7 (Lenney, 1956).

Inactivating activity against tryptophan synthase was not observed in extracts of yeast cells in the early exponential phase of growth (Katsunuma *et al.*, 1972). It appeared near the end of exponential growth and reached a high level in the stationary phase. Both proteases were more active on the apo-form than the holo-form of tryptophan synthase (Schott and Holzer, 1974).

B. GROUP SPECIFIC PROTEASES IN ANIMAL TISSUES

Inactivating enzymes have been described in rat tissues which inactivate the apo-form of either pyridoxal-dependent, NAD-dependent or FAD-dependent enzymes (Katsunuma *et al.*, 1972). It was established that they each had a proteolytic action and four distinct proteases acting on pyridoxal-dependent enzymes have now been described (Katunuma *et al.*, 1975). The pyridoxal group specific proteases have their maximum activity in the alkaline pH range e.g. for inactivation of ornithine transaminase the optimum pH was 8·6–9·0. They are inhibited by DFP and are active with ATEE but not BAEE (see Section IIIc).

The group specific proteases are considered to act by causing a limited proteolysis of their enzyme substrate (Kominami *et al.*, 1975). When highly purified skeletal muscle protease and rat liver ornithine transaminase were incubated (Fig. 1) the loss of activity of the latter was correlated with the release of ninhydrin positive substances. No significant decrease occurred in the amount of protein (ornithine transaminase) recovered by precipitation with 5% trichloroacetic acid. It was shown by paper chromatography that only a few peptides were released during the inactivation reaction. The main component of inactive ornithine transaminase remaining showed no significant alteration in tertiary structure from its native form. It had the same characteristics on polyacrylamide gel, the same antigenic reaction, it retained its capacity to bind the pyridoxal coenzyme and there was no significant alteration in its molecular weight (Kominami *et al.*, 1975). Further degradation of the large inactive fragment was very slow. However, when native

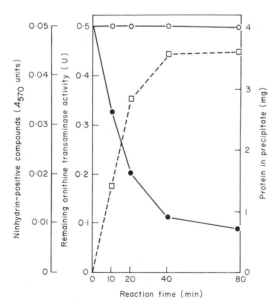

FIG. 1. Kinetics of proteolysis of ornithine transaminase. Reaction mixtures containing 2 U purified skeletal muscle protease and 4 mg ornithine transaminase apoenzyme II in a final volume of 1 ml were incubated at 37°C. At the indicated times aliquots were removed for assay of ornithine transaminase activity (●). Then trichloracetic acid was added at a final concentration of 5%, and the protein concentration in the precipitate (○) and ninhydrin-positive substances in the supernatant (□----□) were analysed. From Kominami *et al.* (1975).

ornithine transaminase was denatured by treatment with 8 M urea at 4°C for 60 min it was then rapidly and extensively degraded by its group specific protease.

The fact that the group specific proteases in rat tissues are not active on the holoenzyme suggest that either they attack the coenzyme binding site or that the conformation of the apoenzyme is more susceptible to attack. The proteases increase under various dietary conditions: during starvation, on a protein free or protein rich diet and under conditions of vitamin deficiency e.g. vitamin B_6 in the case of pyridoxal group proteases (Katunuma, 1973).

C. NITRATE REDUCTASE INACTIVATING ENZYME IN THE MAIZE ROOT

Nitrate reductase from the root tip (0–2 cm of the primary root) of 3- to 4-day maize seedlings is much more stable *in vitro* than the enzyme from the remainder of the primary root referred to as the mature root (Wallace, 1973). A fraction isolated from the mature root was shown to be responsible for this *in vitro* inactivation of nitrate reductase. It was acid and heat labile and had a Q_{10} 15 to 25°C of 2·2. When assayed with a partially purified sample of nitrate reductase (Fig. 2a) the degree of inactivation increased with time. This pattern of inactivation and the fact that the inhibition is irreversible indicates that the inactivating factor was not just binding to the nitrate reductase like the potato invertase inhibitor (Pressey, 1966). It has thus been referred to as an inactivating enzyme like the enzymes described above in yeast and rat tissues.

The maize root inactivating enzyme promotes an exponential loss of nitrate reductase activity (Fig. 2b). It is assayed with a nitrate reductase sample (40% $(NH_4)_2SO_4$ precipitate) from either the root tips or scutella of 3-day maize seedlings (Wallace, 1974). As shown in Fig. 2b the nitrate reductase sample is stable for 90 min at 25°C. The loss of nitrate reductase activity in the presence of the inactivating enzyme allows an estimate to be made of the activity of the latter. This is usually given as units nitrate reductase inactivated, h^{-1}. An increase in the level of the inactivating enzyme or nitrate reductase, its substrate, results in an increase in the amount of nitrate reductase inactivated (Wallace, 1974) but a plot of percentage loss of nitrate reductase activity versus level of inactivating enzyme is linear, indicating that the complex formed is undissociable.

In the purification procedure for the maize root inactivating enzyme (Wallace, 1974) the most significant steps were treatment at pH 4·0 followed by chromatography, of the fraction not precipitated, on a CM-32 cellulose column (Fig. 3). The inactivating enzyme is consistently eluted when approximately 10 volumes of 10 mM acetate pH 5·0 (equilibration buffer) plus 50 mM NaCl have been passed through the column. Its mol. wt was estimated to be 44 000 by gel filtration. With active site inhibitors (Section IIIA) the only significant inhibition was obtained with PMSF. The maize inactivating

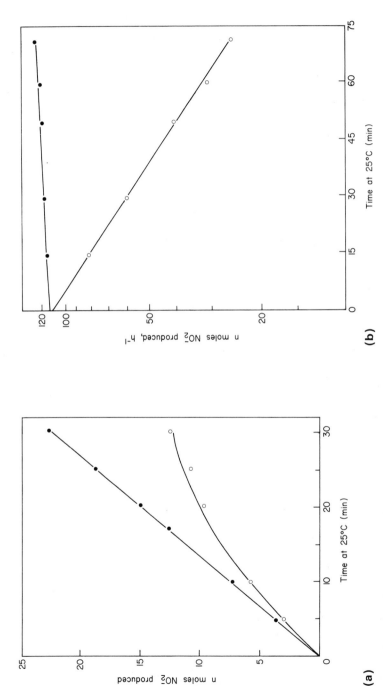

(a)

(b)

Fig. 2. Inactivation of nitrate reductase by a maize root inactivating enzyme (a) The level of nitrite produced was measured in an assay mixture containing scutella nitrate reductase (●——●) or nitrate reductase plus inactivating enzyme (○——○). In the aliquots taken, the protein level of the scutella sample was 0·35 mg and that of the inactivating enzyme 12 µg. (b) Root tip nitrate reductase (3 mg protein) was incubated at 25°C and pH 7·0 with (○——○) and without (●——●) inactivating enzyme (24 µg protein). The total volume was 3·2 ml and aliquots of 0·4 ml were taken at the times indicated and assayed for nitrate reductase activity.

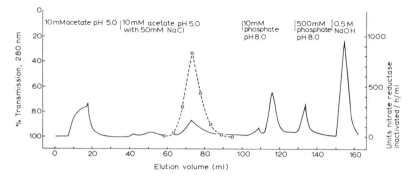

FIG. 3. Chromatography of the nitrate reductase-inactivating enzyme on CM-32 cellulose. A 12 ml sample of the pH 4·0 supernatant (see text), containing 15·5 mg protein, was loaded on a 0·78 cm² × 4·5 cm column equilibrated with 10 mM acetate, pH 5·0, and eluted by the procedure indicated. The percent transmission of the column effluent was monitored (——) and 5 ml fractions were collected for assay of the inactivating enzyme (----) after adjusting the pH to 7·0. From Wallace (1974).

enzyme hydrolysed casein and azocasein but did not have esterase activity (Section IIIC) and peptide hydrolase or amino peptidase activity (Wallace, 1974). A range of enzymes were tested but only nitrate reductase was inactivated (Section IV). Nitrate or NADH did not protect nitrate reductase and FAD and cysteine which stabilize it against heat inactivation were also ineffective. The activity of the inactivating enzyme increased with the age of the maize root and was not influenced by the level of nitrate in the growth medium. It has not been possible to determine accurately the pH optimum of the inactivating enzyme since the stability of nitrate reductase decreased during incubation at pH values below or above 7·0. When the V_{max} of the inactivating enzyme was compared at pH 6·0, 7·0 and 8·0 the highest value was at pH 7·0 (Wallace, 1974).

Nitrate reductase in most plant species is converted to an inactive form by treatment with NADH (Losada, 1974). The inhibition is reversed by oxidation of the enzyme with ferricyanide or nitrate. In the inactive form no NADH-nitrate reductase activity or $FADH_2$-nitrate reductase activity can be detected but the activity of the cytochrome c reductase component of the nitrate reductase complex is unaltered. The maize root inactivating enzyme inactivates all components of the nitrate reductase complex but a preferential inactivation of the NADH cytochrome c reductase component is indicated (Wallace, 1975a). In this case no reactivation of nitrate reductase occurred with ferricyanide treatment.

Nitrate reductase inactivating activity has been observed in the 40–60% $(NH_4)_2SO_4$ precipitate of a range of other maize and plant samples (Wallace, 1975b). The factor involved was non-dialysable and destroyed by heating for 10 min at 70°C. In contrast to the maize root inactivating enzyme, PMSF was either not inhibitory or gave only a low and variable level of inhibition. An inhibitor of nitrate reductase reported in rice seedlings (Kadam *et al.*, 1974) has some similarities to the maize root inactivating enzyme but NADH protected the nitrate reductase against inhibition. In rice and some other plant species (Kadam *et al.*, 1975) the action of NADH differs from that described earlier in that the reduced coenzyme stabilizes or activates nitrate reductase.

D. OTHER EXAMPLES OF SPECIFIC PROTEASE ACTION IN HIGHER PLANTS

A neutral protease has been described in oat shoots (Pike and Briggs, 1972) which degrades phytochrome and a range of other proteins including casein and azocoll (Section IIIB). It was inhibited by PMSF and TLCK (Section IIIA) and by high salt concentration. The oat protease was estimated to have a mol. wt of 61 500, by gel filtration, but there was evidence of autolysis during its storage at 4°C. In the senescence of barley leaves the loss of ribulose 1,5-diphosphate carboxylase protein was correlated with the appearance of a protease which acted on azocasein (Peterson and Huffaker, 1975).

A decrease in the level and stability of alcohol dehydrogenase in the maize scutellum, after 24 h germination, is paralleled by the accumulation of some factor which inhibits the enzyme *in vitro* (Ho and Scandalios, 1975). It was heat labile, partially dialysable but insensitive to trypsin. β-mercapto-ethanol, which protects alcohol dehydrogenase against loss of activity, had no effect on the activity of an endogenous scutellar protease measured with azocasein.

III. CHARACTERIZATION ON INACTIVATING ENZYMES

A. ACTIVE SITE INHIBITORS

Proteases can be divided into four main groups. In addition to those active in the acid pH range there are the thiol, the serine and the metal dependent which can be characterized with active site inhibitors (Table I). Most of the evidence in the vast literature on the use of these inhibitors indicates that they are specific in their action (for references see Shaw, 1970 and Means and Feeney, 1971). It is a good practice with inhibitors such as pCMB (Table I) and metal chelators to establish that the inhibition is reversible e.g. with a thiol and active site metal respectively. It is known that *o*-phenanthroline, a widely used metal chelator, can cause inhibition by oxidizing thiol groups (Kobashi and Horecker, 1967). Alternatively, inhibition by thiols such as cysteine or dithiothreitol may be due to their chelation of essential metal groups (Seifter and Harper, 1971).

PMSF (Table I) has been shown to react specifically with the active site serine in α-chymotrypsin, even when used in large excess (Gold and Fahrney, 1964). However its inhibition of papain, a thiol dependent protease, and the

TABLE I

Reagents used to identify active sites of proteases

Reagent	Labelled residue	Reference
p-chloromercuribenzoate (pCMB)	Thiol	Pontremoli et al. (1965)
Diisopropylfluorophosphate (DFP)	Serine	Kafatos et al. (1967)
Phenylmethylsulphonyl fluoride (PMSF)	Serine	Gold and Fahrney (1964)
α-N-tosyl-L-phenylalanine chloromethyl ketone (TPCK)	Histidine	Schoellmann and Shaw (1963)
α-N-tosyl-L-lysine chloromethyl ketone (TLCK)	Histidine	Pike and Briggs (1972)
N-bromosuccinimide (NBS)	Tryptophan	Garg and Virupaksha (1970)
Metal chelators, e.g. EDTA o-phenanthroline,αα'dipyridyl	Usually zinc	Gracy and Noltmann (1968)

phytochrome degrading protease in oat leaves is prevented by thiol compounds (Pike and Briggs, 1972). The inhibition of the maize-root inactivating enzyme by PMSF is not prevented by either glutathione or mercaptoethanol (25 mM).

Affinity labels are inhibitors designed with substrate like features to localize them at the active centre of selected proteases (Means and Feeney, 1971). The examples given in Table I, TLCK and TPCK, alkylate the active centre histidine in trypsin and α-chymotrypsin respectively. Papain and the oat protease referred to above are also inhibited by TLCK and it has been shown in the case of the former that this is due to the loss of titratable thiol groups (Whitaker and Perez-Villasenor, 1968). This inhibition of the plant proteases could be interpreted as indicating the involvement of an imidazole group at their active centre. However, Shaw (1970) considers that proteases like papain, which have rather broad specificity and are capable of acting on simple trypsin substrates, would be inhibited by compounds like TLCK.

The use of a combination of inhibitors could provide additional information. While the inhibition of leucine amino peptidase by EDTA can be reversed by Mg^{++} the inhibition by EDTA and pCMB is irreversible. This suggests the involvement of both a metal and a thiol group at the active site (Bryce and Rabin, 1964).

B. ESTIMATION OF PROTEASE ACTIVITY

Although the inactivating enzymes are known to be relatively specific in their action (Section IV) they also hydrolyse exogenous proteins like casein and haemoglobin, presumably because of the relatively high concentration at which they are tested. When a purified sample of casein, such as Hammarsten (approximately 0·2% w/v), is incubated with the protease at 37°C the release of low mol. wt amino compounds into the trichloroacetic acid fraction can be followed. Either the increase in the absorbance at 280 nm (Laskowski, 1955) or the ninhydrin reaction is measured (Lee and Takahashi, 1966). With a protein substrate, such as casein, care is necessary in interpreting the effect of pH on the rate of hydrolysis since the degree of its ionization may influence the activity of the protease (Northrop, 1939). Use of a denatured protein as substrate may be more satisfactory and the procedure for the preparation of acid, alkali and urea denatured albumin has been described by Schlamowitz and Peterson (1959).

Protease activity can also be measured with protein-dye complexes like azocoll and azocasein. Azocoll is an insoluble substrate in which hide powder is labelled covalently with remazobrilliant blue (Rinderknecht *et al.*, 1968). The chromophore on azocasein is diazotized sulphanilamide (Charney and Tomarelli, 1947). There is not much information on how these dyes are bound to proteins but it is known that remazobrilliant blue reacts readily with primary and secondary hydroxyl groups and hence binding to the high content of hydroxyproline in hide powder is probably most significant. The coloured derivative of casein is reported to be due to the reaction of diazotized sulphanilamide with histidine and tyrosine residues (Horinishi, 1964). Another reagent available is casein yellow where the chromophore is nitrotyrosine.

In studies with the maize root inactivating enzyme the following procedure was used with azocasein. The reaction mixture was 0·5 ml 0·2 M phosphate pH 7·0, 0·3 ml azocasein (10 mg/ml) and 0·2 ml enzyme solution or equivalent amount of buffer. After incubation for 2 h at 40°C the reaction was stopped by the addition of 2 ml cold 7% perchloric acid. The samples were centrifuged for 10 min at 12 000 × *g*, the supernatant was decanted and made alkaline with 0·3 ml 10 N NaOH and its absorbance at 440 nm determined. Measurement of the absorbance of the acid supernatant at 340 nm (Peterson and Huffaker, 1975) is also a reliable measure of the amount of azocasein degradation.

With these general protease substrates it is necessary to establish that the activity measured relates to that of the proteolytic enzyme under investigation. It is shown in Table II that during purification of the maize root inactivating enzyme its specific activity increases in parallel with the azocasein degrading activity. PMSF inhibited both reactions. Peterson and Huffaker (1975) showed that the loss of ribulose, 1,5-diphosphate carboxylase during senescence of barley leaves was correlated with the appearance of proteolytic activity on azocasein. The application of cycloheximide and kinetin, which delayed senescence and the loss of the carboxylase enzyme, prevented the development of the protease activity.

The oat protease, described by Pike and Briggs (1972), hydrolyses azocoll and most of its characterization was based on the use of this substrate. Moore

TABLE II

Purification of maize inactivating enzyme and azocasein degrading activity

Fraction	Total protein (mg)	Nitrate reductase inactivation[a]			Azocasein degradation[b]		
		Total	Specific activity	Purification	Total	Specific activity	Purification
40–60% $(NH_4)_2SO_4$ precipitate	26·2	1152	44	1	523	20	1
pH 4·0 supernatant	8·7	2820	324	7	799	92	5
CM-32 fraction	1·3	1879	1445	33	601	463	23

[a] Units nitrate reductase inactivated, h^{-1}.
[b] Absorbance units at 440 nm ($\times 10^2$), h^{-1}.

(1969) showed that although azocoll was insoluble it could be used reliably to measure proteolytic activity as a function of time. In yeast, a kinetic study of azocoll hydrolysis was used in the characterization of the inhibition of protease B by its inhibitor (Betz *et al.*, 1974). Azocasein used at the level given for the maize root study was in solution.

C. ESTERASE ACTIVITY

Most proteases are known to hydrolyse other types of bonds in addition to the peptide bond. With an ester compound the reaction is as follows:

$$X-NH-\overset{\overset{\displaystyle R}{|}}{CH}-CO-OR^1 + H_2O \longrightarrow$$

$$X-NH-\overset{\overset{\displaystyle R}{|}}{CH}-COO^- + HOR^1 + H^+$$

X is a blocking group such as acetyl, benzoyl, tosyl or benzyloxycarbonyl, R is a side chain which meets the specificity requirement of the enzyme and R^1 may be a methyl, ethyl or *p*-nitrophenyl group.

The esterase reaction can be monitored by the increase in absorbance due to the formation of the carboxylate group. With α-N-benzoyl-L-arginine ethyl ester (BAEE) the absorbance is measured at 253 nm. If the pH of the reaction is above the pK of the carboxylate group the proton liberated in the reaction can be detected with a pH-stat unit (Saheki and Holzer, 1974). Using these synthetic substrates, where only one bond is susceptible to hydrolysis, the kinetic parameters of a protease may be more readily defined.

The proteases involved in enzyme inactivation in yeast and animal cells (Section IIB) have esterase activity. The maize root inactivating enzyme and the oat protease which degrades phytochrome were found not to have esterase activity with α-N-benzoyl-L-arginine ethyl ester (BAEE) and N-acetyl-L-tyrosine ethyl ester (ATEE), substrates for trypsin and chymotrypsin respectively. Papain is also active on BAEE. In their study on barley leaves Peterson and Huffaker (1975) measured esterase activity with N-carbobenzoxy-L-tyrosine-*p*-nitrophenyl ester but the activity was not correlated with that against azocasein (Section IIIB). It is possible that the absence of esterase activity associated with these plant proteases is due to the use of ineffective substrates. (For further references on esterase activity see Boyer, 1971.)

IV. SPECIFICITY OF INACTIVATING ENZYMES

In the initial characterization of most inactivating enzymes a relative specificity for one or a limited group of enzymes was noted. The maize root inactivating enzyme inactivated the NADH-dependent nitrate reductase from a range of plant tissues and also the *Neurospora* NADPH-dependent enzyme. It had no effect on the following: xanthine oxidase, a similar molybdoflavoprotein, nitrite reductase, and isocitrate lyase from the maize scutella, or invertase and glutamate dehydrogenase from the maize root.

TABLE III

Comparison of action of maize root and yeast inactivating enzyme on yeast tryptophan synthase and maize nitrate reductase

Inactivating system	Protein level	Units tryptophan synthase[a] inactivated h^{-1}		Units nitrate reductase inactivated h^{-1}	
		—	with PMSF (1 mM)	—	with PMSF (1 mM)
Maize root	34 µg	38	12	—	—
Maize root	1·4 µg	—	—	23	0
Yeast protease A[b]	0·44 mg	58	58	9	10
Yeast protease B	17 µg	41	11	13	2

[a] Yeast tryptophan synthase was isolated by the procedure of Katsunuma *et al.* (1972) to step 4.
[b] A yeast tryptophan synthase inactivating fraction was separated into two components by hydroxylapatite as described by Katsunuma *et al.* (1972). These are now known to be protease A and B.

It has now been found that the maize inactivating enzyme will inactivate tryptophan synthase from yeast while the yeast proteases, known to inactivate tryptophan synthase, inactivate maize nitrate reductase. It is shown in Table III that the action of the maize inactivating enzyme on tryptophan synthase, like that on nitrate reductase, is inhibited by PMSF. Only yeast protease B is sensitive to this inhibitor. The maize inactivating enzyme and yeast proteases were completely inhibited by a low level of yeast inhibitor prepared by the method of Ferguson *et al.* (1973) (purification to step III). Both yeast proteases when chromatographed on CM-32 cellulose (Fig. 3) were eluted in the same fraction as the maize inactivating enzyme. In contrast to the latter there was considerable loss of yeast protease activity during this purification procedure. Thus, although there was some apparent similarity between yeast protease B and the maize inactivating enzyme the latter differed in not being inhibited by pCMB and not having esterase activity with ATEE or BAEE. While the partially purified yeast inhibitor sample inhibited the maize inactivating enzyme pure samples of protease A and B inhibitors, supplied from Dr Holzer's laboratory, were found to have no effect on the maize inactivating enzyme.

The yeast proteases are now known to activate or inactivate a range of enzymes (Holzer, 1975). They have been shown to inactivate the cytoplasmic malate dehydrogenase without affecting the mitochondrial enzyme and it has been suggested that they mediate the *in vivo* inactivation of the cytoplasmic enzyme when glucose is added to acetate grown cells.

V. Evidence for Inhibitors of Inactivating Enzymes

Heat stable protein molecules which inhibit the inactivating enzymes have also been demonstrated. They have been investigated in most detail in yeast cells where four inhibitor species have been described for protease A and two for protease B (Holzer, 1975). The average mol. wts were 6000 and 10 000 respectively compared to a mol. wt of 42 000 for protease A and 34 000 for protease B. An immuno-precipitation technique has been used for the estimation of inhibitor B level (Matern *et al.*, 1974). Each inhibitor is specific for its own protease and each inhibitor-protease complex is activated by the opposite protease, e.g. the inactive protease B inhibitor complex is activated by protease A and vice-versa (Saheki *et al.*, 1974). The inhibitors increase in parallel with the proteases and highest levels are found in the stationary phase of yeast growth. Since the proteases are located in the vacuole and the inhibitors in the cytoplasm (Section VI) it has been proposed that the role of the inhibitors is to prevent against deleterious protease action in the cytoplasm (Betz, 1975). There is an excess of the inhibitors so that when extracts of yeast are prepared the proteases are largely recovered in an inactive form. If a similar situation occurred in other cells then the level of protease activity estimated in crude extracts would not have much validity. When yeast extracts are incubated at pH 5·0 and 25°C an activation of the proteases can be demonstrated, presumably due to dissociation of the protease inhibitor complex (Saheki and Holzer, 1975). A similar yeast inhibitor complex in *Neurospora* was dissociated with 4 M urea (Yu *et al.*, 1974).

There is evidence for an inhibitor of the nitrate reductase inactivating enzyme in the maize root (Table IV). In the root tip of the 3-day maize seedling no nitrate reductase inactivating activity was found in the crude

TABLE IV

Recovery of the inactivating enzyme in the maize root tip

Fraction	Total protein (mg)	Total activity (units)[a]	Inhibitor of[b] inactivating enzyme (% inhibition)
40–60% $(NH_4)_2SO_4$ precipitate	12·0	0	80
pH 4·0 supernatant	4·0	864	34
CM32 fraction	0·2	624	33

[a] Units nitrate reductase inactivated, h^{-1}.
[b] Inhibitor activity estimated after boiling samples for 10 min.

extract or fraction precipitated by 40–60% saturation with $(NH_4)_2SO_4$. (The inactivating enzyme in the mature root is recovered in this fraction.) Hence the relative stability of the root tip nitrate reductase *in vitro*. It was also found that this root tip sample contained a heat stable factor, precipitable by $(NH_4)_2SO_4$, which inhibited the inactivating enzyme from the mature root (Table IV). After treatment of the root tip sample at pH 4·0 and removal of the precipitate it was found that the supernatant had a high level of nitrate reductase inactivating activity. A significant reduction also occurred in the inhibitor activity. This root tip inactivating enzyme chromatographed on CM-32 cellulose in an identical manner to that from the mature root (Fig. 3).

VI. Intracellular Compartmentation of the Inactivating Enzymes

It has been demonstrated in yeast that the proteases which inactivate tryptophan synthase are located in the vacuole while their inhibitors are found in the cytoplasm (Hasilik *et al.*, 1974). In the isolation procedure used there are three main steps: spheroplast preparation, disruption of the plasma membrane by metabolic lysis and isolation of the vacuoles by Ficoll gradient centrifugation.

The yeast protoplasts were metabolically lysed by incubation with glucose. This treatment (Indge, 1968) results in an increased permeability of the plasma membrane to the osmotic stabilizer, in this case mannitol, and rupture of the protoplast follows. Since the vacuole remains intact it appears that its limiting membrane differs in permeability properties to the plasma membrane. Ficoll is a copolymer of sucrose and epichlorohydrin and because of its large molecular size membranes are much less permeable to it than sucrose. When the yeast lysate is centrifuged on a gradient of 7·2–8·2% Ficoll, the vacuoles have a density lower than the medium and are recovered in a layer on the surface (Hasilik *et al.*, 1974). Mitochondria were found in the pellet and the remaining supernatant fraction was taken as the soluble cytosolic component of the yeast cell (Table V). Alcohol dehydrogenase and succinate dehydrogenase were used as marker enzymes for the soluble and mitochondrial fraction respectively.

The tryptophan synthase and inhibitor of tryptophan synthase inactivating activity from the lysate were recovered in the cytosol (Table V). No tryptophan synthase inactivating activity was detected in the lysate but a high level of activity was recovered in the vacuole layer. In the lysate sample it appears that there was sufficient inhibitor to complex the inactivating enzyme. Since procedures are now available for the preparation of higher plant protoplasts (Cocking and Evans, 1973), it may be possible to use the fractionation methods described for yeast to determine if protease enzymes like the maize inactivating enzyme are located in the vacuole of the plant cell. Suitable marker enzymes for the plant vacuole would be an acid phosphatase and carboxypeptidase known to occur in the central vacuole of giant algal cells like *Nitella* (Doi *et al.*, 1975).

TABLE V

Enzyme distribution in the fractions after Ficoll-gradient centrifugation of a metabolic lysate from yeast spheroplasts[a]

Fraction	Tryptophan synthase	Tryptophan synthase inactivation	Inhibitor of[b] tryptophan synthase inactivation	Alcohol dehydrogenase	Succinate dehydrogenase
Total lysate	520	0	928	1390	3800
Cytosolic	410	0	700	1380	100
Mitochondrial	5·2	10·4	30	6·4	3330
Vacuolar	1	91	15	0·4	25

Data from Hasilik et al. (1974).

[a] Total activities corresponding to 1 g wet spheroplast are given in International Units.

[b] Results from a separate experiment.

VII. Control of Inactivating Enzymes during Cell Extraction

Although it is not known if an inactivating enzyme like that in the maize root has an *in vivo* role, it does cause serious interference with *in vitro* studies on nitrate reductase. There are two possible ways of preventing such enzyme inactivation. The first involves the use of a specific inhibitor for the inactivating enzyme while the second employs a large excess of exogenous protein also acted on by the inactivating enzyme.

The nitrate reductase inactivating enzyme is inhibited by PMSF and it has been shown (Fig. 4) that the inclusion of this compound in the extraction medium stabilizes the nitrate reductase from the mature root of 3-day maize seedlings. PMSF is less effective in older maize root material (Wallace, 1975b) and it does not stabilize nitrate reductase from other sources where inactivation occurs by a mechanism that is insensitive to this inhibitor. Other protease inhibitors may be effective in these cases.

The use of an exogenous protein such as casein or serum albumin in the extraction medium for nitrate reductase was first reported by Schrader *et al.* (1974). It has been found that casein prevents the action of the maize inactivating enzyme (Wallace, 1975b) and so it appears likely that this is at least

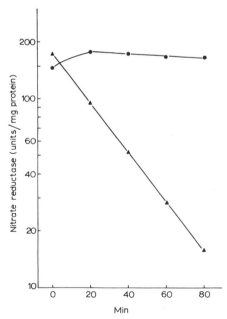

FIG. 4. Use of PMSF to prevent the *in vitro* inactivation of nitrate reductase. A sample of the mature root of 3-day maize seedlings was extracted either with standard extraction medium of 0·05 M phosphate, 0·5 mM EDTA and 5 mM cysteine (▲—▲) or with the addition of 1 mM PMSF (●—●), both at pH 7·5. The crude extracts (27 000 × *g* supernatant) were incubated at 25°C for the times shown. From Wallace (1974).

TABLE VI

Influence of casein in the extraction medium on the recovery of nitrate reductase from maize seedlings

| Organ[a] | n moles NO_2^- produced, h^{-1}, $organ^{-1}$ | |
	No casein	Casein (3% w/v)
Root	69	1802
Scutellum	87	650
Axis	279	458
Leaf	1169	2050

Data from Wallace (1975b).
[a] 7-Day seedlings.

part of the explanation for its stabilization of nitrate reductase *in vitro*. In Table VI it is shown that when casein is used in the extraction medium for 7-day maize seedlings there is a dramatic improvement in the recovery of nitrate reductase, especially from the root sample. As already suggested by the work of Schrader *et al.* (1974) and further exemplified in Table VI, casein is effective in protecting the nitrate reductase from a range of plant tissues.

ACKNOWLEDGEMENT

The investigation in the author's own laboratory was supported by a grant from the Australian Research Grants Committee.

REFERENCES

Betz, H. (1975). *Biochim. biophys. Acta* **404**, 142.
Betz, H., Hinze, H. and Holzer, H. (1974). *J. biol. Chem.* **249**, 4515.
Boyer, P. D. (1971). *In* "The Enzymes", Vol. 3, Academic Press, New York and London.
Bryce, G. F. and Rabin, B. R. (1964). *Biochem J.* **90**, 513.
Charney, J. and Tomarelli, R. M. (1947). *J. biol. Chem.* **171**, 501.
Cocking, E. C. and Evans, P. K. (1973). *In* "Plant Tissue and Cell Culture" (Street, H. E., ed.), Botanical Monographs 11, pp. 100–120 Blackwell Scientific Publications, Oxford.
Doi, E., Ohtsuru, C. and Matoba, T. (1975). *Pl. Cell Physiol.* **16**, 581.
Ferguson, A. R., Katsunuma, T., Betz, H. and Holzer, H. (1973). *Europ. J. Biochem.* **32**, 444.
Garg, G. K. and Virupaksha, T. K. (1970). *Europ. J. Biochem.* **17**, 4.
Gold, A. M. and Fahrney, D. (1964). *Biochem.* **3**, 783.
Gracy, R. W. and Noltmann, E. A. 1968. *J. biol. Chem.* **243**, 4109.
Hasilik, A., Müller, H. and Holzer, H. (1974). *Europ. J. Biochem.* **48**, 111.
Ho, D. T. and Scandalios, J. G. (1975). *Pl. Physiol.* Lancaster **56**, 56.
Holzer, H. (1975). *In* "Mechanism of Action and Regulation of Enzymes" (Keleti, T., ed.), Vol. 32, pp. 181–193. Akadémiai Kiadó, Budapest.

Holzer, H., Betz, H. and Ebner, E. (1975). *In* "Current Topics in Cellular Regulation" (Horecker, B. L. and Stadtman, E. R., eds), Vol. 9, pp. 103–156. Academic Press, New York and London.

Horinishi, H. (1964). *Biochim. biophys. Acta* **86**, 477.

Indge, K. J. (1968). *J. gen. Microbiol.* **51**, 441.

Kadam, S. S., Gandhi, A. P., Sawhney, S. K. and Naik, M. S. (1974). *Biochim biophys. Acta* **350**, 162.

Kadam, S. S., Sawhney, S. K. and Naik, M. S. (1975). *Indian J. Biochem. Biophys.* **12**, 81.

Kafatos, F. C., Law, J. H. and Tartakoff, A. M. (1967). *J. biol. Chem.* **242**, 1488.

Katsunuma, T., Schött, E., Elsässer, S. and Holzer, H. (1972). *Europ. J. Biochem.* **27**, 520.

Katunuma, N. (1973). *In* "Current Topics in Cellular Regulation" (Horecker, B. L. and Stadtman, E. R., eds), Vol. 7, pp. 175–203. Academic Press, New York and London.

Katunuma, N., Kominami, E., Kominami, S. and Kito, T. (1972). *In* "Advances in Enzyme Regulation" (Weber, G., ed.), Vol. 10, pp. 289–306. Pergamon Press, Oxford.

Katunuma, N., Kominami, E., Kobayashi, K., Banno, Y., Suzuki, K., Chichibu, K., Hamaguchi, Y. and Katsunuma, T. (1975). *Europ. J. Biochem.* **52**, 37.

Kobashi, K. and Horecker, B. L. (1967). *Archs Biochem. Biophys.* **121**, 178.

Kominami, E. and Katunuma, N. (1976). *Europ. J. Biochem.* **62**, 425.

Kominami, E., Banno, Y., Chichibu, K., Shiotani, T., Hamaguchi, Y. and Katunuma, N. (1975). *Europ. J. Biochem.* **52**, 51.

Laskowski, M. (1955). *In* "Methods in Enzymology" (Colowick, S. P. and Kaplan, N. O., eds), Vol. 2, pp. 26–36. Academic Press, New York and London.

Lee, Y. P. and Takahashi, T. (1966). *Ann. Biochem.* **14**, 71.

Lenney, J. F. (1956). *J. biol. Chem.* **221**, 919.

Losada, M. (1974). *In* "Metabolic Interconversion of Enzymes" pp. 257–270. Third International Symposium, Seattle. Springer Verlag, Berlin.

Matern, H., Betz, H. and Holzer, H. (1974). *Biochem. biophys. Res. Commun.* **60**, 1051.

Means, G. E. and Feeney, R. E. (1971). *In* "Chemical Modification of Proteins" Holden-Day Inc., San Francisco.

Moore, G. L. (1969). *Analyt. Biochem.* **32**, 122.

Northrop, J. H. (1939). *In* "Crystalline Enzymes" pp. 7–10. Columbia University Press, New York.

Peterson, L. W. and Huffaker, R. C. (1975). *Pl. Physiol.* Lancaster **55**, 1009.

Pike, G. S. and Briggs, W. R. (1972). *Pl. Physiol.* Lancaster **49**, 521.

Pontremoli, S., Luppis, B., Traniello, S., Rippa, M. and Horecker, B. L. (1965). *Archs Biochem. Biophys.* **112**, 7.

Pressey, R. (1966). *Archs Biochem. Biophys.* **113**, 667.

Rinderknecht, H., Geokas, M. C., Silverman, P. and Haverback, B. J. (1968). *Clin. chim. Acta* **21**, 197.

Saheki, T. and Holzer, H. (1974). *Europ. J. Biochem.* **42**, 621.

Saheki, T. and Holzer, H. (1975). *Biochim. biophys. Acta* **384**, 203.

Saheki, T., Matsuda, Y. and Holzer, H. (1974). *Europ. J. Biochem.* **47**, 325.

Schlamowitz, M. and Peterson, L. V. (1959). *J. biol. Chem.* **234**, 3137.

Schoellmann, G. and Shaw, E. (1963). *Biochemistry* **2**, 252.

Schött, E. H. and Holzer, H. (1974). *Europ. J. Biochem.* **42**, 61.

Schrader, L. E., Cataldo, D. A. and Peterson, D. M. (1974). *Pl. Physiol.* Lancaster **53**, 688.

Seifter, S. and Harper, E. (1971). *In* "The Enzymes" (Boyer, P. D., ed.), Vol. 3, pp. 649–697. Academic Press, New York and London.
Shaw, E. (1970). *Physiol. Rev.* **50**, 244.
Wallace, W. (1973). *Pl. Physiol.* Lancaster **52**, 197.
Wallace, W. (1974). *Biochim. biophys. Acta* **341**, 265.
Wallace, W. (1975a). *Biochim. biophys. Acta* **377**, 239.
Wallace, W. (1975b). *Pl. Physiol.* **55**, 774.
Whitaker, J. R. and Perez-Villasenor, J. (1968). *Archs Biochem. Biophys.* **124**, 70.
Yu, P. H., Siepen, D., Kula, M. R. and Tsai, H. (1974). *FEBS Letts.* **42**, 227.

CHAPTER 10

Immunochemical Approaches to Questions Concerning Enzyme Regulation in Plants

J. DAUSSANT, C. LAURIÈRE, N. CARFANTAN AND A. SKAKOUN

Physiologie des Organes Végétaux après récolte, Meudon, France

I.	Introduction	.	.	.	.	.	.	.	.	.	198
II.	Principles of immunochemical Methods Based on the Specific Precipitation between Antigens and Antibodies	.	.	.	.	.	.	198
	A. Immune Sera: Biological Reagents for Identifying Proteins	.	.	198
	B. Antigen-antibody Specific Precipitation	.	.	.	.	.	199
	C. Techniques of Precipitation in Gel	.	.	.	.	.	199
	D. Remarks	.	.	.	.	.	.	.	.	200
	E. Principles of Absorption Techniques	.	.	.	.	.	201
	F. Preliminary Investigations on the Anti-enzyme Immune Serum	.	202
III.	Detection of "Inactive" Enzymes	.	.	.	.	.	.	202
	A. Search for Inactive Enzymes using Techniques of Specific Precipitation in Gels	.	.	.	.	.	.	.	.	203
	B. Search for Inactive Enzymes using the Techniques of Competitive Absorption	.	.	.	.	.	.	.	.	206
	C. Remarks	.	.	.	.	.	.	.	.	206
IV.	Multiple Forms of Enzymes at Different Physiological Stages	.	.	207
	A. No Enzymatic Characterization Reaction is Available	.	.	208
	B. Enzymatic Characterization Reaction on the Gels is Available	.	210
V.	Identification of Phenomena Involved in the Appearance and Disappearance of Enzyme Activity and in the Modification of Enzyme Physico-chemical Properties	.	.	.	.	.	.	213
	A. Search for the Cause Leading to the Disappearance of α-amylase of Developing Wheat Seeds upon Maturation	.	.	.	.	216
	B. Indirect or Direct Proof for Enzyme Synthesis	.	.	.	218
	C. Changes in Physico-chemical Properties of Wheat β-amylase during Germination	.	.	.	.	.	.	.	219
VI.	Changes in the Specific Activity of Enzymes	.	.	.	220
VII.	Studies on Enzyme Structure and Localization of Enzyme Synthesis using Anti-enzyme Subunits Immune Sera	.	.	.	.	221
VIII.	Conclusions	.	.	.	.	.	.	.	.	.	221
	References	.	.	.	.	.	.	.	.	.	222

I. Introduction

The immunochemical characterization of proteins offers a means of identifying specific proteins and, under certain conditions, of quantitating these proteins in protein mixtures. Furthermore, this characterization may even allow detection of proteins which have undergone modifications resulting in changes of physico-chemical properties. Thus, immunochemical characterization of enzymes ought to be valuable in studies concerned with enzyme regulation in plants.

In fact, several reviews have been published concerning the enzymatic characterization of specific precipitin bands in gels (Uriel, 1971), or dealing with the contribution of immunochemistry of enzymes to different biological studies (Cinader, 1963, 1967; Arnon, 1971, 1973). The usefulness of immunochemical techniques in plant protein studies—including plant enzymes—was also reported recently (Daussant, 1975).

This report is not a review. It aims at showing, with a few examples, the kinds of information immunochemical identification of enzymatic proteins may provide in elucidating different aspects of enzyme regulation studies.

Thus we will describe first the principles of *in vitro* antigen–antibody reactions as well as basic immunochemical methods founded on specific precipitation. More recent and unpublished developments of these techniques will be described throughout the text and the figures.

II. Principles of Immunochemical Methods Based on the Specific Precipitation Between Antigens and Antibodies

A. Immune Sera: Biological Reagents for Identifying Protein

Antibodies are seric proteins induced by injection of foreign substances called "antigens" (proteins are antigens) into rabbits, for example. These antibodies form a family of protein molecules which react specifically each with a particular part of the antigen. These structurally different parts of the antigen are called "antigenic determinants". One may assume that there is about one antigenic determinant per 5000 daltons (Crumpton, 1973).

From a methodological point of view, one may consider an immune serum specific for one protein as a biological reagent containing a series of markers corresponding to several, structurally different, discrete areas on the surface of the protein. Thus, if a protein undergoes modifications, several of the antibodies specific for the initial protein will fail to react with the modified one, whereas some of the antibodies may still react if the corresponding determinants are not altered.

Practically, an immune serum may provide, therefore, a means for recognizing proteins which have been modified in several of their physico-chemical properties.

In this report we refer to an immune serum specific for a single protein as "monospecific". In contrast, an immune serum prepared by injecting a mixture of proteins (a plant organ extract for example) and containing several sets of antibodies, each reacting with a distinct protein, is referred to as "organo-specific immune serum"

B. ANTIGEN–ANTIBODY SPECIFIC PRECIPITATION

A particular type of reaction carried out *in vitro* between antigens and antibodies is the reaction of precipitation. This reaction can be effected by mixing the antigen solution with the immune serum. It can also be performed by disposing antigen and antibodies in a gel and causing the reactants to come together through diffusion or electrophoresis.

When increasing amounts of a protein are added to the same amount of a corresponding immune serum, the precipitates first increase in quantity then decrease progressively. The part of the curve where proportion of antigens to antibodies gives maximum precipitation is called the zone of equivalence.

This type of precipitation carried out in tubes using a rabbit monospecific immune serum may be used for quantitating a specific protein in a protein mixture. The ascending part of the precipitation curve is used and a reference scale is prepared with the purified protein. This technique has been applied, for example, for quantitating the ribulose diphosphate carboxylase enzyme protein (Kleinkopf *et al.*, 1970).

If the immune serum is not monospecific, this type of technique cannot be applied, for the precipitation curve is the result of several antigen–antibody precipitating complexes. Other techniques, antigen–antibody precipitation in gels, obviate this problem for they make it possible to visualize individual precipitates corresponding to different antigen–antibody complexes. Techniques of this type are applicable to qualitative and quantitative analyses.

C. TECHNIQUES OF PRECIPITATION IN GEL

An important aspect of precipitin bands formed in gels between antibodies and enzyme antigens should be recalled: the precipitated enzyme–antibody complex generally remains enzymatically active. Thus, if a coloured enzymatic characterization reaction is known, it is possible directly to identify the precipitin band corresponding to the enzyme (Uriel, 1971). Furthermore, this property makes it possible to use an organospecific immune serum as a monospecific one if only one precipitin band is stained by the characterization reaction employed.

Many of the precipitation techniques in gels are derived from one of the following basic methods: double diffusion (Ouchterlony, 1949), single diffusion (Mancini *et al.*, 1965), immunoelectrophoretic analysis (Grabar and Williams, 1953) and crossed immunoelectrophoresis (Ressler, 1960; Laurell, 1965). The two latter ones are of particular interest in studies of enzyme changes during a physiological process and will, therefore, be briefly recalled.

1. Immunoelectrophoretic Analysis (IEA) (Grabar and Williams, 1953), a qualitative method. Proteins are first separated by electrophoresis in agar gel. After electrophoresis a canal parallel to the axis of electrophoretic migration is cut in the gel and filled with the immune serum. Antibodies and antigens diffuse, meet each other and, in the region of the gel where the ratio between antigens and the corresponding antibodies is optimum, a precipitin band occurs. After washing the gel in order to eliminate proteins which are not included in the precipitin bands, a reaction of enzymatic characterization is carried out (see Fig. 4).

2. Crossed Immunoelectrophoresis (Ressler, 1960; Laurell, 1965), a qualitative as well as quantitative method. After a first electrophoresis in agar gel, used for separating the proteins which are to be analysed, this gel is imbedded in a second one containing antibodies (see Fig. 5) and a second electrophoresis is performed perpendicularly to the first one. At the pH chosen for the second electrophoresis most of the antibodies do not move in the gel. The separated proteins are forced to migrate into the antibodies and form soluble antigen–antibody complexes. In the course of this electrophoresis the complexes become more and more charged with antibodies and when the ratio between both reactants approaches the zone of equivalence, the complexes precipitate and do not move further. For one antigen and a given immune serum the areas under the peaks are proportional to the ratio, antigen concentration: antibody concentration (Clarke and Freeman, 1967). When several samples are analysed using the same immune serum at the same concentration, the areas of the peaks corresponding to the same antigen in different samples provide a means for comparing the amount of this antigen in the different samples.

D. REMARKS

Some remarks concerning particular dispositions of these basic techniques or their application in enzyme studies are worth mentioning.

1. A simplified method of the second type of technique is interesting in immunochemical study of enzymes: this is the rocket immunoelectrophoresis, a quantitative technique (Laurell, 1966). Samples are deposited in the gel containing antibodies without previous electrophoretic separation. This technique is simpler and consumes less serum than crossed immunoelectrophoresis. It can be employed when only one precipitin band with the enzymatic activity is identified on each side of the starting well.

2. In IEA as well as in crossed immunoelectrophoresis the first electrophoresis can be carried out in different gels, chosen for superior resolution

or because of properties of characterizing proteins by particular parameters such as isoelectric point or molecular size. Starch gels (Poulik, 1971) polyacrylamide gels (Johansson and Stenflo, 1971; Giebel and Saechtling, 1973) pore gradient polyacrylamide gels (Daussant and Skakoun, 1975) agarose or polyacrylamide gels containing ampholins (Lebas et al., 1974) have been used. It is noteworthy that a special disposition of rocket immunoelectrophoresis called "fused rocket immunoelectrophoresis" (Svendsen, 1973) is particularly suitable for analysing the multiple fractions provided, for example, by a chromatographic separation (Laurière et al., 1975).

3. Immunochemical quantitation of a protein in different plant extracts implies that the protein actually exists in all extracts with the same antigenic determinant set. Such a technique could not be used for quantitating a given protein in two extracts if the protein is antigenically modified in one or the other. Practically, this may be verified in first approximation by using techniques of precipitation in gel for observing, with the immune serum used, identity reactions between the enzymes from the different extracts.

The result of these quantitations may be expressed in milligrams of enzymatic protein, if the enzyme exists in a pure state, by using a reference scale. If the enzyme does not exist in a pure state, results may be expressed in the per cent of its amount in one of the extracts taken as reference.

4. For studies of labile enzymes, it is interesting to note that the crossed immunoelectrophoresis can be carried out within one day, whereas the techniques in which diffusion moves antigens and antibodies together (IEA) are much more time consuming, requiring one or two days of diffusion and several days of washing to eliminate enzymes which do not participate in the specific precipitate (with crossed immunoelectrophoresis this elimination generally occurs by means of long duration electrophoresis).

E. PRINCIPLES OF ABSORPTION TECHNIQUES

These techniques aim at removing certain antibody sets from an immune serum by adding the corresponding antigens to the serum. After sedimentation and elimination of the precipitate, the supernatant solution is put in the presence of different amounts of the proteins used for absorption. A lack of reaction indicates that all antibodies of the immune serum, specific for the proteins used for the absorption were actually removed. A simplified absorption technique in gel, involving no dilution of the immune serum has been designed (Hill and Djurtoft, 1964). An example of this technique is shown in Fig. 4 (III).

Another application of absorption techniques consists in removing enzymes from a tissue extract by means of anti-enzyme immune serum. In order to investigate whether an immune serum actually reacts with all molecular forms of a given enzymatic activity, increasing amounts of immune serum are added to the enzyme preparation. After precipitation of the complex, the activity is measured on the supernatant solution. A parallel analysis is carried out using the non-immune serum, to be assured that the observed absorption is specifically due to the antibodies. Sometimes seric proteins interact with the enzymatic protein and, therefore mask the specific effect of the antibodies. In this case the immunoglobulins have to be purified from the immune serum and from the non-immune serum. An absorption of this sort is shown in Fig. 3 (Faye, 1975). Enzyme absorption may be also carried out in gels if a coloured reaction of enzymatic characterization is available. The reactive molecules are placed in contact either by diffusion or by electrophoresis (Daussant and Skakoun, 1974; Daussant and Carfantan, 1975).

F. PRELIMINARY INVESTIGATIONS ON THE ANTI-ENZYME IMMUNE SERUM

When an anti-enzyme immune serum is prepared, a preliminary study on this reagent ought to be made to answer the following questions.

1. Does the immune serum actually contain anti-enzyme antibodies?
2. Does the immune serum characterize one or several antigenic forms of the enzyme?
3. Does the immune serum also react with other proteins, in addition to the enzymatic constituents?
4. Does the immune serum characterize all forms of the enzyme studied?

The first three questions can easily be answered by using immunoelectro-phoretic analysis or crossed immunoelectrophoresis followed by a coloured reaction of enzymatic characterization and amido black staining. The fourth question may be answered by using an absorption technique.

III. DETECTION OF "INACTIVE" ENZYMES

When a certain enzymatic activity is found in a tissue previously lacking it, one often wishes to know whether this activity is due to an "activation" of a pre-existing protein. Perhaps because of the lack of a necessary cofactor or because of the presence of inhibitors the protein failed to show activity. The conversion of enzyme precursors, "zymogens" or "proenzymes", into active enzymes is brought about by enzymatic splitting of certain peptide bonds of the zymogen. The hydrolysis may be limited, as is the case for the conversion of trypsinogen to trypsin in which one hexapeptide is removed (Neurath, 1957), or the conversion may be more extensive, as is the case for procarboxypeptidase A in which two-thirds of the molecule are removed to produce the active carboxypeptidase A (Yamasaki et al., 1963).

The existence of such activation processes may be verified by identifica-tion of an inactive form of the enzyme. This aim may be difficult to fulfill, because in many cases, inactive and active enzymes differ in several of their physico-chemical properties.

These difficulties may be overcome by an immunochemical approach. Active and inactive enzymes that would differ in overall external structure may still present identical antigenic determinants. Thus, at least some of the antibodies prepared against the active enzyme may react with the inactive form. Some examples will illustrate this approach and the probability of obtaining a positive result will be discussed thereafter.

A. SEARCH FOR INACTIVE ENZYMES USING TECHNIQUES OF SPECIFIC PRECIPITATION IN GELS

The first example concerns glutamate dehydrogenase extracted from tulip tepals. The enzymatic activity which was null or very low in young tepals, dramatically increased during ageing (Carfantan and Daussant, 1974).

If an immune serum monospecific for the enzyme had been available, the search for an inactive enzyme in the extract of young tepals would have been practicable by using a simple technique; double diffusion (Ouchterlony, 1949) for example. If an inactive enzyme were detected, the comparison between antigenic specificities of the inactive and active enzymes could have been carried out with the same technique. In case of identity reaction between the two forms, one could have compared the amount of enzymatic proteins in both extracts by using a quantitative technique.

In the present case the immune serum was not monospecific, so that an enzymatic characterization reaction had to be carried out for identifying the precipitate corresponding to the enzyme. In order to identify also the precipitin band formed by the anti-enzyme antibodies and the inactive enzyme, a technique combining line immunoelectrophoresis (Krøll, 1973a, b), rocket immunoelectrophoresis (Laurell, 1966) and an enzymatic characterization reaction was applied. The disposition and the principle of the technique are described in Fig. 1. In this case, the inactive enzyme was detected in the extract of young tepals (Fig. 1). If there is a reaction of identity between active and inactive enzyme, a comparison between the amounts of enzymatic proteins in the two extracts may be made.

Combination of line immunoelectrophoresis with crossed immuno-electrophoresis (Krøll, 1973c) in which the enzymes are first separated by electrophoresis in appropriate gels offers more general applications. Figure 2 represents one of these techniques used in analysis of alcohol dehydrogenase of barley seeds. The first electrophoresis was carried out in a pore gradient polyacrylamide gel for estimating the molecular size of the enzyme (Margolis and Kenrick, 1968). After the long-duration electrophoresis which is involved in this technique, no activity could be detected on the gel, either because the size of the enzyme was too small and therefore not retained in the gel, or because of enzyme inactivation. By combining crossed immunoelectrophoresis (first electrophoresis in pore gradient polyacrylamide gel) and line immunoelectrophoresis as described in Fig. 2, the inactivated enzyme could be identified by its antigenicity (Skakoun *et al.*, 1976). An interesting advantage of this type of technique (providing an immunochemical identification of the inactive enzyme) is that it gives further characterization of the inactive enzyme: electrophoretic mobility, or pHi, or molecular size according to the conditions chosen for the first electrophoresis.

GDH

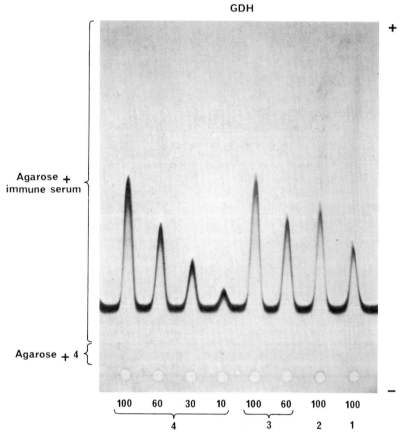

Fig. 1. Combination of rocket immunoelectrophoresis, line immunoelectrophoresis and a reaction of enzymatic characterization. The technique aims at detecting inactive enzymes: glutamate dehydrogenase from tulip tepals 1. Extract of young tepals containing no active enzyme; 2, 3. Extracts of tepals at intermediary ontogenical stages; 4. Extract of ageing tepals containing active enzyme. The undiluted extracts were called 100. Different concentrations of the extracts (100, 60, 30, 10) were submitted to analysis.

During electrophoresis, the active enzyme in the gel strip migrates in the gel containing antibodies, forms complexes then precipitates producing a line perpendicular to the migration axis, stainable with the enzymatic characterization reaction. The addition of antigenic enzyme (active or inactive) moved by electrophoresis from the wells, results in the deformation of the specific precipitin line with the apparition of peaks, the size of which is in relation with the amount of added antigen (see the scale 4, 100, 60, 30, 10).

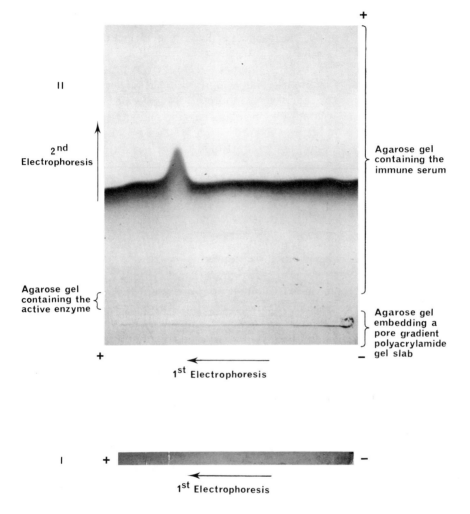

FIG. 2. Combination of electrophoresis, crossed agarose gel immunoelectrophoresis, line immunoelectrophoresis and a reaction of enzymatic characterization. The technique aims at detecting inactive enzymes after a first electrophoretical separation (alcohol dehydrogenase from barley seed).

I. Electrophoresis in pore gradient polyacrylamide gel. The enzyme activity was not detectable after the electrophoresis.

II. Combination of the preceeding gel electrophoresis with the immunochemical analysis. By the second electrophoresis, the active enzyme in the agarose gel strip migrates in the gel containing antibodies, forms complexes then precipitates producing a line perpendicular to the migration axis stainable with the enzymatic characterization reaction. The addition of inactive alcohol dehydrogenase antigen also extracted by the second electrophoresis from the polyacrylamide gel strip results in the deformation of the specific precipitin line with the apparition of a peak which localizes the position of the inactive enzyme in the polyacrylamide gel strip after the first electrophoresis.

In a study on the genetic control of alcohol dehydrogenase in maize, Schwartz (1972) developed a particular two-dimensional immunoelectrophoresis in starch gel; this sensitive technique aimed at analysing allelic isoenzymes at the proteic level and particularly at detecting the inactive ones. Electrophoresis of the seed extract was performed in a starch gel band. Afterwards, active enzyme and immune serum were deposited, each in a slit on each side of the starch gel band, parallel to the migration axis of the first electrophoresis. A second electrophoresis was performed at a right angle to the first one under conditions bringing enzyme and antibodies together. A stainable precipitin line was formed which presented peaks in the regions of the gel where the enzymatic constituents were separated by the first electrophoresis.

B. SEARCH FOR INACTIVE ENZYMES USING THE TECHNIQUE OF COMPETITIVE ABSORPTION

When no enzymatic characterization reaction can be carried out, this type of technique offers, nevertheless, a further possibility of investigation. Its principle is simple: in a preliminary enzyme absorption by antibodies, the quantity of immune serum necessary for absorbing almost the total amount of a given enzymatic activity is determined. In a second experiment the procedure is carried out with the same quantity of immune serum, with the same enzymatic solution, but with the addition of a solution in which an inactive form of the enzyme is suspected. In this case antibodies will react with the two forms of the enzyme and less active enzyme will be absorbed; the remaining enzymatic activity will be greater than in the control absorption. For increasing the sensitivity of the technique, one may determine the amount of immune serum which absorbs only 50% of the given enzymatic activity. As already mentioned, this type of technique requires parallel experiments with non-immune serum and sometimes it is necessary to use immunoglobulins purified from the sera.

This technique was recently applied for determining whether the phenylalanine ammonia-lyase activity induced by far-red illumination in cotyledons of radish results from *de novo* synthesis of the enzyme or from an activating system. A protein antigenically related to the active enzyme was detected in extracts of non-illuminated cotyledons thus indicating the possibility of an activating mechanism in this enzymatic induction (Faye, 1975).

C. REMARKS

As already mentioned, the possibility that the anti-enzyme immune serum will react with the inactive enzyme assumes that there is sufficient antigenical relationship between the active enzyme and its inactive form.

There are now a number of examples indicating reactivity of anti-enzyme immune sera with the inactive enzymes (Arnon, 1973).

Several cases have been reported where anti-enzyme antibodies reacted with an enzyme associated with an inactivator (Arnon, 1973). This is the case for chymotrypsin inhibited by a polypeptide inhibitor (Bucci and Bowman, 1970), for trypsin bound either to diisopropylfluorophosphate or to a pancreatic inhibitor (Arnon and Schechter, 1966), and also for barley β-amylase inhibited by mercuric chloride (Daussant, 1966). Concerning the antigenical relationship between enzymes and the corresponding apo-enzymes, for which the prosthetic group was an organic molecule or a metal ion, it was reported that in most cases the anti-enzyme antibodies appear to react with the apo-enzyme and that no difference between them was detected by using the precipitation techniques (Arnon, 1973). Genetic alteration may also result in enzymic inactivation. There are examples which indicate that such inactivated enzymes retain the capacity of reacting with the anti-active enzyme immune serum. Such is the case for β-galactosidase from *E. coli*: in many mutants of *E. coli* there is no enzyme activity but there is a protein which cross-reacts with the anti-enzyme immune serum (Perrin, 1963). A similar case was also found for tryptophan synthetase from *E. coli* (Suskind *et al.*, 1963). Finally, taking into account the great variability of structural differences existing between an enzyme and its precursor (depending on the enzyme studied), the possibility that anti-enzyme immune serum will recognize the corresponding precursor remains questionable. Thus with anti-enzyme antibodies a positive reaction was observed with chymotrypsinogen (Bucci and Bowman, 1970; Gundlach, 1970), trypsinogen (Barett *et al.*, 1967), proplasmin (Robbins and Summaria, 1966), procccoonase (Berger and Kafatos, 1971; Berger *et al.*, 1971) and sometimes identity reactions were also obtained between the enzyme and its precursor. However, with pepsin (Seastone and Herriott, 1937; Arnon and Perlmann, 1963; Gerstein *et al.*, 1964) and carboxypeptidase A (Barett, 1965; Amiraian and Plummer, 1971; Lehrer and Van Vunakis, 1965) the reactions between the anti-enzyme antibodies and the zymogen was very low and sometimes unobservable. No reaction was observed between antithrombin antibodies and the prothrombin (Shapiro, 1968).

In conclusion, it seems that the identification of inactive enzyme using anti-enzyme antibodies is worth attempting. Nevertheless, a lack of reaction between the anti-enzyme immune serum and the proteic solution in which the inactive enzyme is suspected to exist, provides only a strong assumption but not a proof of the absence of this inactive form.

IV. MULTIPLE FORMS OF ENZYMES AT DIFFERENT PHYSIOLOGICAL STAGES

Multiple forms of enzymes raise questions about the nature of their polymorphism, questions of great importance in studies on enzyme regulation. Enzyme polymorphism in relation to genetic or physiological problems in plants has been dealt with in several recent reviews (Shannon, 1968; Scandalios, 1974; See also Scandalios, Chapter 7). For the elucidation of the nature of an enzyme polymorphism, the antigenic specificity of protein revealed by immunochemical techniques provides comparative data on the structure of enzymatic constituents present in a protein mixture. Examples of such antigenical comparison between different enzymatic forms in plant have already been reported (Daussant, 1975). However, when structural relationships between constituents are investigated there is a methodological aspect worth pointing out. It is interesting to combine the immunochemical identification with a fractionation technique which characterizes proteins by a parameter, which may indicate structural relationship between the constituents, i.e. according to their molecular size. For example, combination of exclusion chromatography and fused rocket immunoelectrophoresis was used for comparing molecular weight and antigenicity of malate dehydrogenases extracted from wheat leaves (Laurière et al., 1975). Combination of pore gradient polyacrylamide gel electrophoresis, which characterizes proteins by their molecular size, and crossed agarose gel immunoelectrophoresis was used for comparing β-amylase constituents extracted from barley seeds (Daussant and Skakoun, 1975).

The identification of enzymatic constituents becomes increasingly interesting when it aims at investigating whether the activity occurring at different physiological steps involves the same, or distinct enzymatic constituents. Multiple forms of enzymes can be detected by using electrophoretic techniques when a characterization reaction is available. Immunochemical techniques may offer original data and complete the information gained by the electrophoretic analysis or even provide an approach to the problem when no characterization reaction is available.

A. NO ENZYMATIC CHARACTERIZATION REACTION IS AVAILABLE

The enzyme extracts corresponding to the different physiological stages are diluted to the same enzymatic activity. Increasing amounts of the anti-enzyme immune serum are added to aliquots of the extracts, and the remaining activity is measured on the supernatant solution. Differences in the absorption curves corresponding to the several extracts, reflect changes in the composition of the enzymatic population.

One example concerns phenylalanine ammonia-lyase (PAL) (Faye, 1975; Fig. 3). As already described, this enzyme activity increases when radish cotyledons, soaked in the dark for 36 h, are submitted to far-red light. This increase is due, at least in part, to a mechanism which renders pre-existing enzyme active. Under prolonged illumination, the activity continues to increase then, finally, decreases. An immune serum was prepared from a purified preparation of the enzyme extracted from the cotyledons at maximum activity (36 h illumination). Preliminary studies of this serum showed that it reacted specifically with all enzyme forms in the purified fraction, but only with part of the enzyme present in the crude extract. The immune serum, therefore, characterized two types of enzymes, one which reacted with the antibodies and one which did not react with the antibodies. The changes in the relative proportions of the two forms were then studied over the course of light exposure: extracts were made: 1. before the PAL activity reached its maximum (18 h exposure); 2. at maximum activity (36 h exposure) and 3. after this maximum (48 h exposure). Immunoabsorption was carried out on the three extracts after adjusting them to the same activity by adding increasing amounts of purified IgG from the immune serum. Results indicated that the proportion of the enzyme which reacted with the immune serum (about 60% of the activity in the first extract, 18 h light exposure) became greater in the second extract (36 h light exposure) and constituted the bulk of the activity in the last extract (48 h light exposure). Thus, by using antibodies in this absorption technique, it was possible to show a shift in the proportion of the two PAL constituents during the exposure to light.

Another result of the absorption is interesting. Absorption is more efficient with extract of 36 h than with extracts of 18 h or 48 h. In other words, the same amount of antibodies absorbs more enzymatic activity in the 36 h extract than in either of the other two. This result indicates that for an equivalent activity there are more antigenic enzymes in extracts of 18 h

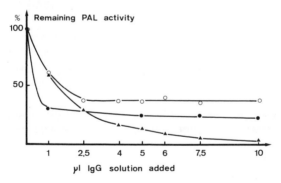

FIG. 3. Immunoabsorption of phenylalanine ammonia-lyase (PAL) extracted from cotyledons of soaked radish, first kept in the dark for 36 h, then exposed to red light for respectively 18 h (extract 18), 36 h (extract 36), 48 h (extract 48), with IgG purified from an anti-PAL immune serum. The immune serum was prepared by immunizing rabbits with an enzyme purified from extract 36. Absorption was carried out from the three extracts previously adjusted to the same activity. Results not reported on the figure, concerning the treatment of the PAL with similar amounts of IgG purified from the serum taken before immunization indicated that no reduction of the activity occurred. The diminution of activity reported here therefore indicated an absorption due to the antibodies.

Note that the absorption is partial for extracts 18 and 36 but is almost total for extract 48 (see text). Note that 1 µl IgG absorbs more activity in extract 36 than in extracts 18 or 48 (see text). Extract 18 O—O—O. Extract 36 ●—●—●. Extract 48 ▲—▲—▲ (Faye, 1975).

and of 48 h, than in extract of 36 h. One may hypothesize that in addition to the active enzyme in the 18 h extract, there are also pre-existing enzymes which have not yet been activated. One may also speculate that in the 48 h extract the loss of activity is due to an inactivating system or to molecular ageing of the enzymatic proteins; these remain in the extract but become less active.

B. ENZYMATIC CHARACTERIZATION REACTION ON THE GELS IS AVAILABLE

In this case various immunochemical techniques of specific precipitation in gels are practicable. An example will illustrate that results obtained by these techniques may complete those obtained by electrophoresis. α-Amylase in developing and germinating wheat seeds is chosen to illustrate this.

The activity of α-amylase present at an early stage of development in wheat, decreases and practically disappears as maturation proceeds (Olered, 1964). However, the activity increases again significantly, during germination (Kneen, 1944). Electrophoretic studies indicated that α-amylase constituents in developing seeds were identical to α-amylase constituents in germinating seeds (Kruger, 1972b), whereas immunochemical absorption using an anti-α-amylase immune serum from germinated seeds, indicated that the greatest part of the activity in germinated seeds was due to constituents antigenically distinct from the α-amylase of developing seeds (Daussant and Renard, 1972). Combining both electrophoretic and immunochemical studies, similarities and differences between enzymatic constituents in developing and germinating seeds could be further defined. By using agarose gel electrophoresis (Fig. 4), one constituent of germinating seeds (G1) seems identical to the anodic constituent of developing seeds (D1). Nevertheless, the main activity in germinating seeds is borne by constituents characteristic of germinating seeds (G2). Moreover, a cathodic constituent is unique to the developing seeds (D2).

Immunoelectrophoresis was carried out with both extracts using the anti-α-amylase immune serums specific for: 1. enzyme from developing seeds (anti-α-D), Fig. 4 (II) and 2. enzyme from germinated seeds (anti-α-G).

1. The Immunochemical Analysis Confirms the Electrophoretic Data

a. D2 is antigenically unique to developing seeds, since the immuno-absorption of the anti-α-D immune serum with the proteins of germinating seeds does not prevent the immune serum from reacting with D2, Fig. 4 (III).

b. In the same way G2 was shown to be characteristic for germinating seeds: the immuno-absorption of the anti-α-G immune serum with the proteins of developing seeds did not prevent the immune serum from reacting with G2.

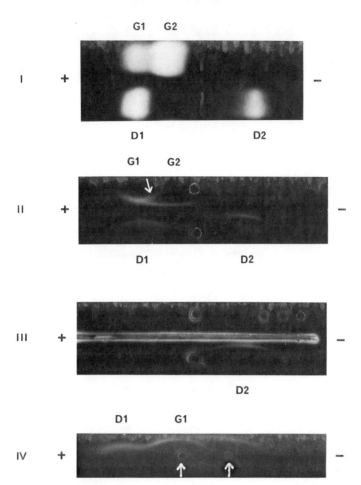

FIG. 4. Electrophoretical and immunoelectrophoretical comparisons between α-amylase extracted from 7 days germinated wheat seeds (for each slide upper well) and α-amylase extracted from developing seeds taken 15 days after anthesis (for each slide under well). A reaction of α-amylase characterization was carried out on the gels after electrophoresis and immunoelectrophoresis. G_1 and G_2, D_1 and D_2 design respectively 2 α-amylase groups in extracts of germinated seeds, and of developing seeds.

I. Agarose gel electrophoresis.

II. Immunoelectrophoresis using an immune serum specific for α-amylase from developing seeds. The arrow indicates a semi-identity reaction between G_1 and G_2.

III. Immunoelectrophoresis using the same immune serum after absorption with the extract of germinated seeds. (After electrophoresis, the three wells on the right above the trough were filled with different dilutions of the extract of germinated seeds.) There is no more reaction between the immune serum and the extract of germinated seeds; the absorbed immune serum still reacts with constituent D_2 from developing seed extract.

IV. Tandem immunoelectrophoresis using the immune serum specific for α-amylase of developing seeds. Two wells cut on the same side of the trough (indicated by the arrows) were respectively filled, from the anode to the cathode, by extract of developing and of germinating seeds. Note that the precipitin bands corresponding to D_1 and G_1 form a single line without any spur.

212 J. DAUSSANT ET AL.

2. *The Immunochemical Analysis Completes the Information Given by Electrophoresis*

Structural relationship is suggested between G1 and G2 by using both immune serums.

With the anti-α-D immune serum, the precipitin band corresponding to G1 crosses the precipitin band corresponding to G2 but is not crossed by this latter one (Fig. 4 (II)). With the anti-α-G immune serum the reverse result was obtained: the precipitin band corresponding to G2 crossed the precipitin band corresponding to G1 but was not crossed by this latter one. The semi identity reactions are schematized in Table I.

TABLE I

Antigenical relationship suggested between G1 and G2 by using
the two types of immune serums;
A, B, C, D, define groups of
antigenic determinants

	G1	G2
anti-α-D	AB	A
anti-α-G	C	CD

3. *The Immunochemical Analysis Confirms and Completes Data Provided by Electrophoresis*

D1 and G1 are electrophoretically identical. With the anti-α-D immune serum, D1 and G1 appear not only antigenically related (G1 reacts with the immune serum) Fig. 4 (II), but also antigenically identical (the absorption of the immune serum by the amylases of germinated seeds results in removing all antibodies from the immune serum which reacted with D1) Fig. 4 (III). Another way of indicating that D1 and G1 are antigenically identical is shown on Fig. 4 (IV). Precipitin bands corresponding to D1 and G1 form a continuous line without any spur.

By using the same techniques, the antigenic identity between D1 and G1 was also found with the anti-α-G immune serum.

Further information on the comparison between G1 and D1 is provided by using a prolonged electrophoresis for better separation and the corresponding cross immunoelectrophoresis. In Fig. 5 (I), where only anodic constituents are characterized, G1 and D1 contain each three constituents, whereas G2 includes four constituents. The results are in accordance with electrophoretic studies carried out on polyacrylamide gel (Kruger, 1972a, b).

Using the anti-α-D immune serum under conditions adequate for studying D1 and G1, the three constituents of D1 as well as of G1 form three peaks without any spur (Fig. 5 (II, III)). The immune serum indicates great structural analogy between these constituents and no differences.

Furthermore, since there are identity reactions between the three constituents of D1 and the three constituents of G1, a quantitative evaluation of the enzymatic proteins in the course of the development and the germination will be possible. For example, a quantitative comparison carried out as shown in Fig. 5 (II and III), indicated that similar amounts of D1 and G1 were detected in developing seed extracts and in similar extracts of germinating seeds after a 40-fold dilution. In other words, for the same weight of freeze-dried developing seeds (20 days after anthesis) and of freeze-dried germinating seeds (7 days after steeping) there is 40-fold less D1 protein than G1 protein. In the same number of seeds the common anodic α-amylase constituent is about 80-fold more abundant in germinating seeds than in developing seeds.

Thus, the immunochemical approach may add some structural comparative data and provide quantitative information concerning amounts of enzymatic protein.

V. Identification of Phenomena Involved in the Appearance and Disappearance of Enzymatic Activity and in the Modification of Enzyme Physico-chemical Properties

As already mentioned, the occurrence of an enzymatic activity raises questions on the level at which the enzyme activity originates; by means of a system activating pre-existing enzyme or by synthesis *de novo*? The immunochemical means for detecting the existence of inactive enzyme has already been discussed. Here it will be shown how immunochemical methods, directly or indirectly, may indicate a *de novo* synthesis of enzymatic constituents. The two examples concern α-amylase in germinating wheat seeds and ribulose diphosphate carboxylase in barley leaves.

Similar questions arise concerning the disappearance of an enzymatic activity; degradation, inactivation, insolubilization? In the two latter cases the enzymatic protein must still be present in a soluble or insoluble form. This aspect will be dealt with by taking as example the disappearance of α-amylase activity in wheat seed during maturation.

Finally, changes in physico-chemical properties of enzymes or isoenzymatic systems may be brought about by factors which act on the already existing proteins. It will be indicated how immunochemical techniques may contribute to evidence on such events by taking β-amylase in germinating seeds as example.

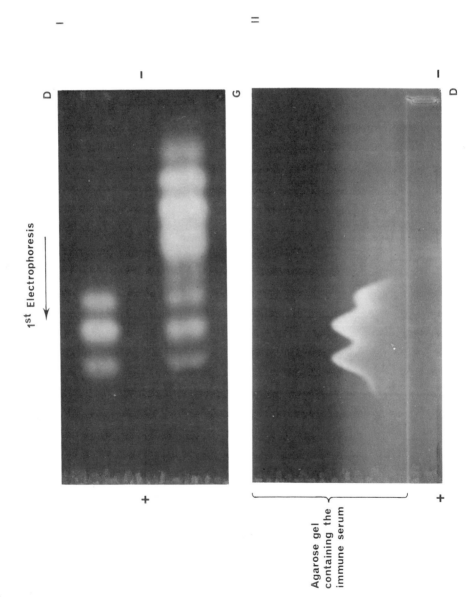

1st Electrophoresis

Agarose gel containing the immune serum

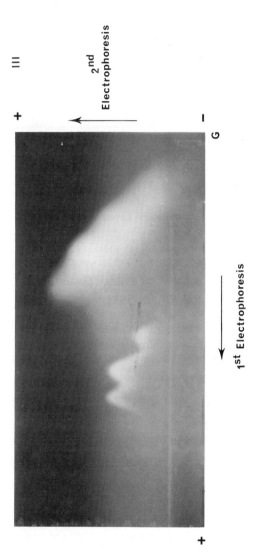

III

+

2nd Electrophoresis

−

G

1st Electrophoresis

+

FIG. 5. Comparison by prolonged agarose gel electrophoresis and crossed agarose gel immunoelectrophoresis of anodic α-amylases from developing wheat seeds taken 15 days after anthesis (D) and from 7 days' germinated wheat seeds (G). Starting wells (on the right of the figures) were filled with the extracts, and when the solutions were sucked in by the gel they were filled again with agarose gel. A reaction of α-amylase characterization was carried out on the gels after electrophoresis and cross immunoelectrophoresis.

I. Prolonged agarose gel electrophoresis (5 V/cm at 4°C for 6,5 h). Groups G_1 and D_1 of Fig. 4 are separated, each into three constituents, group G_2 into four constituents.

II, III. Crossed agarose gel immunoelectrophoresis. After the first electrophoresis (I) the agarose gel strips were sealed against another agarose gel containing an immune serum specific for α-amylase from developing seeds. During the second electrophoresis perpendicularly to the first one, the separated constituents form complexes with antibodies then precipitate producing peaks stainable by the enzyme characterization reaction. The three constituents of D_1, (II), form a continuous precipitin band without spur. Same observation for the three constituents of G_1 (III).

A. SEARCH FOR THE CAUSE LEADING TO THE DISAPPEARANCE OF α-AMYLASE OF
DEVELOPING WHEAT SEEDS UPON MATURATION

In a first attempt to detect inactive forms of the cathodic and anodic α-amylases in mature seeds, a combination of line rocket immunoelectrophoresis and enzymatic characterization reaction (as shown in Fig. 1 for evidence of inactive glutamate dehydrogenase) was applied using different extracts of the mature seeds. The different extracts were made to solubilize different categories of proteins, and the following solutions were used: 1. saline solutions extract albumins and globulins (active α-amylases belong to such protein types); 2, 3 M or 8 M urea solutions in presence of a reducing agent which aimed at extracting enzymes not solubilized by the preceeding solutions. The technique failed to detect in extracts proteins antigenically related to the anodic and cathodic α-amylases, but it remains possible that the lack of reaction of the immune serum with protein extracted with urea, resulted from an irreversible denaturation of the enzymes and loss of their original antigenical structure.

In order to complete this information another approach involving immunoabsorption of the immune serum directly with the seed flour was attempted. A fine flour of mature seeds was added to the immune serum specific for α-amylase of developing seeds and slowly stirred at 20°C for 36 h. Under these conditions it was expected that no artificial denaturation of the enzyme would occur and that during the stirring, antibodies would complex with the enzymes in the flour. A quantitation of the antibodies would detect such an immunoabsorption. Practically, flour from 10 mature seeds was added to 6 ml aliquot of 20-fold diluted immune serum; the flour of 3 developing seeds and the flour of one developing seed were added, respectively, to 2 other aliquots of the diluted immune serum; an aliquot of the diluted immune serum was also submitted to the same stirring at the same temperature and served as a control. After centrifuging the immune sera, agarose gel strips, each containing exactly 6% of the different aliquots of the diluted and treated immune serum, were prepared. The same quantity of the extract of developing seeds was deposited into wells cut in the centre of each gel strip, and electrophoresis was started. During electrophoresis, anodic and cathodic constituents move into the gel, form complexes with the corresponding antibodies and precipitate as two peaks, one towards the anode, one towards the cathode, each with the α-amylase activity. For each anodic and cathodic constituent, the surface of the peaks characterizes a certain ratio between the amounts of antigen and the corresponding antibodies. Since the same amount of antigen was deposited, an increase in the dimension of the peaks in comparison to the peaks of the control would reflect a diminution in the amount of antibodies. Absorption with flour from 3 developing seeds resulted in a clear cut diminution of antibodies corresponding to the anodic as well as to the cathodic constituents (compare Fig. 6 (I) and Fig. 6 (IV)). The absorption was also perceptible when only the flour corresponding to one developing seed was used (compare Fig. 6 (I) and Fig. 6 (III)). Absorption with the flour of 10 mature seeds did not significantly reduce the antibody titre in the immune serum (compare Fig. 6 (I) and Fig. 6 (II)). Even if there is a reduction of the antibodies corres-

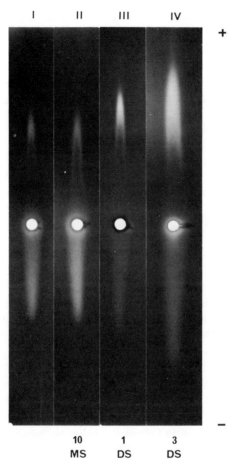

Fig. 6. Search for proteins antigenically related to α-amylase of developing wheat seeds in mature seeds by using an absorption technique. Absorption was carried out on the immune serum specific for α-amylase from developing seeds with seed flour (see Section VA). The analysis of the unabsorbed and absorbed immune serum samples is represented on the figure. Each agarose gel strip contains the same percentage of the different immune serum samples.

I. Unabsorbed immune serum;

II. immune serum absorbed with flour from 10 mature seeds (10 MS);

III. immune serum absorbed with flour from 1 developing seed (1 DS);

IV. immune serum absorbed with flour from 3 developing seeds (3 DS). The same amount of a same extract of developing seeds was deposited in each well. Thus, in comparison to the control experiment (I) an increase in peak size indicates a reduction of the antibodies titre in the immune serum samples (see text).

ponding to the cathodic constituent, the absorption caused with 10 mature seeds remained inferior to the absorption caused by one developing seed (compare Fig. 6 (I), Fig. 6 (II) and Fig. 6 (III)).

These results indicate that if the α-amylases characterized in developing seeds still exist in mature seeds, they are probably not in a soluble and inactivated form, but merely in an insoluble form. In any case the external structure of these constituents would be quite modified or profoundly hidden under other molecular species, but these results suggest rather that the bulk of the enzyme molecules might have been degraded during maturation.

B. INDIRECT OR DIRECT PROOF FOR ENZYME SYNTHESIS

Results reported in Section IV concerning the amount of α-amylases antigenically and electrophoretically identical in developing and germinating wheat seeds already provided indirect evidence for *de novo* synthesis of these constituents during germination. The results indicated that the amount of these constituents was approximately 80-fold greater in germinating seeds than in developing seeds. These results suggested that if there was an activation of pre-existing enzyme during germination, this might concern only a small proportion of the enzymatic constituents in germinating seeds. The negative results in the search for these constituents in mature seeds supports this view. In the same way, a lack of reaction between the immune serum specific for α-amylases of germinated seeds and proteins from mature seeds constituted indirect evidence supporting the view that the occurrence of the corresponding antigenic enzymes upon germination resulted from *de novo* synthesis in barley seeds (Grabar and Daussant, 1964) and in wheat seeds (Daussant and Corvazier, 1970).

More directly, a quantitative increase in enzymatic proteins which parallels an increase in enzymatic activity supports the view for *de novo* synthesis of the enzyme. Thus, using a monospecific immune serum prepared against ribulose diphosphate carboxylase, it was shown that the increase in the enzymatic activity which occurred in etiolated barley leaves subsequently exposed to light, corresponded to an accumulation of this enzymatic protein (Kleinkopf *et al.*, 1970).

Direct proofs of *de novo* enzyme synthesis may be provided by combining the *in vivo* incorporation of labelled amino acid into proteins with the antigenical identification of proteins.

Such an approach was used in the same study (Kleinkopf *et al.*, 1970)· *in vivo* incorporation of labelled leucine into proteins was carried out with the excised etiolated leaves put in the light; radioactivity was found in the enzyme precipitated from the leaf extracts using the anti-enzyme monospecific immune serum. This approach was also used in a study concerning α-amylase isoenzymes of germinated wheat seeds. Seeds were germinated in water containing labelled amino acids. By immunoelectrophoresis of the extracts of germinated seeds with an immune serum specific for α-amylases of germinated seeds, the two precipitin bands corresponding to the enzyme became labelled. This result indicated that *de novo* synthesis upon germination gave rise to both antigenic forms of α-amylase (Daussant and Corvazier, 1970).

This type of approach, using quantitative techniques of antigen determination and quantitative measurement of radioactivity on specific antigen-antibody precipitates, may provide original information on enzymatic protein turnover. An example of this approach was provided by studies on ribulose diphosphate carboxylase extracted from barley leaves (Peterson *et al.*, 1973, 1975). After *in vivo* ^{14}C incorporation, extracts of leaves were made at different times; radioactivity was measured on specific precipitates obtained from the different extracts by using the anti-enzyme immune serum. A decrease in the specific radioactivity (radioactivity per given amount of specific precipitate) indicates that the constant amount of enzymatic proteins resulted from degradation of enzyme molecules balanced by synthesis of new ones. A constant specific radioactivity indicates that the population of enzymatic molecules was not renewed.

C. CHANGES IN PHYSICO-CHEMICAL PROPERTIES OF WHEAT β-AMYLASE DURING GERMINATION

Upon germination, the electrophoretic mobilities of β-amylase constituents are modified although the enzyme keeps its antigenic specificity (Kruger, 1972b; Daussant and Corvazier, 1970). This could be caused by *in vivo* modification of the pre-existing β-amylase molecules or by a disappearance of certain β-amylase constituents and synthesis of new ones. Wheat seeds were germinated in water containing labelled amino acids. Autoradiography carried out on the immunoelectrophoregram of the seed extracts using an anti-β-amylase monospecific immune serum did not detect any radioactivity in the β-amylase precipitin band. Thus it appeared that the bulk of the antigenic β-amylase with the new electrophoretic mobility was due to modifications of the pre-existing β-amylase rather than to synthesis of new molecules (Daussant and Corvazier, 1970).

These examples aimed at illustrating the originality of the information which may be provided by immunochemical quantitation of enzymatic proteins from crude extracts; they also aimed at indicating how the significance of data obtained with *in vivo* incorporation of labelled molecules may be enhanced when combined with a qualitative or quantitative immunochemical determination of enzymatic proteins.

VI. Changes in the Specific Activity of Enzymes

Immunochemical techniques for evaluating the quantitative changes in enzymatic proteins are particularly interesting when carried out together with a quantitative determination of changes in enzymatic activity; together these data provide means for evaluating the evolution of the specific activity of the enzymes.

When an enzymatic activity results from a single molecular species, the activity may be measured spectrophotometrically and the amount of enzymatic protein may be evaluated by a technique of precipitation in tube or in gel, depending on the monospecificity of the immune serum. When several isoenzymes are present, the specific enzymatic activity may be obtained only for constituents for which antibodies are available. The activity of the enzyme constituents may be evaluated, under certain conditions by electrophoretic separation, enzymatic characterization reaction and densitometry on the electrophoregrams, or by spectrophotometric measurement after elution of the enzymes from the gel. The enzymatic protein amount may be evaluated by combining the same electrophoretic separation with a crossed immunoelectrophoresis (Freeling, 1973; Skakoun and Daussant, 1975).

In a study of enzymes extracted from banana fruits during maturation, alcohol dehydrogenase was characterized as a single antigen, and two antigenic malate dehydrogenases were identified as two distinct molecular species. During ripening an increase of the alcohol dehydrogenase activity was paralleled by an equal increase in the amount of antigen. The enzymatic increase could thus be ascribed to an accumulation of enzymatic proteins. The situation was slightly different for each of the antigenically defined malate dehydrogenases. Here an increase in the activity corresponded also to an increase in the amount of antigen, but to a lesser extent. So in this case the activity increase was probably due to an accumulation of enzymatic proteins and also to an activating system at the enzymatic protein level (Skakoun and Daussant, 1975).

In studies on ribulose diphosphate carboxylase from barley leaves exposed to different light conditions the activity and the amount of the antigenic enzyme were recorded for several days. Under constant illumination the specific activity remained first unchanged, then decreased slightly after a certain time. This may suggest a molecular ageing of the enzyme molecules. During a dark period, both the amount of enzymatic protein and the activity decrease, but to a different extent, so that the specific activity of the enzyme increases. The reverse phenomenon was observed during a subsequent period of illumination. Thus, the results which indicate the action of darkness on the loss of enzymatic proteins and the action of subsequent lightening on synthesis of new enzymatic molecules also reveal a curious phenomenon concerning the *in vitro* observed specific activity. This latter discovery certainly calls for further studies (Peterson *et al.*, 1973).

The possibility for investigating *in vitro*, the evolution of the specific activity of enzymes represents probably one of the most interesting aspects of the antigenic determination of enzymatic proteins.

VII. Studies on Enzyme Structure and Localization of Enzyme Synthesis by Using Anti-enzyme Subunits Immune Sera

Further studies concerning the ribulose diphosphate carboxylase have been reported recently. The higher plant multimeric enzyme is known to be formed from several large subunits and several small subunits. Immune sera, specific for the large one and for the small one were used in these studies.

Absorption was performed with each of these immune sera, on ribulose diphosphate carboxylase extracted from bean leaves and activity measurements were made on the resuspended specific precipitates. A marked inhibition was noted with the anti-large subunit immune serum but not with the anti-small subunit immune serum. The result supported the idea that the large subunit has an enzymatic function (Gray and Kekwick, 1974). Similar results were obtained with the enzyme extracted from spinach leaves (Nishimura and Akazawa, 1974). Moreover, the action of the anti-small subunit immune serum resulted in the loss of the ability of Mg^{++} to shift the pH optimum of the enzyme activity. The action of each of the two immune sera was also studied on the ribulose diphosphate oxygenase activity of the same enzyme. It was found that the anti-large subunit immune serum had an inhibiting effect on the enzyme whereas the anti-small subunit immune serum did not affect the activity. Nevertheless, the latter immune serum did shift the pH optimum of the enzyme activity. All together these results indicate that the large subunit has a catalytic role whereas the small one has a regulatory function (Nishimura and Akazawa, 1974).

The important fact that the anti-small subunit immune serum did not react with the large subunit and, conversely, that the anti-large subunit immune serum did not react with the small subunit provided an essential means for gaining insight into the localization of the synthesis of these two subunits: In an *in vitro* protein synthesizing system, the immune sera were used for precipitating nascent chains on the cytoplasmic and/or chloroplastic ribosomes. The results obtained, indicated that chloroplast ribosomes make the large subunits (Gooding *et al.*, 1973) whereas the cytoplasmic ribosomes make the small subunits. Further, evidence was provided that completed large subunits already associate with small subunits while still on the ribosomes (Gooding *et al.*, 1973; Gray and Kekwick, 1974).

VIII. Conclusions

The examples provided in this report aimed at indicating the usefulness of an immunochemical approach in several problems related to plant enzyme regulation and at describing some recent immunochemical techniques adapted to these studies. There are actually other applications of these techniques, useful in other fields. For example, the monospecific anti-enzyme immune sera may serve in microscopy or in electron microscopy for localizing enzymes for which no enzymatic characterization reaction is known, or for localizing different isoenzymes.

There is no doubt that characterizing enzymes at the same time by their function and by their antigenic structure offers a discriminating tool for studies concerned with enzyme regulation. The recent developments of immunochemical techniques provide means for using anti-enzyme antibodies in such studies in many different ways.

REFERENCES

Amiraian, K. and Plummer, T. H., Jr. (1971). *J. Immun.* **107**, 547.

Arnon, R. (1971). *Curr. Top. Microbiol. Immun.* **54**, 47.

Arnon, R. (1973). *In* "The Antigens" (Sela, M., ed.), Vol. 1, pp. 88–159. Academic Press, New York and London.

Arnon, R. and Perlmann, G. E. (1963). *J. biol. Chem.* **238**, 963.

Arnon, R. and Schechter, B. (1966). *Immunochemistry* **3**, 451.

Barett, J. T. (1965). *Immunology* **8**, 129.

Barett, J. T., Nilsson, B. and Ghiron, C. A. (1967). *Int. Arch. Allergy appl. Immun.* **31**, 399.

Berger, E. and Kafatos, F. C. (1971). *Immunochemistry* **8**, 391.

Berger, E., Kafatos, F. C., Felsted, R. L. and Law, J. H. (1971). *J. biol. Chem.* **246**, 4131.

Bucci, E. and Bowman, D. E. (1970). *Immunochemistry* **7**, 289.

Carfantan, N. and Daussant, J. (1974). *Acta Horticulturae* **41**, 31.

Cinader, B. (1963). *Ann. N.Y. Acad. Sci.* **103**, 495.

Cinader, B. (1967). *In* "Antibodies to Biologically Active Molecules" (Cinader, B., ed.), pp. 85–137. Pergamon Press, Oxford.

Clarke, H. G. M. and Freeman, T. (1967). *In* "Protides of the Biological Fluids" (Peeters, H., ed.), Vol. 14 pp. 503–509. Elsevier, Amsterdam.

Crumpton, M. J. (1973). *In* "The Antigens" (Sela, M., ed.), Vol. 1, pp. 1–77. Academic Press, New York and London.

Daussant, J. (1966). *Biotechnique* **4**, 1.

Daussant, J. (1975). *In* "The Chemistry and Biochemistry of Plant Proteins" (Harborne, J. B. and Van Sumere, C. F., eds), pp. 31–69. Academic Press, New York and London.

Daussant, J. and Carfantan, N. (1975). *J. Immun. Methods* **8**, 373.

Daussant, J. and Corvazier, P. (1970). *FEBS Letts.* **7**, 191.

Daussant, J. and Renard, M. (1972). *FEBS Letts.* **22**, 301.

Daussant, J. and Skakoun, A. (1974). *J. Immun. Methods* **4**, 127.

Daussant, J. and Skakoun, A. (1975). *J. Immun. Methods* **7**, 39.

Faye, L. (1975). Thesis, University of Rouen, France.

Freeling, M. (1973). *Molec. gen. Genet.* **127**, 215.

Gerstein, J. F., Levine, L. and Van Vunakis, H. (1964). *Immunochemistry* **1**, 3.

Giebel, W. and Saechtling, H. (1973). *Hoppe–Seyler's Z. Physiol. Chem.* **354**, 673.

Gooding, L. R., Roy, H. and Jagendorf, A. T. (1973). *Archs Biochem. Biophys.* **159**, 324.
Grabar, P. and Daussant, J. (1964). *Cereal Chem.* **41**, 523.
Grabar, P. and Williams, C. A. (1953). *Biochim. biophys. Acta* **10**, 193.
Gray, J. C. and Kekwick, R.G.O. (1974). *Europ. J. Biochem.* **44**, 481.
Gundlach, H. G. (1970). *Hoppe–Seyler's Z. Physiol. Chem.* **351**, 696.
Hill, R. J. and Djurtoft, R. (1964). *J. Inst. Brew.* **70**, 416.
Johansson, B. G. and Stenflo, J. (1971). *Ann. Biochem.* **40**, 232.
Kleinkopf, G. E., Huffaker, R. C. and Matheson, A. (1970). *Pl. Physiol.* **46**, 416.
Kneen, E. (1944). *Cereal Chem.* **21**, 304.
Krøll, J. (1973a). *Scand. J. Immun.* **2**, *Suppl.* 1, 61.
Krøll, J. (1973b). *Scand. J. Immun.* **2**, *Suppl.* 1, 79.
Krøll, J. (1973c). *Scand. J. Immun.* **2**, *Suppl.* 1, 83.
Kruger, J. E. (1972a). *Cereal Chem.* **49**, 379.
Kruger, J. E. (1972b). *Cereal Chem.* **49**, 391.
Laurell, C. B. (1965). *Ann. Biochem.* **10**, 358.
Laurell, C. B. (1966). *Ann. Biochem.* **15**, 45.
Laurière, C., Skakoun, A. and Daussant, J. (1975). *Physiol. Vég.* **13**, 467.
Lebas, J., Hayem, A. and Martin, J. P. (1974). *C.r. hebd. Séanc. Acad. Sci.* Paris **278D**, 2359.
Lehrer, H. I. and Van Vunakis, H. (1965). *Immunochemistry* **2**, 255.
Mancini, G., Carbonara, A. O. and Heremans, J. F. (1965). *In* "Immunochemistry" pp. 235–254. Pergamon Press, Oxford.
Margolis, J. and Kenrick, K. G. (1968). *Ann. Biochem.* **25**, 347.
Neurath, H. (1957). *Adv. Protein Chem.* **12**, 319.
Nishimura, N. and Akazawa, T. (1974). *Biochemistry* **13**, 2277.
Olered, R. (1964). *Arkh. Kemi.* **22**, 175.
Ouchterlony, O. (1949). *Acta path. microbiol. scand.* **26**, 507.
Perrin, D. (1963). *Ann. N.Y. Acad. Sci.* **103**, 1058.
Peterson, L. W. and Huffaker, R. C. (1975). *Pl. Physiol.* **55**, 1009.
Peterson, L. W., Kleinkopf, G. E. and Huffaker, R. C. (1973). *Pl. Physiol.* **51**, 1042.
Poulik, M. D. (1971). *In* "Methods in Immunology and Immunochemistry" (Williams, C. A. and Chase, M. W., eds.), Vol. 3, pp. 279–294. Academic Press, New York and London.
Ressler, N. (1960). *Clin. Chim. Acta* **5**, 795.
Robbins, K. C. and Summaria, L. (1966). *Immunochemistry* **3**, 29.
Scandalios, J. G. (1974). *A. Rev. Pl. Physiol.* **25**, 225.
Schwartz, D. (1972). *J. Chromat.* **67**, 385.
Seastone, C. V. and Herriott, R. M. (1937). *J. gen. Physiol.* **20**, 797.
Shannon, L. M. (1968). *A. Rev. Pl. Physiol.* **19**, 187.
Shapiro, S. S. (1968). *Science* **162**, 127.
Skakoun, A. and Daussant, J. (1975). *In* "Facteurs de la Régulation et de la Maturation des Fruits" Vol. 238, pp. 281–289. Coll. Int. CNRS., Paris.
Skakoun, A., Niku-Paavola, M. L. and Daussant, J. (1976). *Bios.* **13**.
Suskind, S. R., Wickham, M. L. and Carsiotis, M. (1963). *Ann. N.Y. Acad. Sci.* **103**, 1106.
Svendsen, P. J. (1973). *Scand. J. Immun.* **2**, *suppl.* 1, 69.
Uriel, J. (1971). *In* "Methods in Immunology and Immunochemistry" (Williams, C. A. and Chase, M. W., eds.), Vol. 3, pp. 294–321. Academic Press, New York and London.
Yamasaki, M., Brown, J. R., Cox, D. J., Greenshields, R. N., Wade, R. N. and Neurath, H. (1963). *Biochemistry* **2**, 859.

CHAPTER 11

The Use of Density Labelling Techniques in Investigations into the Control of Enzyme Levels

C. B. JOHNSON

Department of Physiology and Environmental Studies, University of Notting-ham, School of Agriculture, Sutton Bonington, England

I. Introduction	225
II. The Methodology of Density Labelling	225
A. Density Labels	226
B. Solutes for Density Gradient Centrifugation	228
C. Centrifugation Conditions	232
III. Density Labelling as an Experimental Technique	233
A. For the Demonstration of *De Novo* Protein Synthesis . . .	233
B. For the Measurement of Enzyme Turnover	236
C. For Investigations into the Control of Enzyme Activity . . .	238
D. Measurement of Incorporation of Density Labels into Amino Acid Pools	240
IV. Conclusion	242
Acknowledgements	242
References	242

I. INTRODUCTION

This review falls conveniently into two parts. In the first I have attempted to survey the most recent advances in the density labelling technique itself. The second part is devoted to an evaluation of density labelling as a tool for investigating problems concerning enzyme control. In this section, I have placed most emphasis on the design and interpretation of experiments aimed at measuring or comparing rates of enzyme turnover and enzyme synthesis, on which topics the most density labelling work is currently being pursued.

II. THE METHODOLOGY OF DENSITY LABELLING

The principle of density labelling is a simple one. The incorporation into an enzyme of atoms of carbon, hydrogen, oxygen or nitrogen of densities greater than those found in nature, leads to increases in the densities of the enzyme molecules. These increases can be detected by isopycnic centrifugation in a suitable solute. Because individual enzymes can be detected, after fractionation of the gradients, merely by measurement of the enzyme activity in each fraction, the technique circumvents the need for purification of the enzyme. Purification is a prerequisite for the demonstration of synthesis by all the other available methods.

The demonstration of the occurrence of enzyme synthesis *per se* is rarely a noteworthy event. Except in specialized situations, for example, germination, the problem facing the investigator is not to establish whether or not enzyme synthesis is occurring but, rather, whether a particular stimulus alters the rate of enzyme synthesis, enzyme degradation (turnover) or enzyme activation/inactivation. Thus, quantitative measurements must be made of the fate of a density label applied to a tissue. As the problems tackled become more intricate the demands on the technique are increased and it therefore seems appropriate to begin by establishing the conditions which will lead to the maximum possible resolution of molecules differing by only small amounts in density. Because many proteins are labile *in vitro*, and especially so in the solutes used for density gradients, methods whereby the time needed for ultracentrifugation can be reduced are also given some attention.

A. DENSITY LABELS

The maximum density increases which can be introduced into proteins of typical amino acid composition are shown in Table I. It may be noted, however, that these are indeed maximum values and the relative incorporation of these labels may be very different, depending on the experimental conditions. For example, C^{13} is normally applied in the form of labelled amino acids, whereas N^{15} may be applied also as inorganic nitrogen, usually in the form of nitrate or ammonia. Consequently, the relative incorporation of these

TABLE I

Density increases introducible into protein molecules by incorporation of different density labels

Density label	Maximum density increase $kg\, l^{-1}$
N^{15}	0·013
C^{13}	0·050
H^2	0·060[a]
O^{18}	0·030

From Hu *et al.* (1962).
[a] Assuming that H^2 replaces all carbon-bound H^1.

labels could depend on the contributions of amino acid reutilization and amino acid biosynthesis to protein synthesis in the tissue. These contributions may, themselves, be altered by effectors of enzyme activity. The situation is even more complex when labelled water is used. Both O^{18} and H^2 are then incorporated both via amino acid biosynthesis and during amino acid reutilization. However, whereas the carbonyl oxygen atoms of peptide

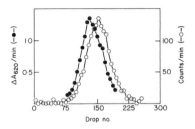

FIG. 1a. Isopycnic centrifugation in CsCl of α-amylase extracted from barley aleurone layers. Closed circles represent enzyme activity of H_2O^{18} labelled, gibberellic acid treated enzyme, open circles represent radioactivity associated with native, purified enzyme. (From Filner and Varner, 1967.)

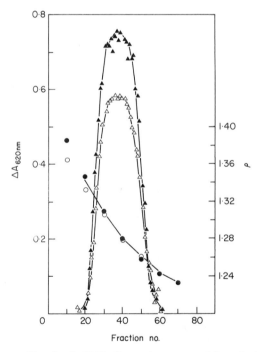

FIG. 1b. Isopycnic centrifugation in CsCl of α-amylase extracted from barley aleurone layers, treated with 10^{-6} M gibberellic acid. Open symbols refer to unlabelled enzyme, closed symbols to H_2^2O labelled enzyme. Enzyme activity is indicated by triangles, buoyant density determinations by circles.

chains synthesized via reutilization of amino acids will retain the labelled isotope, deuterium atoms incorporated in this manner are freely exchangeable with the hydrogen atoms in the surrounding medium. Only carbon-bound hydrogen is permanently replaced by deuterium. This is strikingly illustrated in Fig. 1. The synthesis of α-amylase in barley aleurone layers is largely at the expense of amino acids derived from storage proteins (Varner and Ram Chandra, 1964). Filner and Varner (1967) used H_2O^{18} to label α-amylase synthesized in response to the application of gibberellic acid (Fig. 1a). With H_2^2O as density label, on the other hand, there is no incorporation into the enzyme (Fig. 1b). Furthermore it should not be forgotten that a stimulus which increases the activity of an enzyme may well alter the comparative contributions of amino acid synthesis and amino acid reutilization. Gibberellic acid is a case in point (Varner and Ram Chandra, 1964).

Another factor which may affect the choice of density label is the biological damage likely to be caused. Most evidence indicates that deuterium causes rather more marked biological abnormalities than the other isotopes, presumably because of the large change in mass of the deuterium atom compared with hydrogen. Nevertheless H_2^2O is the density label which has been used most widely in density labelling studies, largely, one imagines, on account of its comparative cheapness.

B. SOLUTES FOR DENSITY GRADIENT CENTRIFUGATION

For density gradient centrifugation of proteins, solutions of mean density about 1.3 kg l^{-1} are required. There are only a small number of solute/solvent systems in which this density can be attained without excessive viscosity and without causing damage to proteins and large losses in enzyme activity. The density gradient work (Meselson et al., 1957; Meselson and Stahl, 1958) was carried out using caesium chloride in aqueous solution. Caesium salts are well suited to nucleic acid work since they are soluble at high concentrations (e.g. 63%:CsCl, $\rho = 1.9 \text{ kg l}^{-1}$) and nucleic acids are stable in solutions of them.

Most workers in the field of enzyme control (the present author included) have followed the example of their colleagues studying nucleic acids for earliest density labelling work and have, similarly, used CsCl solutions. However, it is becoming increasingly clear that CsCl is far from ideal for this purpose. Indeed, in their elegant exposition of the density labelling technique, Hu et al. (1962) used rubidium chloride rather than caesium chloride to great advantage. In the same paper they included details of a number of other useful solutes/solvents.

It seems pertinent to consider here the factors which are of most importance in the selection of a solute.

1. *The Gradients Formed by the Solutes during Centrifugation*

The comparative slopes of the gradients formed by some solutes/solvents at densities of about $1\cdot3$ kg l^{-1} are given in Table II. It can be seen that CsCl forms the steepest gradient of all. This inevitably means that the separation between two macromolecular species (e.g. enzymes) will be less in CsCl gradients than in those formed by the other salts listed. Ideally the gradient should be such that banding of all the molecular species under investigation should only just be possible. This is rarely the case with CsCl gradients.

TABLE II

A comparison of the slopes of gradients formed when various solutes are expressed as percentages of that obtained with caesium chloride and apply to solutions of density, approximately $1\cdot3$ kg l^{-1}

Solute used	Comparative slope of gradient
Rubidium chloride	55
Potassium bromide	40
Lithium bromide	12
Potassium acetate	11

2. *The Distribution of Each Macromolecular Species within the Gradient i.e. the Bandwidth*

Since

$$\sigma^2 = \frac{RT}{M\overline{V}\left(\dfrac{d\rho}{dr}\right)_r \omega^2 r}$$

where R = the gas constant, T = the absolute temperature, M = the mol. wt of the macromolecular species, $\overline{V}$ = the apparent specific volume of the macromolecular species, ρ = the density of the salt solution, ω = the angular velocity, r = the radius in the centrifuge at which the macromolecule is isopycnic, σ = the distribution of the macromolecule at equilibrium, (Hu *et al.*; 1962), the distribution of any macromolecular species is, other conditions being constant, inversely proportional to the square root of the slope of the gradient, whereas the separation is inversely proportional to the slope. Thus the resolving power of any gradient (i.e. the ratio separation:distribution) is greater with solutes which form shallower gradients. CsCl solutions offer the worst resolution of all those tested by Hu *et al.*

3. *The Time Required for Equilibration of the Macromolecule with the Gradient*

Other factors being constant, the time required for a macromolecule to equilibrate with the gradient in which it is being centrifuged is inversely proportional to the square root of the angular velocity (Hu *et al.*, 1962). Thus, for example, in a salt which under similar conditions forms a gradient of 50% the slope of a CsCl gradient, two macromolecules will have the same separation as in CsCl, when centrifuged at $\sqrt{2}$ times the speed. Under these conditions resolution will still be better by a factor of $\sqrt{2}$ and equilibrium achieved in half the time. The limit to the improvement in resolution and centrifugation time which can be achieved by the use of those solutes, such as LiBr, which form very shallow gradients, is set, at present, by the maximum centrifugation speeds available. A comparison of some results obtained using CsCl and KBr gradients is shown in Fig. 2. The improvement is strikingly illustrated by the effect of mixing density labelled and unlabelled extracts in the same tube. The heterogeneous distribution of such a mixture is, of course, not reduced in the shallower KBr gradient, but, because of the narrower distribution of either species alone, the increase in bandwidth produced by the mixture is 86% compared with 16% in the CsCl gradient. This improvement is obviously of great value when bandwidth determinations play an important part in the interpretation of density labelling experiments.

4. *The Stability of the Enzyme in the Solute*

None of these solutes, especially at the high concentration required, is ideal as regards enzyme stability. There seems to be no evidence of any advantage of CsCl over some of the other salts, especially KBr, KĀc and RbCl.

A compromise must normally be made between the opposing parameters of maximum separation between two macromolecules and the time required for them to reach equilibrium. Recently, however there have been reports of the use of metrizamide as a solute for density labelling (Rickwood and Birnie, 1975; Hüttermann and Guntermann, 1975). The key factor associated with the use of such gradients is that preformed metrizamide gradients do not alter significantly in profile during centrifugation. This contrasts with the gradients with the salts discussed above, which form spontaneously during centrifugation. Thus, preformed, shallow gradients, which would normally have formed only at low centrifugation speeds, can be maintained at high speeds, when equilibration with the macromolecule occurs more rapidly. The gradient can be tailored to the requirements of the experiment whilst centrifugation time is kept permanently at a minimum. Furthermore Hüttermann and Guntermann report that a number of enzymes which are

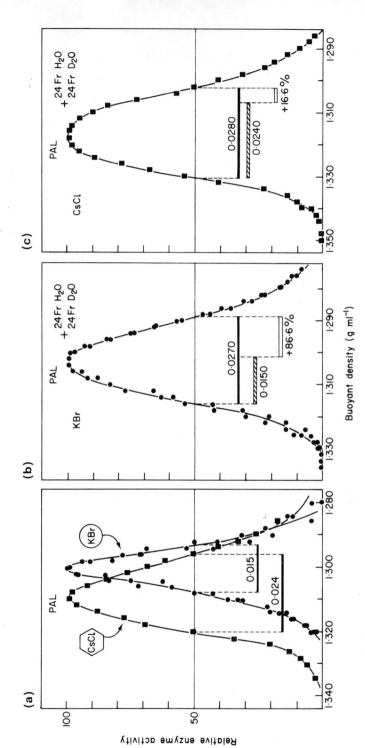

FIG. 2. Isopycnic centrifugation in CsCl and KBr gradients of phenylalanine ammonia-lyase. 2a. A comparison of the distribution of native lyase in CsCl and KBr gradients centrifuged under otherwise similar conditions. 2b and 2c. A comparison of the percentage increase in bandwidth (compared with the native enzyme) obtained when a mixture of equal activities of native and density labelled lyase is centrifuged in the same tube in gradients of KBr (Fig. 2b) and CsCl (Fig. 2c) under otherwise similar conditions.

labile in CsCl or RbCl solutions, are stable in metrizamide. The only disadvantages seem to be the need to construct the gradients prior to centrifugation and, (at present), the comparative cost of, and difficulty in obtaining,
metrizamide.

C. CENTRIFUGATION CONDITIONS

Whilst density labelling workers seem to have been the slaves of tradition
in their choice of solutes, they have, until recently, ignored worthwhile information obtained from nucleic acid density gradient work concerning the
choice of centrifugation conditions. As long ago as 1966, Flamm *et al.*
pointed out that fixed-angle rotors offered advantages over swinging-buckets
rotors, in respect of both resolution and loading capacity. Previously,
swinging-buckets rotors had been used exclusively for density gradient work,
but since 1966 nucleic acid workers have routinely used fixed-angle rotors.
By contrast, most density labelling work with proteins has, until recently,
been carried out using swinging-buckets rotors. An extensive range of
experiments carried out in this laboratory (Johnson *et al.*, 1973) has shown
that the use of the fixed-angle rotor at comparable angular velocities leads
to improved resolution and a reduction in equilibrium times in proportion
to the sine of the angle θ (Fig. 3). It can be seen that, effectively, gradients in
fixed-angle rotors are shallower than those obtained in swinging-buckets
rotors and the effective column length is shorter. The extra bonus of higher
loading capacity may also be important if the enzyme under investigation
represents only a small proportion of the total protein. In the Spinco Type

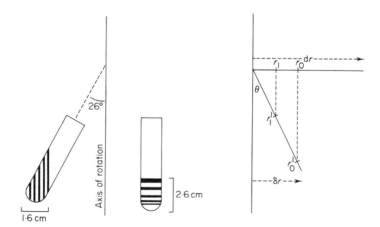

FIG. 3. The distribution of macromolecules during and after isopycnic centrifugation in a
fixed-angle rotor. Because of the reorientation of the gradient in the tube following centrifugation
molecules which band at r_0 and r_1 during centrifugation are separated by the distance $r'_0-r'_1$
when the rotor has stopped. The increase in separation and the effective column length during
centrifugation are both a function of sine θ.

75 Ti angle rotor the separation, loading capacity, and equilibration time are each improved by a factor of 2·3 compared with a swinging-buckets rotor. A comparison of the distribution of phenylalanine ammonia-lyase centrifuged in the two types of rotors is shown in Fig. 4.

Where enzyme decay *in vitro* is a major problem, centrifugation times can be further reduced by the use of step gradients (Brunk and Leick, 1969). In such gradients the lower phase contains a high concentration of the solute, the upper phase a low one. Consequently, the enzyme extract migrates rapidly to the interface which, if the gradient has been correctly prepared, is close to the final equilibrium position. Figure 5 shows that there is no difference in distribution between an extract which has been centrifuged for 24 h in the normal way (in the fixed-angle rotor; in a swinging-buckets rotor 60 h, would have been required) and one centrifuged for 5 h in a two step gradient.

By using a combination of high centrifugation speeds, conditions (solutes and rotors) which lead to the formation of shallow gradients, and step gradients there is no reason why resolution superior to that of most work published to date should not be achieved with runs of 5–8 h duration, about one-tenth of the average time originally required. This surely gives the lie to the argument that density labelling is an uneconomical technique. It is the wrong choice of solutes and rotors which has, in the past, led to variability and wasted time; the technique itself is not to blame.

III. DENSITY LABELLING AS AN EXPERIMENTAL TECHNIQUE

A. FOR THE DEMONSTRATION OF DE NOVO PROTEIN SYNTHESIS

The incorporation of density label into an enzyme provides unequivocal proof of synthesis of all or part of the protein during the labelling period. The extent of the increase in density enables an assessment to be made of the minimum extent of synthesis. In Fig. 6 the labelling of acid phosphatase in the cotyledons of germinating pea seedlings is shown. On the unlikely basis that all H^1 atoms could be replaced by H^2 in this experiment, the incorporation of label into the enzyme suggests that a minimum of 45% of the activity is due to newly synthesized protein.

By contrast it has been shown (Goatly and Smith, 1975) that the increase in activity of phospho-enol pyruvate carboxylase which occurs in sugar-cane leaves exposed to the light occurs without concomitant synthesis of the enzyme protein. Similarly, Ho and Scandalios (1975) report an increase in the activity of alcohol dehydrogenase in the scutellum of maize seedlings, again without incorporation of density label. An important point is that, in each case, label is incorporated into a marker enzyme, thus excluding artefacts due to lack of incorporation of the label into amino acids.

Examples where the simple, qualitative, distinction between synthesis and activation can be made are, however, rare. Most investigations into the

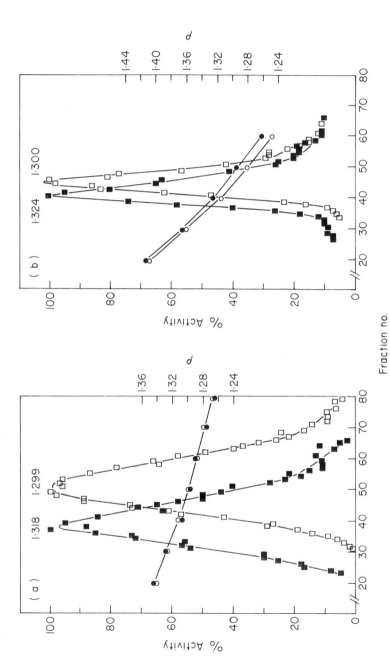

FIG. 4. Ultracentrifugation of native (open squares) and H²-labelled (closed squares) phenylalanine ammonia-lyase in separate tubes in a, a Spinco Type 75 Ti fixed-angle rotor at 45 000 rev/min for 24 h and b, a Spinco SW 50·1 swinging-buckets rotor at 40 000 rev/min for 60 h. (From Johnson et al., 1973.)

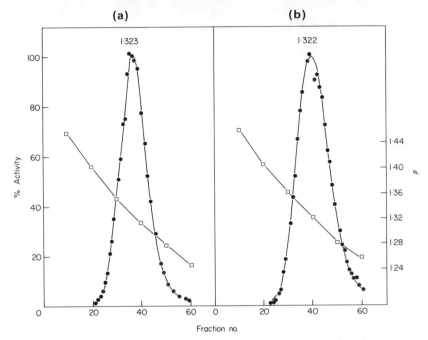

FIG. 5. Ultracentrifugation at 65 000 rev/min in a Spinco Type 75 Ti fixed-angle rotor for; a, 24 h and b, 5 h using a two-step gradient of native acid phosphatase. (From Johnson *et al.*, 1973.)

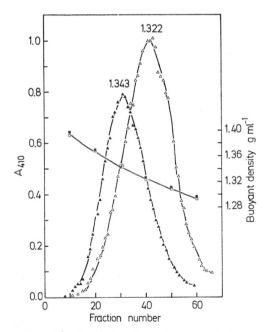

FIG. 6. Isopycnic centrifugation of acid phosphatase from germinating pea cotyledons. Seedlings were germinated in H_2O for 4 days (open symbols) or in 80% H_2^2O for 8 days. Enzyme extracted was centrifuged in CsCl for 40 h in a Spinco Type 75 Ti fixed-angle rotor at 45 000 rev/min and 4°C. (From Johnson *et al.*, 1973.)

control of enzyme activity involve changes in activity promoted by, for example, substrates, hormones or light. The task before the investigator is to compare the rates of enzyme synthesis and turnover in the stimulated and unstimulated tissue (Attridge *et al.*, 1974). The extent and the reliability of determination of these variables by density labelling methods is the subject of the remainder of this review.

B. FOR THE MEASUREMENT OF ENZYME TURNOVER

A satisfactory definition of enzyme turnover is elusive. From the results of a private opinion poll carried out by Koch (cited by Boudet *et al.*, 1975) it would seem that most biochemists equate protein turnover with enzyme degradation. However, most biologists also estimate turnover of individual enzymes by measuring the half-life of catalytic activity. Between these two definitions the distinction between enzyme degradation and enzyme inactivation is lost. The importance of this distinction is obvious and it is as well to consider just what is being measured in density labelling experiments aimed at determining rates of enzyme turnover. Such experiments have been carried out by a number of workers, notably Williams and Neidhardt (1969), Zielke and Filner (1971), Acton *et al.* (1974) and Acton and Schopfer (1974, 1975). The method generally used is based on the fact that the distribution of protein molecules on a density gradient (the bandwidth) is maximal when the population is half native and half labelled. Thus, under steady-state conditions, the time taken for the maximum bandwidth to be attained following the addition of a density label is taken as an estimate of the half-life of an enzyme.

The extent to which such measurements genuinely reflect turnover depends on the way in which activity is being lost. Clearly, if activity is simply the resultant of synthesis and degradation then the half-life measured in this way will be a true reflection of the rate of turnover. On the other hand if enzyme inactivation is occurring, thus:

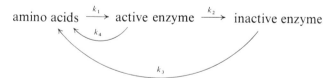

then the half-life calculated from the distribution of enzyme activity in a density gradient (or by any other method based on enzyme activity) need give no indication of k_4, the rate of turnover of active enzyme, nor indeed of $k_3 + k_4$, the turnover rate of all the enzyme, active and inactive, but only the sum of k_2 (the rate of inactivation) and k_4 (the rate of degradation of active enzyme). The extent to which the estimate reflects inactivation rate or turnover depends on the relative values of k_2, k_3 and k_4. It is interesting to note that two enzymes which are known to be subject to inactivation, nitrate reductase (Wallace, 1973; Solomonson et al., 1973) and phenylalanine ammonia-lyase (Attridge and Smith, 1973) both have very high "apparent turnover" rates, as measured by density labelling.

Thus, on balance, it must be said that density labelling is not a good method for estimating enzyme turnover unless enzyme inactivation can be excluded. The same criticism applies also to most other methods based on measurement of enzyme activity. There are, however, occasions when some information concerning the rate of enzyme turnover can be obtained despite the simultaneous occurrence of enzyme inactivation. One such opportunity arises when it is possible to re-activate the inactive enzyme. In gherkin hypocotyl tissue the enzyme phenylalanine ammonia-lyase can be activated *in vivo* by cycloheximide treatment (Attridge and Smith, 1973). In untreated control tissue following the introduction of density label, the enzyme is labelled within 3 h to about 50% of the maximum attained (Table III). Cycloheximide treatment causes a very large increase in activity with no evidence of density label incorporation. Indeed, it can be concluded that some of the enzyme responsible for this increase in activity must have been synthesized more than 12 h prior to the cycloheximide treatment since 12 h pretreatment with H_2^2O is insufficient to increase the density of the activated enzyme to that of the untreated control. Whilst such an approach may not yield an accurate estimate of either the half-life of enzyme inactivation or of enzyme degradation it does give an indication of the possible extent of the difference between the two.

There is at present no convenient method for the determination of either turnover or inactivation rate of individual enzymes. All the available methods have a common prerequisite for purification to homogeneity of the enzymes

TABLE III

The effect of cycloheximide on the incorporation of density label into PAL

Treatment	Buoyant density of PAL kg l^{-1}
96 h H_2O: 3 h H_2O	1·294
96 h H_2O: 3 h H_2^2O	1·300
96 h H_2O: 3 h H_2^2O CH	1·295
84 h H_2O:15 h H_2^2O	1·302
84 h H_2O:15 h H_2^2O CH	1·299
72 h H_2O:27 h H_2^2O	1·306
72 h H_2O:27 h H_2^2O CH	1·304

From Attridge and Smith (1974).
Treatments suffixed CH terminated with 3 h in the presence of 100 µg ml cycloheximide. After the treatments listed, the lyase was extracted from gherkin hypocotyls and centrifuged to equilibrium in CsCl gradients.

concerned, which fact is sufficient in itself to account for the dearth of reliable information on this subject.

The factors discussed in the previous section impinge directly on the interpretation of density labelling experiments concerning mechanisms of control of enzyme activity, to which problem the majority of the published density labelling work has been addressed. The control of enzyme activity is normally quantitative: the stimulus causing an increase in the activity of an enzyme already present rather than evoking the appearance of a totally new activity. Thus, unless the contrary is proved, it must be presumed that enzyme turnover is occurring in both control and stimulated tissue. This fact complicates the interpretation of experiments concerning enzyme control. Nevertheless, whilst the rate of turnover has a major effect on the density distributions obtained even in the absence of a stimulus, some conclusions concerning possible mechanisms of enzyme control can be reached by comparing the profiles observed with enzyme extracted from stimulated and unstimulated tissue during the period following introduction of a density label. The density gradient profile may be altered by the stimulus in a number of ways and, depending on which pattern is observed, conclusions, differing in precision, can be drawn (Attridge et al., 1974). Recently, a very thorough theoretical analysis of the conclusions which can be reached as a consequence of different comparative distributions has been made by Lamb and Rubery (1976).

The profiles expected if different control mechanisms are operating may be summarized thus:

In situations where the stimulus causes an increase in the rate of enzyme synthesis, the newly synthesized enzyme responsible for the increase in activity will be density labelled and thus of high buoyant density. The density of the enzyme extracted from the control tissue will be a function of the half-life of enzyme activity (not necessarily of the rate of turnover, nor of inactivation). The shorter this time, the more rapidly labelled will the enzyme be. Thus an increased rate of synthesis will lead to a comparative increase in density only if the period of the response is not significantly more than the half-life of the enzyme activity.

Where the stimulus causes activation of existing inactive enzyme two patterns are possible. In one case, thus:

$$\text{amino acids} \xrightarrow{k_1} \text{active enzyme} \xrightarrow{k_2} \text{inactive enzyme}$$
$$\xleftarrow{\quad\text{stimulus}\quad}$$

the activation of previously synthesized inactive enzyme will lead in the stimulated tissue, to a pool of active enzyme of predominantly low buoyant density. Once again the amount of label incorporated into the enzyme in the control tissue depends on the half-life of enzyme activity. The shorter the half-life the greater the incorporation will be.

By contrast, if activation involves the activation of an inactive precursor, thus

$$\text{amino acids} \xrightarrow{k_1} \text{inactive precursor} \xrightarrow{k_2} \text{active enzyme}$$

$$\xrightarrow{\text{stimulus}}$$

the profile observed will be similar to that observed if the stimulus modulates the rate of synthesis. Indeed, if k_2 is close to zero, there need, in theory at least, be little or no incorporation of label into enzyme during a prolonged period of incubation. There is, however, at present no demonstrated example of activation by such a mechanism in a higher plant.

A third possibility is that the stimulus causes enzyme activity to increase as a result of a reduction in the rate of enzyme degradation. This will, inevitably, result in a lowering of the buoyant density of enzyme extracted from stimulated tissue, compared with the control.

Thus, a comparative increase in density in response to the stimulus (e.g. the phytochrome controlled increase in ascorbic acid oxidase in mustard cotyledons, Fig. 7a) is most readily explained by an increased rate of synthesis, but the possibility of activation of an inactive precursor should be borne in mind. A comparative decrease in density (e.g. the phytochrome mediated increase in phenylalanine ammonia-lyase in the same tissue Fig. 7b) will reflect activation of previously inactivated enzyme or a reduction in the rate of turnover, but eliminates the possibility of control of the rate of synthesis. The distinction between the remaining alternatives must usually be made on experimental evidence other than density labelling.

Some workers, particularly Acton (Acton and Schopfer, 1974; Acton et al., 1974; Acton and Schopfer, 1975) have placed much emphasis on the use of bandwidth measurements in investigations into enzyme control. However, as has already been pointed out, the estimation of half-life from measurements of enzyme activity distributions, even under steady-state conditions need be neither an estimate of turnover nor of inactivation rate. Furthermore, such measurements are meaningful only if the tissue represents a steady-state. The assumption that, for example, a germinating seedling represents a steady-state system is clearly not valid. It may be noted that the comparison of density changes between control and stimulated tissue outlined above, does not rely in any way on the assumption that either situation approximates to a steady-state. In practice too, with CsCl gradients the experimental variation in bandwidths between different extracts of enzyme are often of the same order of magnitude as the expected differences due to heterogeneity of enzyme populations (see, for example Acton and Schopfer, 1975). These experimental problems, but not of course, the conceptual ones, can be overcome by the use of more suitable solutes.

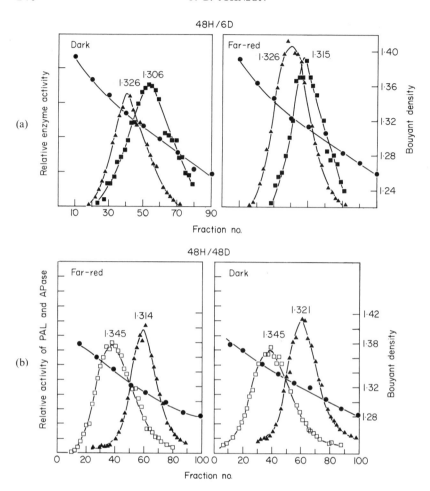

FIG. 7a. Isopycnic centrifugation in CsCl of ascorbate oxidase (squares) and acid phosphatase (triangles) extracted from the cotyledons of mustard seedlings grown in the dark on H_2O for 48 h and then transferred to 100% H_2^2O. After a further 6 h in the dark, or in far-red light the seedlings were harvested. The density of native ascorbate oxidase is 1·302 kg l and of acid phosphatase 1·325 kg l. (From Attridge, 1974.)

FIG. 7b. Isopycnic centrifugation in CsCl of phenylalanine ammonia-lyase (triangles) and acid phosphatase (squares) extracted from the cotyledons of mustard seedlings grown in the dark on H_2O for 48 h and then transferred to 100% H_2^2O. The seedlings were harvested after a further 48 h in the dark, or in far-red light. The density of native phenylalanine ammonia-lyase is 1·295 kg l^{-1}. (From Attridge et al., 1974.)

D. MEASUREMENT OF INCORPORATION OF DENSITY LABELS INTO AMINO ACID POOLS

An assumption which has been made in the previous section is that the stimulus responsible for modulating the activity of an enzyme does not also alter the proportion of density label in the amino acid pools from which the enzyme is synthesized. This assumption needs to be verified in each experiment. There are, indeed, many situations in which it may not be valid. Gibberellic acid, for example, is known to enhance the rate of mobilization of storage protein reserves in the barley aleurone layer (Varner and Ram Chandra, 1964). Far-red light treatment of mustard cotyledons, in addition to stimulating various enzymes, alters the rate of storage protein mobilization (Häcker, 1969) as well as of protein synthesis (Jakobs and Mohr, 1966) and nitrate reduction (Johnson, 1976). The overall effect of these changes on the incorporation of label into proteins is almost impossible to predict. In order to safeguard against artefactual results arising from possible differences in the incorporation of density label an internal control is normally used. Most workers (e.g. Attridge *et al.*, 1974, Acton and Schopfer, 1974) have used another enzyme in the same tissue, whose activity is not under the control of the particular stimulus. However, it is not always easy to find a suitable marker enzyme. To be suitable, such a marker should ideally have a half-life similar to or shorter than that of the enzyme under investigation, otherwise changes in labelling of the amino acid pool might not be reflected rapidly enough.

The ideal measurement to make is of the labelling of the amino acid pool itself. Amino acids do not themselves form discrete bands in density gradients but they can be made to do so after they have been used to synthesize protein. We have made use of this approach in this laboratory (Johnson and Smith, 1977) using *Escherichia coli* as the biological tool for synthesizing proteins from amino acids extracted from experimental plant material. The bacterial proteins made, are then extracted and centrifuged in density gradients and

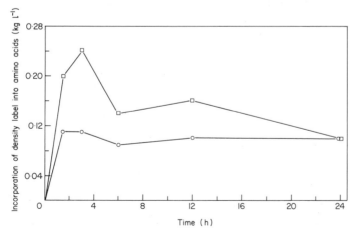

FIG. 8. Incorporation of density label into amino acids extracted from the cotyledons of mustard seedlings (for method, see text). Seedlings were grown on H_2O in the dark for 36 h and then transferred to 80% H_2^2O (and where appropriate to continuous far-red light) for the times indicated. (Squares represent far-red light treated seedlings; circles, dark grown seedlings.)

the increase in density, compared with native protein, measured. The results of such an experiment are shown in Fig. 8. It can be seen that in previously dark-grown mustard cotyledons, far-red light stimulates the uptake of H_2^2O into the amino acid pool. Thus proteins synthesized in light-treated cotyledons would be expected to be heavier than those in the dark controls. Because the effect is only apparent for the first 6–12 h of irradiation, only proteins with fairly short half-lives would reflect these differences. These data, as well as helping to prevent erroneous conclusions being reached from comparative density labelling experiments, incidentally indicate that, as expected from the phytochrome induction of nitrate reductase, far-red light stimulates amino acid synthesis.

IV. Conclusions

It will be obvious to the reader that great care is needed in the interpretation of many density labelling experiments concerned with enzyme turnover and the modulation of enzyme activity. Such interpretive problems are not, in fact, exclusive to density labelling experiments. Indeed, even when all the essential control experiments have been carried out the difficulties of performing and analysing such experiments are considerably less than those found with any of the other procedures currently available. The reader will, perhaps, note that in work presented in four other chapters in this volume, (those of Scandalios, Smith, Varner and Yeoman) density labelling has been one of the key methods employed in studies on the control of enzyme activity. If the improvements in the technique outlined here are used to the full there is no reason why suitable density labelling experiments should not yield much more information concerning enzyme control.

ACKNOWLEDGEMENTS

I would like to take this opportunity of thanking Drs Terry Attridge and Pierre Fourcroy for their contributions towards the work described here and Professor Harry Smith for his encouragement and for much helpful discussion.

REFERENCES

Acton, G. J. and Schopfer, P. (1974). *Biochem. J.* **142**, 449.
Acton, G. J. and Schopfer, P. (1975). *Biochim. biophys. Acta* **404**, 231.
Acton, G. J., Drumm, H. and Mohr, H. (1974). *Planta* **121**, 39.
Attridge, T. H. and Smith, H. (1973). *Phytochemistry* **12**, 1569.
Attridge, T. H., Johnson, C. B. and Smith, H. (1974). *Biochim. biophys. Acta* **343**, 440.
Boudet, A., Humphrey, T. J. and Davies, D. D. (1975). *Biochem. J.* **152**, 409.
Brunk, C. F. and Leick, V. (1969). *Biochim. biophys. Acta.* **179**, 136.

Filner, P. and Varner, J. E. (1967). *Proc. natn. Acad. Sci. U.S.A.* **58**, 1520.
Flamm, W. G., Bond, H. E. and Burr, H. E. (1966). *Biochim. biophys. Acta* **129**, 310.
Fourcroy, P. and Johnson, C. B. (1976). *Planta* (in press).
Goatly, M. B. G. and Smith, H. (1975). *Planta* **117**, 67.
Häcker, M. (1969). *Planta* **76**, 309.
Ho, D. and Scandalios, J. G. (1975). *Pl. Physiol.* **56**, 56.
Hu, A. S. L., Bock, R. M. and Halvorson, H. O. (1962). *Ann. Biochem.* **4**, 489.
Hüttermann, A. and Guntermann, U. (1975). *Ann. Biochem.* **64**, 360.
Jakobs, M. and Mohr, H. (1966). *Planta* **69**, 187.
Johnson, C. B. (1976). *Planta* **128**, 127.
Johnson, C. B. and Smith, H. (1977). *Phytochemistry* (in press).
Johnson, C. B., Attridge, T. H. and Smith, H. (1973). *Biochim. biophys. Acta* **317**, 219.
Lamb, C. J. and Rubery, P. H. (1976). *Biochim. biophys. Acta* **421**, 308.
Meselson, M. and Stahl, F. W. (1958). *Proc. natn. Acad. Sci. U.S.A.* **44**, 671.
Meselson, M., Stahl, F. W. and Vinograd, J. (1957). *Proc. natn. Acad. Sci. U.S.A.* **43**, 581.
Rickwood, D. and Birnie, G. D. (1975). *FEBS Letts.* **50**, (2), 102.
Solomonson, L. P., Jetschmann, K. and Vennesland, B. (1973). *Biochim. biophys. Acta* **309**, 32.
Varner, J. E. and Ram Chandra, G. (1964). *Proc. natn. Acad. Sci. U.S.A.* **52**, 100.
Wallace, W. (1973). *Pl. Physiol.* **52**, 197.
Williams, L. S. and Neidhardt, F. C. (1969). *J. molec. Biol.* **43**, 529.
Zielke, H. R. and Filner, P. (1971). *J. biol. Chem.* **246**, 1772.

CHAPTER 12

The Extraction and Purification of Enzymes from Plant Tissues

M. J. C. RHODES

Agricultural Research Council, Food Research Institute, Norwich, England

  I. Introduction 245
 II. General Problems of Extraction and Stabilization of Enzymes from
     Higher Plant Tissues 247
     A. General Considerations 247
     B. Problems due to Phenolic Compounds and Tannins . . . 248
     C. Some Factors Affecting the Stability of Isolated Enzymes . . 256
III. Techniques for Purification of Enzymes from Higher Plant Tissues . 258
     A. Solubilization of Membrane Bound Enzymes 259
     B. Concentration of Enzyme Extracts 261
     C. Desalting of Enzyme Extracts 262
     D. Chromatographic Methods 262
     E. Electrophoretic Procedures 265
 IV. Criteria for Assessing the Purity of Isolated Enzymes . . . 266
  V. Conclusion 267
References 267

I. INTRODUCTION

Much of our knowledge of the molecular biology of enzyme structure and function has come from a study of isolated enzymes purified to homogeneity. Enzymes are proteins which may bear a specific prosthetic group and may be associated with carbohydrate or lipid. Modern work has established that many enzymes are composed of subunits which may be associated with either the catalytic or regulatory functions of the enzyme (Koshland and Neet, 1968). The existence of multiple molecular forms (isoenzymes) of many enzymes, which may be determined genetically or may result from secondary modifications of the enzyme during extraction, has been described (Scandalios, 1974).

The reasons for undertaking the purification of an enzyme may vary from situation to situation. Often it may be necessary to purify an enzyme to some extent even to obtain a reliable measurement of enzyme activity. A higher degree of purification is required before kinetic studies of enzyme activity and inhibition can be undertaken and it is necessary to purify to homogeneity in order to study the chemical structure of enzymes and the molecular mechanism of their catalytic activity.

It is important to consider the intracellular location of enzymes before discussing the methods of their extraction and purification. Enzymes may be held within a membrane enclosed space, form an integral part of a cell membrane (or be associated with its surface) or be bound to the cell wall. Many enzymes isolated by the standard methods of homogenization appear in the so-called "soluble" fraction which is defined as that cell fraction which is not sedimented by forces of $100\,000 \times g$ applied for one hour or more. There is good reason to believe that this does not represent the *in vivo* situation of these enzymes (Ginsburg and Stadtman, 1970) and that it is to some extent an artefact of the extraction methods. These so called "soluble" enzymes could be bound within delicate lipoprotein-bound vesicles and there are a number of examples of plant enzymes originally classified as soluble which on subsequent study have been shown to be associated with cell organelles (Tolbert, 1971). The isolation of soluble multienzyme complexes such as the fatty acid synthetase of yeast (Lynen, 1961) may be indicative of the existence of protein; protein interactions between neighbouring enzymes in a given pathway may be important in maintaining spatial relationships between enzymes in the cytosol *in vivo* (Frieden, 1971). It may well be that these protein interactions are weak and normally proteins bearing single activities are dissociated in aqueous extracts.

The approach to the purification of proteins is generally an empirical one since often there is very little information on the chemical structure of the enzyme protein under investigation. The specificity of enzyme activity allows a single enzyme protein to be detected in a complex mixture. Increasing purification of an active fraction is assessed by an increase in the specific activity (activity/mg total protein) over that of the initial extract. As activity is retained and inactive protein is removed during the various procedures the specific activity increases to reach a constant value at homogeneity. Jakoby (1971) has said that "advances in the tactics of protein purification have been such that one can reasonably expect that any protein of a given order of stability may be purified to currently acceptable standards of homogeneity." This statement is true of plant as well as animal enzymes but the problems of stability are probably greater in the case of plant enzymes.

Plant tissues generally have much lower protein contents than animal tissues. For instance, protein contents range from $3 \cdot 2\%$ of the fresh weight for spinach to $0 \cdot 2\%$ for apples compared with $19 \cdot 9\%$ in bovine liver (Walt and Merrill, 1963). In spite of this, some plant tissues may be rich in a particular enzyme and this is an important consideration in the choice of material for study. The enzyme concentration, as a proportion of the total protein has an important bearing on the difficulty of purification if the enzyme is to be purified to homogeneity. For instance, Fraction 1 protein which bears the ribulose 1:6 diphosphate carboxylase/oxygenase activities accounts for over 50% of the total soluble protein of tobacco leaves and thus it is only necessary to raise the specific activity by about two to achieve homogeneity (Kung, 1976). Fraction 1 protein can be purified by a simple method from crude extracts owing to its insolubility in the absence of substrate, on addition of bicarbonate and magnesium ions (Chan *et al.*, 1972). In general, the increase in specific activity required for complete purification is much greater than this and will vary from tissue to tissue and from enzyme to enzyme. Other factors such as the presence of specific inhibitors of the enzyme or of general protein inhibitors such as tannins may also have a bearing on the choice of material. It is interesting but not surprising that much of our knowledge of plant enzymology has been derived from studies on a relatively small number of species of food plants which are characterized by low levels of endogenous tannins (Bate-Smith, 1972). The problems of isolation of plant enzymes including both the general problems of stability of enzymes and the particular problems due to the presence of phenolic compounds and tannins in plant tissue will be considered prior to a discussion of the application of some of the classical purification procedures to plant proteins.

II. General Problems in the Extraction and Stabilization of Enzymes from Higher Plant Tissues

A. GENERAL CONSIDERATIONS

The conditions under which plant homogenates are prepared can have a crucial bearing on the successful extraction of an enzyme. The forces required to rupture plant cells with their thick cellulose walls are greater than those required with many animal cells and this makes the isolation of plant cell organelles difficult. It is advantageous to use organelles as the starting point for the purification of enzymes associated with that organelle, particularly if there are several isoenzymic forms of the enzymes housed in different cell compartments. Methods for the homogenization of plant tissues under mild conditions, adapted for the isolation of organelles have recently been reviewed (Price, 1974) but for "soluble" enzymes more vigorous methods of homogenization may be employed.

The composition of the extraction medium used, is an important factor and under unfavourable circumstances may lead to preparations with low or even no activity (Anderson, 1968). Owing to the acidic pH of the sap of many plants it is usually necessary to include a buffer in the extraction medium to maintain the pH of the extract within the range of stability of the enzyme. The choice of pH for homogenization varies from enzyme to enzyme. The pH for optimum activity of the enzyme is often used in view of the greater stability of many enzymes at this pH. However, the solubilization of proteins tends to be greater at higher pHs, and pH can be an important factor in determining the apparent distribution of activity between cell organelles and the supernatant. For many enzymes, more activity is associated with sedimentable fractions at low pH (5·5–6·0) than at higher pH (Chin and Weston, 1973; Ricardo and Ap Rees, 1970). In order to minimize phenolic oxidation, pHs close to or slightly acid of neutral are desirable (Loomis, 1974). Thus the choice of pH for homogenization must be a compromise between a number of conflicting factors. The choice of buffering ion depends on the particular activity under study. Tris and phosphate are widely used but the range of buffers covering the physiological range has been greatly extended by the introduction of the zwitterion buffers of Good et al. (1966). These buffers have lower metal complexing characteristics and lower temperature coefficients of Ka than Tris and have less tendency to inhibit some enzyme activities (Good et al., 1966). The ratio of buffer to tissue during homogenization is important since at high ratios the extraction of the tissue is more efficient and any potential inhibitors are diluted. The ratio is limited at the upper level by the inconvenience of handling large volumes of extract and by the instability of some enzyme systems in very dilute solutions (e.g. Takeda and Hizukuri, 1972). Ratios of 2–5:1 are commonly used in practice. In addition to a buffer, extraction media commonly include an osmoticum such as sucrose or mannitol, particularly if organelles are to be isolated and a metal chelator, such as EDTA to minimize the possibility of inhibition of the isolated enzymes by metal ions.

Inert proteins are often included, particularly bovine serum albumin (BSA) which binds fatty acids and may protect mitochondria and membrane bound enzymes from inactivation (Dalgardo and Birt, 1962). BSA also protects particular enzymes from endogenous inhibitors, e.g. p-coumarate CoA ligase of swede roots (Rhodes and Wooltorton, 1973). BSA protects nitrate reductase from inactivation in extracts of oat and tobacco leaves (Schrader et al., 1974). Inclusion of BSA or casein in the extraction medium led to a large increase in the extractable nitrate reductase especially in extracts of older leaf tissue. Wallace (1975) showed that 3% casein increased the extractable level of nitrate reductase from maize roots. Casein in this case is thought to act by inhibiting a nitrate reductase inactivating system which may involve proteolytic attack on the enzyme (see Chapter 9). The addition of inert proteins can be valuable in the quantitative measurement of enzyme activities in tissues, but has obvious difficulties if the enzyme protein is to be further purified.

B. PROBLEMS DUE TO THE PRESENCE OF PHENOLIC COMPOUNDS AND TANNINS

It is usual to include in extraction media for use with plant tissues, compounds which protect enzymes against inhibition by phenolic compounds and tannins. Failure to include effective protective agents can lead to failure to detect activity in the extract. There are two related problems caused by phenolic compounds; one due to the presence of preformed tannins, and the other to the oxidation of endogenous low mol. wt phenolic compounds to form quinones and their subsequent polymerization to form brown products with tannin-like properties. Tannins have been defined by Swain and Bate-Smith (1962) as "naturally occurring substances with chemical and physical properties akin to those converting raw hide to leather. These are water soluble, polyphenolic compounds with molecular weights in the range 500–3000". Two types of tannins have been described; namely, the hydrolysable tannins which yield either gallic acid or a mixture of gallic and ellagic acids on acid hydrolysis, and the condensed tannins which are polymers of catechins (flavan-3-ols) or leucoanthocyanins (flavan 3:4 diols) (Ribbereau-Gayon, 1972). The leucoanthocyanins are the most widely distributed and representative of the tannins (Bate-Smith and Lerner, 1954) and are typically polymers of 2–10 monomeric units. Tannins link to proteins by two types of bond; hydrogen bonds between the phenolic hydroxyl groups of the tannin and the peptide bond of the protein probably through the peptide oxygen, and covalent linkages between quinone groups in the tannin and reactive groups in the protein (Ribbereau-Gayon, 1972). The water solubility, molecular size and the number of hydroxyl groups on the tannin capable of external hydrogen bonding are important in determining whether a given tannin will react with protein. There are degrees of interaction with only tannins within a defined range of molecular size being able to precipitate proteins; low mol. wt tannins may form soluble complexes while very high mol. wt tannins do not precipitate proteins as they are insoluble in water and are unable to orientate themselves to form stable linkages to protein.

The other aspect of the phenolic problem is the oxidation of released monomeric phenols by the copper containing enzyme, o-diphenol:O_2 oxidoreductase (phenolase). The intracellular localization of phenolase and of its substrates, and for that matter the tannins, is a matter for conjecture, although it is usually thought that most phenolics, which are generally present as esters or glycosides, are held within the vacuole and are thus physically separated from the cytoplasmic phenolase. On cell rupture, mixing of substrates and enzymes occurs, leading to hydrolysis of the phenolic glycosides and oxidation of the released phenolic compounds by phenolase to form quinones. Quinones, particularly o-quinone are very reactive oxidizing agents (Pierpoint, 1970; Synge, 1975; Van Sumere et al. 1975) which can form covalent linkages with amine, α-amino, imino and thiol groups in proteins (see Van Sumere et al., 1975). Such quinones can undergo non-enzymic polymerization reactions either in situ on the protein or free in solution to form polymers which have tannin-like properties and can precipitate proteins. These polymers are brown in colour and are products of the process known as enzymic browning.

The strategy to overcome these problems in practice is to prevent oxidation of simple phenols and to prevent interaction between isolated proteins and tannins and quinones during homogenization. This necessitates including effective protective agents, allowing good access of the medium to the tissue and minimizing conditions which stimulate oxidative reactions during homogenization. Tissues vary widely in their content and composition of phenolase substrates and tannins and in the activity of phenolase. Thus the practical problems vary from tissue to tissue and enzyme to enzyme; many enzymes show different susceptibilities to inhibition by phenolics. For instance, Slack (1966) found that extracts of sugar-cane, prepared under conditions in which enzymic browning was allowed to proceed, had no sucrose synthetase activity but relatively high invertase activity. The addition of phenolase inhibitors during extraction led to a large rise in sucrose synthetase activity while that of invertase was unchanged. Enzymes which have quinone binding sites close to their active sites may be very susceptible to inhibition by oxidized phenols while the activity of other enzymes which do not bind quinones at positions close to their active sites may be unaffected. However it is likely that some quinone attachment to regions of the protein remote from the active site occurs in both cases and this modifies their structure and properties. Pierpoint (1973a) has shown that virus proteins may have markedly altered properties following exposure to products of chlorogenic acid oxidation but that the infectivity of the virus is largely unaffected. It is desirable to prevent enzymic browning during extraction whether or not the enzyme activity under consideration is affected if the isolation of proteins in an unaltered state is sought. The procedures developed to overcome the problems due to phenols and tannins will be discussed in the next two sections.

1. *Use of Polymeric Protective Agents*

The use of polymeric agents for adsorbing tannins during tissue extraction has been reviewed by Loomis and Battaile (1966), Anderson (1968) and Loomis (1974). What is clear from these reviews and from much of the more recent work is that there is no universal remedy to this problem but a particular polymer may be best suited to a particular tissue or particular enzyme system. Inert proteins such as hide powder and casein (Haard and Hultin, 1968) and various forms of powdered nylon (such as ultramid, perlon) have been used with varying degrees of success (Jones *et al.*, 1965). More recently, polyethylene glycol (PEG) and polyvinylpyrrolidone (PVP) have been introduced and offer particular advantages over previous materials. PEG is used in some tissues (Doronton and Hawker, 1973) but PVP has found a much wider usefulness. PVP is available in water soluble forms with average mol. wts ranging from 10 000 to 360 000 and in a highly cross-linked insoluble form sold as Polyclar AT. Although the soluble forms offer advantages with some tissues (Jones *et al.*, 1965), Polyclar AT has become the material of choice mainly because it is easily separated by centrifugation from soluble enzymes (Loomis and Battaile, 1966; Loomis, 1974). PVP acts by forming both water soluble and water insoluble complexes with tannins, by hydrogen bonding between phenolic OH groups of the tannins and the oxygen attached to the pyrrolidone ring of PVP. PVP is more efficient at adsorbing phenolics at pHs on the acid side of neutral and forms more stable complexes with high mol. wt tannins than with low mol. wt phenolic compounds such as chlorogenic acid (Andersen and Sowers, 1968; Anderson, 1968). PVP is able to dissociate tannin–protein complexes and to some extent reverse the inhibition of activity (Goldstein and Swain, 1965) and in addition PVP may have a secondary effect as an inhibitor of phenolase (Walker and Hulme, 1965).

Other polymeric materials have been introduced for binding phenols during enzyme extractions. Lam and Shaw (1970) showed that Dowex-1-Cl′ (200–400 mesh) used at 10% w/v was more effective than insoluble PVP, both in reducing the level of phenolics and in preserving peroxidase activity in extracts of flax cotyledons. Haissig and Schipper (1971) have pointed out the dangers of using anion exchange resins such as Dowex-1 under low salt conditions when adsorption of proteins bearing a high net negative charge is possible. Anion exchange resins have however been successfully used in recent work (Fieldes and Tyson, 1973; Gross *et al.*, 1975).

Loomis (1974) suggested in his review the potential value of the use of non-ionic macroreticular adsorbents in enzyme extraction. These materials are in the form of highly porous beads of polymeric materials produced by fusion of small microspheres. The commercially available adsorbents are based on polystyrene (Amberlite XAD 2 and 4) or polyacrylic esters (Amberlite XAD 7) and have a very high adsorptive surface (equivalent to up to 750 m^2/g dry weight). Such materials have been used for adsorption of tannins from tannery effluents (Briggs *et al.*, 1973) and are thought to act by interactions between the polymer surface and hydrophobic portions of the aromatic molecule. We have shown that these adsorbents remove both monomeric phenols such as chlorogenic acid and tannins such as tannic acid from solution. These results seem to be very encouraging as they appeared to be equally effective in absorbing small and large mol. wt phenolics and might prove useful in the isolation of enzymes.

We have tested a number of the newly introduced polymeric agents with rather variable success. The extraction of NADP-malic enzyme from apple peel was used as this tissue combines a high phenolase content with a high level of endogenous tannins. In past work it was shown that PVP was the only available agent giving protection during this extraction; PEG, various forms of nylon, chelating agents such as DIECA and mercaptobenzothiazole and reducing agents such as metabisulphite were ineffective. Table I shows that in the absence of protective agents the extract was a dark brown colour and was inactive. Im DTE completely prevents enzymic browning but does not lead to the release of activity while the further addition of soluble PVP

TABLE I

Extraction of NADP-malic enzyme from apple peel tissue

	Colour of extract	NADP malic enzyme activity (μ moles/min/10 g fwt)
No additions	Dark brown	0
+ DTE (1 mM)	Light green	0
+ DTE + soluble PVP (1%)	Dark green	19·3
+ DTE + 5 g Amberlite XAD 2	Pale green	0·2
+ DTE + 7 g Amberlite XAD 4	Pale green	0·3
+ DTE + 5 g Dowex 1-Cl × 8 200–400 mesh	Pale green	0
+ DTE + 5 g Amberlite IRA 904 Cl'	Almost colourless, cloudy	1·3
+ DTE + 5 g Amberlite IRA 938 Cl'	Almost colourless, cloudy	2·1
+ DTE + 10 g Amberlite IRA 938 Cl'	Almost colourless, cloudy	5·1
+ DTE + 5 g Amberlite IR 93	Almost colourless, cloudy	0·4
+ DTE + 5 g Amberlyst 15	Green-brown	0·1
+ DTE + 5 g Amberlyst 26 Cl'	Colourless	1·7

5 g of frozen apple peel powder was extracted in 35 ml of medium (0·1 M Hepes–0·01 M EDTA pH 7·8 containing the additions shown above) in a Turrax homogenizer. The extract was filtered through miracloth and the residue reextracted with 15 ml of the medium. The combined extract and washings were clarified by centrifugation at 40 000 × g for 45 min and made to volume. Aliquots of the extract were taken and the activity of malic enzyme measured using standard spectrophotometric techniques.

at 1% w/v led to a dark green extract with a high malic enzyme activity. In other experiments insoluble PVP was shown to be less effective than soluble PVP giving at most 70% of the rate with soluble PVP.

The macroreticular adsorbents XAD 2 and XAD 4 gave only marginal protection to malic enzyme but their use led to the production of almost colourless extracts. In the absence of DTE, these macroreticular adsorbents were ineffective in preventing enzymic browning. Dowex-1-Cl which has been used successfully in other systems was also ineffective in protecting the malic enzyme. We have also tested a range of macroreticular ion exchange resins which are similar to XAD 2 and XAD 4 in being based on styrene-divinyl benzene polymers but have in addition charged groups either strongly acidic (Amberlyst 15), strongly basic (Amberlites IRA 904, IRA 938 and Amberlyst 26A) or weakly basic groups (Amberlite IRA 93). These combine the adsorptive capacity of the macroreticular adsorbents with the ion exchange properties of protective agents such as Dowex-1. All of these with the exception of the acidic exchanger offer some degree of protection to the enzyme. Amberlite IRA 938 was the most effective giving a maximum yield of about 25% of that given by soluble PVP and giving turbid extracts which were almost colourless. In the absence of DTE, IRA 938 completely prevented enzymic browning of apple extracts and gives the same yield of malic enzyme activity as in the presence of DTE. IRA 938 (6·25% w/v) also prevented enzymic browning in extracts of potato tubers (var. Redskin) and led to the production of almost colourless extracts in the absence of reducing agents or thiols. Under these conditions there is no significant inhibition of phenolase and thus the effect of the macroreticular ion exchange must be in removing endogenous phenolics from solution.

We have used Dowex-1 in the extraction of hydroxycinnamate CoA ligase from swede roots and XAD 2 and XAD 4 during the extraction of this enzyme from pea seedlings with some success. But in these cases it may well be that the adsorbents are removing an endogenous inhibitor rather than protecting against tannins or phenolic oxidation products. Croteau et al. (1973) have used XAD 4 to adsorb interfering endogenous monoterpene substrates during the extraction of an enzyme from peppermint catalysing the cyclization of neryl pyrophosphate to α-terpineol. Our preliminary studies suggest that macroreticular ion exchange resins rather than the macroreticular adsorbents may find a wider usefulness and add another tool in the armoury of techniques to protect plant enzymes from inhibition by tannins and quinones during extraction.

2. Use of Low Molecular Weight Protective Agents

Extracted plant enzymes can generally be protected from the products of enzymic browning by the presence of soluble reducing agents and specific inhibitors of phenolase (Anderson, 1968). The use of anaerobic conditions for the extraction of enzymes to prevent oxidation of phenolic substrates has generally proved to be inconvenient in practice but is used by some workers (Loomis, 1974). It is interesting that the activity of pulegone reductase extracted from peppermint leaves was stimulated if homogenization was carried out in a nitrogen atmosphere even in the presence of both polyclar AT and a reducing agent, metabisulphite (Loomis, 1974). Phenolase is inhibited by copper chelators such as diethyldithiocarbamate (DIECA) and mercaptobenzothiazole, by thiols such as thioglycollate, mercaptoethanol and dithiothreitol and by strong reducing agents such as metabisulphite and dithionite. These compounds have been used successfully either singly or in combination in the extraction of enzymes. For instance the presence of DIECA (10 mM) prevents enzymic browning and is essential for the extraction of active sucrose synthetase from sugar-cane (Slack, 1966). Mercaptobenzothiazole was shown by Palmer and Roberts (1967) to inhibit phenolase at very low concentrations (10^{-4} M) and it is thought to combine with copper at the active site of phenolase forming an inactive complex. Mercaptobenzothiazole (100 μM) has recently been used in the extraction of lipolytic acyl hydrolase from potato tissue (Galliard and Dennis, 1974).

Thiols have been widely used for the protection of enzymes during extraction from plants. The effects of thiols on phenolic oxidation should be distinguished from their effects on the longer term stability of enzymes during subsequent steps in a purification procedure. Thiols such as thioglycollic acid, mercaptoethanol and cysteine are effective inhibitors of phenolase and of enzymic browning. Thioglycollic acid was particularly effective as an inhibitor of sugar-cane phenolase (Coombs et al., 1974). The phenolase of this tissue was separated into two fractions which differed in their sensitivity to thioglycollate. 10^{-5} M thioglycollate gave a 50% inhibition of form 1 (PPO 1) while 10^{-4} M was required in the case of form 2. 10^{-3} M mercaptoethanol was required for 50% inhibition with either form of the enzyme and concentrations of greater than 10^{-3} M were required in the case of cysteine and dithiothreitol to achieve this degree of inhibition. Montgomery and Sgarbieri (1975) showed that mercaptoethanol completely inhibited phenolase from the pulp of banana fruits at 10^{-3} M. There seem to be quite wide variations between the sensitivities of phenolases from different sources to thiols and this may explain the differences in the effectiveness of different thiols in protecting enzymes in different tissues.

Reducing agents such as metabisulphite and dithionite which are very effective inhibitors of phenolase were shown by Anderson and Rowan (1967) to be more efficient than thiols in protecting peptidase during extraction from tobacco leaf tissue. The inhibition was permanent but no studies on the mechanism of inhibition were made. Metabisulphite at this concentration has proved very effective in the isolation of pulegone reductase from peppermint leaves (Loomis, 1974) and of p-coumarate CoA ligase from potato tubers (Rhodes and Wooltorton, 1975). Metabisulphite in solution is in equilibrium with bisulphite, sulphite ions, sulphurous acid and SO_2 and the equilibrium mixture is dependent on factors such as pH, temperature and concentration, so it is not clear which species is responsible for the observed effects. The inhibitory effect of sulphites on some plant enzymes have been described (Luttge *et al.*, 1972). Nevertheless, many enzymes appear resistant to inhibition by sulphites and thus metabisulphite can be useful in the protection of such enzymes.

The inhibition of phenolase by these protective agents during homogenization is probably never complete and, except possibly in the case of metabisulphite, the formation of small amounts of *o*-quinone cannot be prevented. An important part of the effectiveness of thiols in preventing enzyme inhibition depends on their capacity to react with *o*-quinones and to prevent both the attachment of quinones to protein and their polymerization. These reactions are of two types; one in which the thiol reduces the *o*-quinone back to the *o*-diphenol and compounds active in this include thioglycollate and mercaptobenzothiazole (Anderson, 1968), and the other in which the thiol reacts with the quinone to form a product which is not further oxidized and does not inhibit enzymes. This second group of thiols includes DIECA and cysteine. The importance of these reactions in removing quinones is illustrated by the fact that cyanide, although an effective inhibitor of phenolase, is not very effective as a protective agent for enzymes during extraction partially because it is unable to react with quinones (Anderson, 1968). Pierpoint (1973b) has used benzene sulphinic acid, a very effective agent for trapping quinones (Pierpoint, 1970) in the extraction of an N-acylamino acid acylase from tobacco leaves.

It can be seen from this brief description that a wide range of materials can protect plant extracts and it is clear from the literature that no single compound or combination of compounds is universally applicable. This, perhaps, is not surprising in view of the wide differences in the qualitative and quantitative composition of phenolic compounds between plant tissues. It is possible to distinguish between two types of problems. With tissues with high levels of endogenous tannins PVP and the other polymers may well be effective, while in tissues in which the tannin content is low but where there are high levels of phenolase and phenolic substrates low mol. wt phenolase inhibitors such as metabisulphite and thiols may be most effective. In tissues with both high tannins and an active phenolase a combination of polymer protectors together with metabisulphites or thiols may be required.

In spite of all the advances in this field it is likely that even under favourable conditions most of the protein preparations made from plant tissues contain some attached phenolic or quinone molecules which may lead to instability during subsequent steps in a purification procedure and that such attachment may well modify the behaviour of the protein molecule even when enzymic activity is not impaired (Fieldes and Tyson, 1973; Pierpoint, 1973a).

C. FACTORS AFFECTING THE STABILITY OF ISOLATED ENZYME PROTEINS

The problem of tannins and quinones is peculiar to work with higher plants. The problem of the stability of enzymic activity in isolated systems is however, common to all biological systems. There are a number of factors such as heat, extremes of pH etc. which can cause denaturation of proteins and thus inactivate enzymes, but here I would like to consider more subtle changes than this which can lead to the loss of enzymic activity. Changes in the molecular configuration of the enzyme or modification of groups close to the active site of the enzyme which are involved in the binding of substrate or cofactor can lead to loss of activity. For instance, the oxidation of —SH bonds, close to the active site may be important in this and the effectiveness of thiols (such as EtSH and especially DTT), in maintaining —SH groups in the reduced state may be important in preventing loss of activity by the formation of —S—S— bonds and subsequent changes in enzyme conformation (Cleland, 1964). There are numerous examples in the literature of plant enzymes being stabilized by DTT and EtSH.

Changes in the molecular configuration of enzymic proteins during storage of extracts which may or may not be associated with loss of activity have been described by a number of workers. Changes in the electrophoretic mobility of catechol oxidase isoenzymes on storage has been ascribed to dissociation of the initially isolated enzyme into subunits (Harel *et al.*, 1973). Pridham and Dey (1974) have described the conversion of an isoenzyme of α-galactosidase into a high mol. wt form on storage. Gerbrandy *et al.* (1975) have shown a change in electrophoretic properties of potato phosphorylase on storage which they ascribe to partial proteolysis of the molecule. There is a loss of regulatory properties of leaf phosphofructokinase on storage at 4° which may result from dissociation into subunits (Dennis and Coultate, 1967). Changes in molecular conformation probably account for the cold lability of some enzymes i.e. prolyl t-RNA synthetase (Norris and Fowden, 1974) and the loss of activity on dilution of enzymes such as potato β-amylase (Takeda and Hizukuri, 1972). Both these changes are reversible; that of the cold lability on warming and that of dilution on addition of non-ionic detergents such as Triton X-100.

Proteolytic attack may be important in causing instability of isolated proteins particularly where specific attack occurs near the active site. General proteolytic attack can occur with little effect on activity although the properties of the protein back-bone may well be severely altered. In an animal system, a rapid decline in activity of ornithine aminotransferase (Kominani and Katunuma, 1976) on storage at 37° was shown to be due to the irreversible inactivation of the enzyme by failure of the cofactor, pyridoxyl phosphate to attach to the apo-enzyme. Addition of protease inhibitors such as soyabean trypsin inhibitor, chymostatin or phenyl methyl sulphonyl fluoride (PMSF) or the presence of the cofactor, almost completely prevented inactivation of the enzyme. Other enzymes in the extract including aspartate aminotransferase, which has a non-dissociable coenzyme, were stable during this period. It seems likely that a group-specific protease attacks the apo-enzyme of ornithine aminotransferase and prevents attachment of the cofactor to form active enzyme. The scope for proteolysis playing a part in loss of activity seems to be important particularly in plant tissues with high levels of protease. Wallace (1975) showed that PMSF promoted the extrac-tion of nitrate reductase from maize roots although casein at 3% w/v was much more effective.

There has been some recent work on the use of substrates and cofactors to protect plant enzyme activities. For example, the problems of protecting dehydrogenases have been described by Haissig and Schipper in a series of papers on the extraction of enzymes from both woody and herbaceous tissues (Haissig and Schipper, 1972; 1975a; 1975b). Woody tissues offer a particu-larly difficult "phenolic" problem which affects the extractability and longer term stability of these relatively sensitive enzymes (Haissig and Schipper, 1972, 1975a). Insoluble PVP together with either EtSH or DTT was used to overcome these problems but the concentrations of EtSH (or DTT) required for optimum extraction of activity was found to be more than 10 times that required to prevent enzymic browning of the extracts. The authors took this to mean that the thiol was protecting the active site and that substrates or cofactors, because of their higher affinities for the binding site, might be more effective in protecting the enzyme. Haissig and Schipper (1975b) showed that the inclusion of substrate (0·25–0·75 mM) in the extraction medium led to a large increase in the extractable activity of glyceraldehyde-3-phosphate dehydrogenase and glucose-6-phosphate dehydrogenase. Table II shows the effect of EtSH (100 mM) and NAD (1 mM) on the extractable levels of glyceraldehyde-3-phosphate dehydrogenase from 10 different species and their stability over a period of storage of the extracts at 0°C. The results show that in nearly all cases the cofactor stimulates the extraction of the dehydro-genase and increases its stability during subsequent storage. The exceptions to this are barberry and tomato where there are no increases in the amount of initially extracted enzyme, but even in these two cases the loss of activity on subsequent storage is greatly reduced in the presence of the cofactor. Similar data on the effect of pyridine nucleotides on the extraction and stabilization of malate and glucose-6-phosphate dehydrogenases is also presented in this study. The use of substrate to stabilize enzyme activity has also been employed in work on asparagine synthetase where substrate binding leads to the forma-tion of a stable dimeric form (Rognes, 1975).

Enzymes can be stabilized on addition of a specific inhibitor. For example, aspartokinase is stabilized by its inhibitor, threonine (Aarnes and Rognes, 1974). Threonine (at 1 mM) as well as mercaptoethanol (14 mM) was included in the extraction medium and in the elution buffers during the 85–fold purification of aspartokinase from pea seedlings. The enzyme fractions were

TABLE II

The influence of mercaptoethanol (100 mM) and NAD (1 mM) on the extraction and subsequent decay of glyceraldehyde-3-phosphate dehydrogenase activity

Species	Time, h	Activity (mean of three replications) Units/g. dry wt		
		Control	EtSH	EtSH + NAD
Sugar maple	0	1·35	3·56	10·72
	72	0	0	3·23
Red pine	0	0·21	0·93	3·14
	72	0	0	1·50
Walnut	0	0	9·3	10·7
	72	0	1·0	5·1
Aspen	0	16·0	17·73	30·87
	72	0	2·55	14·93
Barberry	0	18·0	16·53	10·84
	72	0	1·40	8·51
White spruce	0	0	1·83	3·28
	72	0	1·24	3·38
Barley	0	10·49	31·84	113·05
	72	0	0	10·10
Spinach	0	175·16	210·0	224·09
	72	0	101·13	138·58
Tomato	0	0·59	60·97	43·35
	72	0	0·58	34·35
Sweet potato	0	0·38	42·36	57·47
	72	0·13	11·40	48·05

Modified from Haissig and Schipper (1975b).

assayed after either dilution to a final threonine concentration of 0·2 mM (a concentration giving only a low degree of inhibition), or chromatography on Sephadex G25 against buffer lacking threonine. Threonine also inhibits and stabilizes homoserine dehydrogenase (Di Marco and Grego, 1975).

The use of substrate, cofactor, or inhibitor to stabilize a particular form of an enzyme may be an important consideration for the future in enzyme purification. During a purification procedure an enzyme preparation will experience a wide range of different conditions and it is important to recognize stages in the procedure in which the enzyme may become unstable and to develop methods to combat this instability.

III. TECHNIQUES FOR THE PURIFICATION OF ENZYMES FROM HIGHER PLANTS

Jakoby (1971) has stated that "the trend in the development of techniques for the purification of enzymes has gradually evolved from a type of trial and error application to that of an art form and more recently from an art to a form approximating a reasonably scientific basis". In attempting the purification of a new enzyme activity one is aware of the possibilities in terms of the presence of subunits, of isoenzymic forms, of more than one enzymic activity associated with a single protein, and of the presence of soluble multienzyme complexes, but often one has very little information of the chemical and physical properties of the protein entity under study. A certain amount of information on properties is obtained from the behaviour of the protein during the purification steps and as the chemical and physical bases of the separation methods are increasingly understood a more scientific approach to the resolution of complex mixtures of proteins can be undertaken.

In order to apply the classical procedures of protein purification it is nearly always necessary to have the protein in true solution An exception is in the separation of dispersions of complex lipoproteins for which special techniques have been developed. In the case of "soluble" enzymes the protein is in true solution in the homogenate. In the case of enzymes bound to membranous structures it is necessary as a preliminary to develop methods for their "solubilization" prior to purification.

A. SOLUBILIZATION OF MEMBRANE BOUND ENZYMES

Some of the techniques for the "solubilization" of membrane bound enzymes are set out in Table III. In the case of matrix enzymes such as the stroma proteins of chloroplasts all that is required is to rupture the membrane

TABLE III

Methods for the "solubilization" of enzymes associated with lipoprotein bound cell compartments

1. Matrix enzymes
   a. Physical methods for disruption e.g. mechanical or osmotic shock

2. Membrane associated enzymes
   a. Physical methods
      Mechanical disruption
      Sonic disintegration
   b. Chemical methods involving the use of:
      Salts at low ionic strength
      Concentrated salts (chaotropic agents)
      Alkaline pH
      Metal chelating agents
      Organic solvents
      Surfactants
   c. Enzymic methods using
      Lipases and phospholipases
      Proteolytic enzymes

and to dissolve out the enzymes. However, for enzymes bound to membranes a range of more drastic techniques have been developed (Penefsky and Tzagoloff, 1971). Some membrane bound enzymes acquire the characteristics of classical water soluble proteins on release from membranes. Others, however, are released as a dispersion of very small lipoprotein fragments which do not sediment under forces of $100\,000 \times g$ applied for one hour or more. It is often possible to distinguish between these two conditions since lipoprotein dispersions tend to be precipitated on adjustment of the pH to 5–6 or on addition of salt to 1–2%.

The available methods (see Table III) range from physical methods such as mechanical or sonic disruption to chemical and enzymic methods. Mild chemical methods such as low ionic strength buffers have been used for the solubilization of cytochrome c from mitochondrial membranes (Jacobs and Sanadi, 1960) while EDTA has been employed in the solubilization of ATPase from spinach chloroplasts (McCarty and Racker, 1966). Recently the use of the "chaotropic salts" such as chlorate and thiocyanate (Hatefi and Hanstein, 1974) has been introduced. These salts seem to decrease hydrophobic interactions within membranes and lead to the solubilization of particular protein constituents. Sodium chlorate has been used successfully for the isolation and purification of succinnic dehydrogenase from bovine heart mitochondria (Davis and Hatefi, 1972). Chaotropic agents are powerful denaturants of proteins and can inactivate enzymes. However, in many cases the inactivation is reversible on removal of the chaotropic agent. Organic solvents such as acetone and butanol have been used to extract proteins from membranes. For example, Morton (1950) used butanol for the solubilization of alkaline phosphatases from membranes of animal cells. Surfactants such as bile salts, cholate and deoxycholate and non-ionic detergents such as Triton X-100 are used for the solubilization of membrane bound enzymes. Deoxycholate has been used to solubilize NADP cytochrome c reductase (Hill, 1975) and Triton X-100 to solubilize cinnamic acid-4-hydroxylase from swede microsomes (Hill and Rhodes, 1975). Maslowski and Komoszynski (1974) used a combination of Triton X-100 and deoxycholate to solubilize ATPase from *Zea mays* microsomes. Enzymic methods involve the use of phospholipases (e.g. the commercially available snake venom phospholipase A) or pancreatic lipase to attack lipid components of membranes and proteolytic enzymes such as trypsin or pronase to attack protein components, thus freeing enzymic components into solution.

For the solubilization of lipoprotein complexes by bile salts, differential methods of extraction with bile salts, extractions at different pH in the presence or absence of salt are used. $(NH_4)_2SO_4$ may be useful for their further purification especially if cholate rather than deoxycholate is used as the dispersing agent (Penefsky and Tzagaloff, 1971). With synthetic detergents, nonionic detergents such as Triton X-100 are widely used as they cause minimal denaturation and the use of 0·2–1·0 M KCl is often recommended as an aid to solubilization. Such detergent dispersed lipoprotein complexes can be separated by ion exchange chromatography using standard techniques provided sufficient detergent is maintained in the elution buffers to preserve the stability of the lipoprotein complexes. An example of the successful use of this technique is in the resolution of the microsomal hydroxylating system of liver cells which involves flavoprotein and cytochrome P450 (Lu and Coon, 1968). An analogous plant system is the cinnamic acid-4-hydroxylase which is also microsomal and requires cytochrome P450 for its activity (Russell, 1971; Potts *et al.*, 1974). This system has been "solubilized" by Triton X-100 treatment (Hill and Rhodes, 1975) and partially resolved into its components by chromatography on DEAE cellulose in the presence of Triton X-100 (Hill, 1975).

B. CONCENTRATION OF PLANT EXTRACTS

As crude extracts of plant tissues are generally rather dilute in protein, a concentration step is usually required at an early stage in the purification prior to application of chromatographic or electrophoretic procedures. Concentration steps are also used following a chromatographic separation when the subsequent step requires a small sample volume. Table IV sets out some of the more important concentration steps. It is clearly not possible to distinguish these completely from normal purification steps since each of the steps described can involve some degree of purification. For instance, differential precipitation with salts such as $(NH_4)_2SO_4$ or Mn^{++} ions or solvents such as acetone or ethanol gives some degree of purification but taking up the resultant precipitate in a minimal volume of buffer can lead to

TABLE IV

Methods for the concentration of enzyme extracts

1. Precipitation methods
   Ultracentrifugation
   Salt/solvent
   Precipitation at the isoelectric point

2. Solvent removal methods
   Lyophilization
   Freezing/thawing methods

3. Partition methods
   Gel exclusion
   Dialysis against solute of high mol. wt
   Adsorption chromatography
   Ultrafiltration
   a. Centrifugally accelerated ultrafiltration
   b. Pressure accelerated ultrafiltration

the concentration of the enzyme fraction. Similar concentration and purification can be achieved by precipitation at the isoelectric point of the protein. Attempts at solvent removal are generally not very successful but lyophilization and freeze/thaw methods can be useful in some circumstances. Other procedures such as gel exclusion using dry dextran gels to absorb aqueous phase while concentrating protein and dialysis against a high mol. wt solute such as polyethyleneglycol are used but are not very convenient. A very useful method of concentrating enzymes is to adsorb the protein from dilute solution onto a suitable adsorbent (ion exchange cellulose, calcium phosphate gel etc.) and to elute in a small volume (see Rhodes and Wooltorton, 1973). In the past few years, the use of ultrafiltration through membranes of defined porosity has been introduced using either centrifugal force or pressure to force the water and salts of low mol. wt through the membrane leaving a concentrated protein solution behind. The commercially available ultrafiltration systems can also be used for dialysis and seem to offer useful prospects for the rapid dialysis of crude extracts followed by concentration of the desalted extract. Caution should be exercised if concentration of the protein solution is taken beyond a certain point owing to a phenomenon known as concentration polarization. In this the protein concentrates at the membrane surface, which reduces the solute flux and makes recovery of the protein fraction difficult. Systems which have a stirrer or agitator mounted just above the membrane to minimize this effect are available commercially.

C. DESALTING OF EXTRACTS

Another important step prior to chromatography or electrophoresis is to remove salt and other low mol. wt solutes from the enzyme extract. This is often also required between chromatographic procedures. As mentioned above, pressure dialysis is a convenient method to achieve this. Standard dialysis techniques are still used but are increasingly being replaced by desalting on highly cross-linked gels of dextrans or polyacrylamide in view of the greater speed and convenience of this technique. With plant extracts this technique has the advantage that many phenolic materials adsorb strongly on the gels and are separated from protein. This adsorption effect is shown more dramatically by dextrans than by polyacrylamides and thus dextrans are the materials of choice for plant tissues. With highly cross-linked dextrans such as Sephadex G25, polyphenols are markedly retarded and sometimes bound irreversibly (Coombs et al., 1974). With all these desalting procedures there is some degree of dilution of the extract.

D. CHROMATOGRAPHIC METHODS

The most important properties of proteins exploited by chromatographic methods are the mol. wt and the ionic properties. The more important chromatographic methods will now be outlined.

1. *Gel Permeation Chromatography*

The principle of gel permeation chromatography was introduced by Porath and Flodin (1959). The chromatographic matrix consists of beads of cross-linked gel which are composed of pores of a sufficient size for molecules below a certain critical size to enter. Compounds of mol. wts above this "exclusion limit" do not interact with the gel and are not retarded. However, smaller molecules which can enter the pores are progressively retarded by the column and elute in order of decreasing mol. wt. This gel permeation chromatography has proved to be a powerful tool for the resolution of complex mixtures of proteins on the basis of their mol. wts. Gels composed of dextrans, polyacrylamide and agarose are available and the resolution of proteins of up to 400 000 is possible on polyacrylamides, up to 800 000 on dextrans and up to 1.5×10^8 on agarose. The principles of gel filtration have been reviewed (Reiland, 1971) and it is clear that although gel permeation methods depend principally on mol. wt and have been used for mol. wt determinations (Andrews, 1964), other factors, particularly molecular shape, also play a role. The commercially available dextrans have up to 10–20 micro equivalents of free carboxyl groups/g dry gel and thus ionic interactions can occur in low ionic strength buffers. Adsorption effects with aromatic and heterocyclic compounds, especially with the highly cross-linked small pore gels, also occur.

For optimum resolution on large pore gels in fine beads the sample should be applied at low viscosity, and in the absence of lipoproteins which are adsorbed by the gel. The sample volume is critical and ideally only between 1–4% of the bed volume of the column should be occupied by the sample before elution. Flow rate should be low (Reiland, 1971) for optimum resolution although this is limited at the lower end by the tendency of bands to diffuse with time. The conditions for desalting on small pore gels is not as critical and up to 25% of the bed volume may be taken up by the sample before elution.

2. *Ion Exchange Chromatography*

Ion exchange chromatography has proved a most powerful tool in the purification of proteins since the introduction of the substituted celluloses by Peterson and Sober (1956). Previously only very limited success had been achieved by the application of ion exchange resins for protein separations. A range of substituted celluloses, dextrans and polyacrylamides are now available commercially and have found wide application. The latter two materials have gel permeation properties, in addition to ion exchange properties, but these are only important with relatively low mol. wt proteins. Among the substituted celluloses, DEAE (diethylamino-ethyl-) cellulose and to a lesser extent, CM- (carboxymethyl) cellulose are most widely used. The use of columns of cellulose exchangers has the advantage over the other types in that they do not change in volume over a wide range of the salt concentrations and pHs met in practice. Proteins are adsorbed by the formation of multiple ionic bonds between the charged groups on the protein and the available groups of opposite charge attached to the cellulose. In view of the fact that most plant proteins have an isoelectric point in the acidic region and are thus negatively charged at neutral pH; DEAE cellulose, an anionic exchanger, is most widely used (Boman, 1963). However, in spite of this, some proteins with a net negative charge will adsorb onto CM-cellulose. This is presumably due to the charge distribution in the protein moiety being such that positively charged regions of the protein are able to bind to the negatively charged groups on the CM-cellulose.

Pretreatment of cellulose exchangers is important to overcome internal hydrogen bonding and to allow access of all the ionic groups on the cellulose to the bonding protein. Alternate acid and alkali treatments are used to achieve this before equilibration with the appropriate starting buffer. In recent years cellulose exchangers with cellulose in a microcrystalline form and having a high capacity have become available and have largely replaced the older fibrous forms for high resolution work. The sample must be presented to the cellulose column at low ionic strength at the appropriate pH and thus dialysis or desalting are important preludes to ion exchange chromatography. The sample volume is not critical but the column size should be such that less than 10% of the adsorbent bed is occupied by adsorbed protein at the start of elution. Elution is normally carried out using gradients of salt or pH as this elution system is less prone to artefacts than step-wise systems. The slope and shape of the gradient can be relatively easily controlled, using simply constructed or commercially available equipment.

3. Other Chromatographic Methods

Other forms of chromatography applied to protein separations include the use of alumina or various forms of calcium phosphate gel and these are often used in batch form for adsorption and subsequent elution of the activity. Hydroxyapatite, a form of calcium phosphate developed by Tiselius et al. (1956), has been widely used for the final stages of separation of proteins after DEAE-cellulose or gel permeation chromatography (Knobloch and Hahlbrock, 1975). Hydroxyapatite $Ca_{10}(PO_4)_6(OH)_2$ is equilibrated with a dilute solution of phosphate buffer, and the samples are adsorbed and subsequently eluted with a gradient of increasing concentration of phosphate buffer. Hydroxyapatite is amphoteric in nature and the nature of the linkages it makes with protein are not clearly established, but they are thought to involve binding of the free carboxyl groups of proteins to Ca^{2+} on the surface of the hydroxyapatite crystals. The binding of $-NH_2$ groups of proteins is also thought to occur but the nature of the receptor groups on the hydroxyapatite is unknown. The presence of high concentrations of salts such as KCl can influence the chromatographic behaviour of proteins on hydroxyapatite and this can be successfully exploited for protein separations (Bernardi, 1971). Another form of chromatography which has been used with success in the last few years is affinity chromatography and this subject will be reviewed elsewhere in this volume (Chapter 13).

E. ELECTROPHORETIC PROCEDURES

Electrophoretic procedures are the other main methods for separating proteins in addition to the chromatographic ones. These procedures may be based on the net charge on the molecule at a particular pH (electrophoresis) or on the isoelectric point of the protein (isoelectric focusing), namely the pH at which the protein does not move in an electric field.

1. *Electrophoresis*

Two methods of electrophoretic separation are widely used for the purification of proteins. In conventional electrophoresis (Tiselius, 1957) the protein mixture is applied to a column of a stabilizing medium such as cellulose, starch gel or dextran, which minimizes diffusion of the sample during electrophoresis. A voltage is applied and components of the mixture separated on the basis of their net charge at constant pH and ionic strength. The separation is allowed to continue and the fractions collected in an elution chamber fitted into the lower end of the column and washed out with elution buffer.

Another method which has been used in enzyme purification is preparative disc electrophoresis (Shuster, 1971) in which polyacrylamide gel is used as the supporting medium with a discontinuous buffer system. During electrophoresis a specific combination of ions forms a moving front with the leading ion, followed by a trailing ion with the protein sample, bracketed between them. This leads to a concentrating of the protein at the origin during the early stages of electrophoresis and enhanced resolution during later migration in the gel. The use of polyacrylamide gel as an anticonvection medium leads to a molecular-sieving effect and thus, molecular size and shape, as well as net charge at the selected pH, affect the degree of resolution. A more recent technique, isotachophoresis, uses the same principles which are involved in the concentration and stacking which occur during the first stages of disc electrophoresis (Svendsen and Rose, 1970). Here the protein bands are stacked in the moving front between a leading and trailing ion which are arranged to have mobilities in the field respectively greater and less than that of all the protein species. Spacer ions such as carrier ampholytes are included in the protein fraction and stack between the protein bands thus greatly enhancing the separation between bands. When separation is complete the separate protein bands and the spacer ions migrate with the same velocity down the column of polyacrylamide gel. With both disc electrophoresis and isotachophoresis, migration of the protein bands is allowed to proceed to the bottom of the gel columns and fractions collected with buffer from an elution chamber.

2. *Isoelectric Focusing*

In isoelectric focusing, a stable pH gradient is set up between the anode and cathode of an electric field. Proteins migrate in this field until they reach a point in the pH gradient equal to their isoelectric point (where they have no net charge) and thus focus at that point (Vesterberg, 1971). The pH gradient is formed by electrophoresis of a mixture of carrier ampholytes having iso-electric points over the desired pH range. These carrier ampholytes are isomers and homologues of aliphatic polyamino-polycarboxylic acids. In order to minimize convective mixing, the pH gradient is usually set up in a sucrose density gradient. The enzyme fraction to be separated can be intro-duced at any point or throughout the gradient so sample volume is not critical, as it can be in normal electrophoresis. After focusing the sucrose gradient is separated into fractions and assayed for enzyme activity. A dis-advantage of this method is that some proteins precipitate at their isoelectric point and this can lead to inactivation. The separations obtained by iso-electric focusing show a high degree of resolution and separations not possible by other methods can often be effected.

IV. CRITERIA FOR ASSESSING THE PURITY OF ISOLATED ENZYMES

Having developed a number of techniques for maintaining the activity of an enzyme preparation and by a number of steps brought this preparation to a high state of purification it is then necessary to assess the degree of purity of the fraction. Over the years a number of tests of homogeneity have been applied (O'Donnell and Woods, 1962) and none is entirely satisfactory mainly because they give evidence of an essentially negative character (i.e. on the absence of contaminants). Crystallization of an enzyme protein was once thought of as a good criterion of purity but when it was shown that mixed protein crystals can form, doubt was cast on its validity. The production of a single sharp moving boundary in the ultracentrifuge with no evidence of additional peaks or broadening of the main band during centrifugation has also been used. However, these methods have been largely replaced by high resolution analytical electrophoretic separations particularly on rods or flat beds of polyacrylamide gels or high resolution isoelectric focusing also using flat beds of polyacrylamide. These electrophoretic separations are used in conjunction with specific stains for proteins such as Coomassie blue or amido black and specific tests to locate enzyme activities. It is still true that no single procedure can definitely establish the homogeneity of a preparation but polyacrylamide gel electrophoresis at a range of pHs together with iso-electric focusing (Boulter and Ramshaw, 1975) are the accepted standard at present. Further evidence can be provided from the numbers of N- and C-terminal amino acids in relation to the number of polypeptide chains in the protein molecule, if this is known.

V. Conclusion

The most difficult aspect of the purification of enzymes from plants is to isolate the enzyme in a stable unchanged form and to maintain it in a stable condition over a number of days so that the standard methods of purification can be applied. The application of these standard methods to plants can generally overcome the problems of resolution of closely related species but there is need for a considerable amount of exploratory work to determine the correct conditions for optimum resolution at each stage in the procedure.

REFERENCES

Aarnes, H. and Rognes, S. E. (1974). *Phytochemistry* **13**, 2717.
Andersen, R. A. and Sowers, J. A. (1968). *Phytochemistry* **7**, 293.
Anderson, J. W. (1968). *Phytochemistry* **7**, 1973.
Anderson, J. W. and Rowan, K. S. (1967). *Phytochemistry* **6**, 1047.
Andrews, P. (1964). *Biochem. J.* **91**, 222.
Bate-Smith, E. G. (1972). *In* "Phytochemical Ecology" (Harborne, J. B., ed.), pp. 45–56. Academic Press, London and New York.
Bate-Smith, E. G. and Lerner, N. H. (1954). *Biochem. J.* **58**, 126.
Bernardi, G. (1971). *In* "Methods in Enzymology" (Jakoby, W. B., ed.), Vol. 22, pp. 325–339. Academic Press, New York and London.
Boman, H. G. (1963). *In* "Modern Methods of Plant Analysis" (Linskens, H. F. and Tracy, M. V., eds), Vol. 6, pp. 393–416. Springer Verlag, Berlin.
Boulter, D. and Ramshaw, J. A. M. (1975). *In* "The Chemistry and Biochemistry of Plant Proteins" (Harborne, J. B. and van Sumere, C. F., eds), pp. 1–30. Academic Press, London and New York.
Briggs, T. M., Hauck, R. A. and Eye, J. D. (1973). *J. Am. Leath. Chem. Ass.* **68**, 176.
Chan, P. H., Sakano, K., Singh, S. and Wildman, S. G. (1972). *Science*, **176**, 1145.
Chin, C. K. and Weston, G. D. (1973). *Phytochemistry* **12**, 1229.
Cleland, W. (1964). *Biochemistry* **3**, 480.
Coombs, J., Baldry, C., Bucke, C. and Long, S. P. (1974). *Phytochemistry* **13**, 2703.
Croteau, R., Burbolt, A. J. and Loomis, W. D. (1973). *Biochem. biophys. Res. Commun.* **50**, 1006.
Dalgardo, L. and Birt, L. M. (1962). *Biochem. J.* **83**, 195.
Davis, K. A. and Hatefi, Y. (1972). *Archs Biochem. Biophys.* **149**, 505.
Dennis, D. T. and Coultate, T. P. (1967). *Biochim. biophys. Acta* **146**, 129.
Di Marco, G. and Grego, S. (1975). *Phytochemistry* **14**, 943.
Doronton, W. J. S. and Hawker, J. S. (1973). *Phytochemistry* **12**, 1557.
Fieldes, M. A. and Tyson, H. (1973). *Phytochemistry* **12**, 2133.
Frieden, C. (1971). *A. Rev. Biochem.* **40**, 653.

Galliard, T. and Dennis, S. (1974). *Phytochemistry* **13**, 1731.

Gerbrandy, S. J., Shankar, V., Shivaram, K. N. and Stegemann, H. (1975). *Phytochemistry* **14**, 2331.

Goldstein, J. L. and Swain, T. (1965). *Phytochemistry* **4**, 185.

Good, N. E., Winger, D., Winter, W., Connolly, T. N., Izawa, S. and Singh, R. M. M. (1966). *Biochemistry* **5**, 467.

Ginsburg, A. and Stadtman, E. R. (1970). *A. Rev. Biochem.* **39**, 429.

Gross, G. G., Mansell, R. L. and Zenk, M. H. (1975). *Biochem. Physiol. Pfl.* **168**, 41.

Haard, N. F. and Hultin, H. O. (1968). *Ann. Biochem.* **24**, 299.

Haissig, B. E. and Schipper, A. L. Jr. (1971). *Biochem. biophys. Res. Comm.* **45**, 598.

Haissig, B. E. and Schipper, A. L. Jr. (1972). *Ann. Biochem.* **48**, 129.

Haissig, B. E. and Schipper, A. L., Jr. (1975a). *Phytochemistry* **14**, 345.

Haissig, B. E. and Schipper, A. L., Jr. (1975b). *Physiol. Pl.* **35**, 249.

Harel, E., Mayer, A. M. and Lehman, E. (1973). *Phytochemistry* **12**, 2649.

Hatefi, Y. and Hanstein, W. G. (1974). *In* "Methods in Enzymology" (Fleischer, S. and Packer, L., eds), Vol. 31, pp. 770–790. Academic Press, New York and London.

Hill, A. C. R. (1975). Ph.D. Thesis, University of East Anglia.

Hill, A. C. R. and Rhodes, M. J. C. (1975). *Phytochemistry* **14**, 2387.

Jacobs, E. E. and Sanadi, D. R. (1960). *J. biol. Chem.* **235**, 531.

Jakoby, W. B. (1971). *In* "Methods in Enzymology" (Jakoby, W. B., ed.), Vol. 22, p. 11. Academic Press, New York and London.

Jones, J. D., Hulme, A. C. and Wooltorton, L. S. C. (1965). *Phytochemistry* **4**, 659.

Knobloch, K. H. and Hahlbrock, K. (1975). *Europ. J. Biochem.* **52**, 311.

Kominani, E. and Katunuma, N. (1976). *Europ. J. Biochem.* **62**, 425.

Koshland, D. E., Jr. and Neet, K. E. (1968). *A. Rev. Biochem.* **37**, 359.

Kung, S-d. (1976). *Science* **191**, 429.

Lam, T. H. and Shaw, M. (1970). *Biochem. biophys. Res. Comm.* **39**, 965.

Loomis, W. D. (1974). *In* "Methods of Enzymology" (Fleischer, S. and Packer, L., eds), Vol. 31, pp. 528–544. Academic Press, New York and London.

Loomis, W. D. and Battaile, J. (1966). *Phytochemistry* **5**, 423.

Lu, A. Y. H. and Coon, M. J. (1968). *J. biol. Chem.* **243**, 1331.

Luttge, U., Osmond, C. B., Ball, E., Brinckmann, E. and Kinze, G. (1972). *Pl. Cell Physiol.* **13**, 505.

Lynen, F. (1961). *Fedn. Proc. Fedn. Am. Socs exp. Biol.* **20**, 941.

Maslowski, P. and Komoszynski, M. (1974). *Phytochemistry* **13**, 89.

McCarty, R. E. and Racker, E. (1966). *Brookhaven Symp. Biol.* **19**, 202.

Montgomery, M. W. and Sgarbieri, V. C. (1975). *Phytochemistry* **14**, 1245.

Morton, R. K. (1950). *Nature, Lond.* **166**, 1092.

Norris, R. D. and Fowden, L. (1974). *Phytochemistry* **13**, 1677.

O'Donnell, I. J. and Woods, E. F. (1962). *In* "Modern Methods of Plant Analysis" (Linskens, H. F. and Tracey, M. V. eds.), Vol. 5, pp. 250–324. Springer Verlag, Berlin.

Palmer, J. K. and Roberts, J. B. (1967). *Science* **157**, 200.

Penefsky, H. S. and Tzagoloff, A. (1971). *In* "Methods in Enzymology" (Jakoby, W. B., ed.), Vol. 22, pp. 204–230. Academic Press, New York and London.

Peterson, E. A. and Sober, H. A. (1956). *J. Am. chem. Soc.* **78**, 751.

Pierpoint, W. S. (1970). *A. Rep. Rothampstead exp. Stn.* Part 2, 199.

Pierpoint, W. S. (1973a). *J. gen. Virol.* **19**, 189.

Pierpoint, W. S. (1973b). *Phytochemistry* **12**, 2359.

Porath, J. and Flodin, P. (1959). *Nature, Lond.* **183**, 1657.

Potts, J. R. M., Weklych, R. and Conn, E. E. (1974). *J. biol. Chem.* **249**, 5019.

Price, C. A. (1974). *In* "Methods in Enzymology" (Fleischer, S. and Packer, L., eds), Vol. 31, pp. 501–519. Academic Press, New York and London.

Pridham, J. B. and Dey, P. (1974). *In* "Plant Carbohydrate Biochemistry" (Pridham, J. B., ed.), pp. 83–96. Academic Press, London and New York.

Reiland, J. (1971). *In* "Methods in Enzymology" (Jakoby, W. B., ed.), Vol. 22, pp. 287–321. Academic Press, New York and London.

Rhodes, M. J. C. and Wooltorton, L. S. C. (1973). *Phytochemistry* **12**, 2381.

Rhodes, M. J. C. and Wooltorton, L. S. C. (1974). *Phytochemistry* **13**, 107.

Rhodes, M. J. C. and Wooltorton, L. S. C. (1975). *Phytochemistry* **14**, 2161.

Ribbereau-Gayon, P. (1972). *In* "Plant Phenolics" Oliver and Boyd, Edinburgh.

Ricardo, C. P. P. and apRees, T. (1970). *Phytochemistry* **9**, 239.

Rognes, S. E. (1975). *Phytochemistry* **14**, 1975.

Russell, D. W. (1971. *J. biol. Chem.* **246**, 3870.

Scandalios, J. G. (1974). *A. Rev. Pl. Physiol.* **25**, 225.

Schrader, L. E., Cataldo, D. A. and Peterson, D. M. (1974). *Pl. Physiol.* **53**, 688.

Shuster, L. (1971). *In* "Methods in Enzymology" (Jakoby, W. D. ed.), Vol. 22, pp. 411–433. Academic Press, New York and London.

Slack, C. R. (1966). *Phytochemistry* **5**, 197.

Svendson, P. J. and Rose, C. (1970). *Sci. Tools* **17**, 13.

Swain, T. and Bate-Smith, E. C. (1962). *In* "Comparative Biochemistry" (Florkin, A. M. and Mason, H. S., eds), Vol. 3, pp. 755–809. Academic Press, New York and London.

Synge, R. L. M. (1975). *Qual. Plant.* **24**, 337.

Takeda, Y. and Hizukuri, S. (1972). *Biochim. biophys. Acta,* **268**, 175.

Tiselius, A., Hjebten, S. and Levin, O. (1956). *Archs Biochem. Biophys.* **65**, 132.

Tiselius, A. (1957). *In* "Methods in Enzymology" (Colowick, S. P. and Kaplan, N. O., eds), Vol. 4, pp. 3–20. Academic Press, New York and London.

Tolbert, N. E. (1971). *In* "Methods in Enzymology" (San Pietro, A., ed.), Vol. 23A, pp. 665–682. Academic Press, New York and London.

Van Sumere, C. F., Albrecht, J., Dedonder, A., De Pooter, H. and Pe I. (1975). *In* "The Chemistry and Biochemistry of Plant Proteins" (Harborne, J. B. and van Sumere, C. F., eds), pp. 211–264. Academic Press, London and New York.

Vesterberg, O. (1971). *In* "Methods in Enzymology", (Jakoby, W. B., ed.), Vol. 22, p. 389–412. Academic Press, New York and London.

Walker, J. R. L. and Hulme, A. C. (1965). *Phytochemistry* **4**, 677.

Wallace, W., (1975). *Pl. Physiol.* **55**, 774.

Walt, B. K. and Merrill, A. L. (1963). *In* "Composition of Foods" USDA Agriculture Handbook No. 8, USDA, Washington D.C.

CHAPTER 13

Affinity Chromatography

P. D. G. DEAN AND M. J. HARVEY

Department of Biochemistry, University of Liverpool, England

I. Introduction	. . . .	.	271
II. General Principles	. . . .	.	271
III. Selection of Ligand	. . . .	.	274
IV. Synthesis of Affinity Adsorbents	. . . . . .	.	277
V. Elution Procedures	. . . .	.	280
Acknowledgements	. . . .	.	286
References	. . . .	.	286

I. INTRODUCTION

Affinity chromatography, biospecific adsorption or ligand-specific chromatography are all terms which theoretically relate to the same general principle. This principle, first exploited by Starkenstein (1910), is that the biological specificity of a macromolecule for its complementary biomolecule can be utilized for a variety of applications when one of the components of the biological reaction is immobilized to an inert support. Examples of these applications are listed in Table I together with appropriate literature references for more detailed descriptions.

II. GENERAL PRINCIPLES

At present the major application of affinity chromatography is in the purification of macromolecules, and Fig. 1 illustrates the adsorption and desorption stages of this process. The affinity adsorbent usually comprises a ligand (biologically reactive component) and a spacer arm to hold the ligand clear of the matrix. When a tissue homogenate is applied to the adsorbent the specific biological interaction of the macromolecule with the immobilized ligand results in adsorption, whilst those macromolecules having no affinity for the ligand are eluted in the column void volume. The selectively retained macromolecule is eluted from the adsorbent by a variety of desorption techniques depending on the nature and strength of the macromolecule-ligand interaction (Table II). The adsorbent is then washed and reequilibrated prior to further use. It should be noted at this juncture that the chromatographic process consists of two distinct entities, adsorption and desorption, each of which should be considered quite separately in the design of affinity separations.

TABLE I

Applications of affinity chromatography

Protein purification
  Enzymes   1, 2
  Antibodies and antigens   3, 4
  Binding and transport proteins   5, 6
  Receptor proteins   7, 8
  Complementary and synthetic peptides and proteins   9, 10
  Repressor proteins   11
  Concentration of dilute protein solutions
  Storage of otherwise unstable proteins in immobilized form

Separative procedures
  Cells and cell membranes   12, 13, 14
  Viruses and phages   15, 16, 17
  Nucleic acids, nucleotides and derivatives   18, 19, 20
  Denatured and chemically modified proteins from native protein   21, 22

Analytical
  Enzyme structure and affinity labelled peptides   23, 24
  Enzyme activity probes   25, 26
  Cell surface structure probes   27, 28, 29
  Investigation of kinetic sequences and metabolism   30, 31, 32
  Defined sequence nucleic acids   33, 34, 35

Immobilized enzymes
  Enzyme reactors—degradative and synthetic   36, 37, 38
  Analytical probes   39, 40
  Therapeutic   41, 42

---

1. Cuatrecasas (1972); 2. Wilchek and Jakoby (1974); 3. Neurath et al. (1973); 4. Weintraub (1970); 5. Allen and Majerus (1972); 6. Sica et al. (1973); 7. Olsen et al. (1972); 8. Schmidt and Raftery (1972); 9. Gawronwski and Wold (1972); 10. Hofmann et al. (1966); 11. Tomino and Paigen (1970); 12. Krug et al. (1971); 13. Soderman et al. (1973); 14. Truffa-Bachi and Wofsy (1970); 15. Cuatrecasas and Illiano (1971); Kenyon et al. (1973); 17. Sundberg and Hoglund (1973); 18. Inouye et al. (1973); 19. Shih and Martin (1973); 20. Weith et al. (1970); 21. Massey et al. (1970); 22. Sluyterman and Wijdenes (1970); 23. Wilchek (1970); 24. Wilchek and Givol (1973); 25. Chan (1970); 26. Feldman et al. (1972); 27. Edelman et al. (1971); 28. Katzen and Soderman (1972); 29. Wallach et al. (1972); 30. O'Carra and Barry (1972); 31. Mawal et al. (1971); 32. Trayer and Trayer (1974); 33. Astell and Smith (1972); 34. Fridkin et al. (1974); 35. Jovin and Kornberg (1968); 36. Marshall (1974); 37. Worthington (1974); 38. Messing (1975); 39. Guilbault (1970); 40. Inman and Hornby (1972); 41. Horvath et al. (1973); 42. Sampson et al. (1974).

TABLE II

Methods of elution from affinity matrices

Ligand competition
Inhibitor competition
Allosteric modification
Co-substrate elution
Ionic strength alterations
Solvent changes
Temperature elution
Buffer and/or pH changes
Chaotropic reagents
Electrophoresis

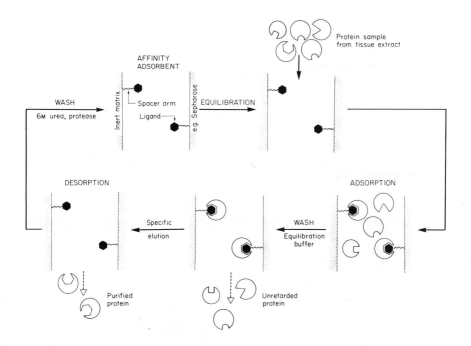

FIG. 1. A diagrammatic representation of the process of affinity chromatography.

Is there a need for affinity chromatography in protein purification? Most conventional protein purifications rely on separations based on small differences between individual proteins. These differences reflect the physical and chemical properties of the protein; molecular charge, size and density, and because each protein is made from the same 20 amino acids these properties vary only slightly. Proteins have a limited stability to extremes of pH and temperature and are sensitive, at all stages of their purification, to trace metals, oxidation and bacterial contamination. The narrow range in ionic charge differences makes separations based on ion-exchange chromatography or zone electrophoresis tedious, whilst precipitation by salt or solvent has limited application. Separations based on protein size, molecular sieving, are only viable in situations where the molecular size of the protein of interest is considerably different from that of the contaminating species. Thus the purification of many proteins by conventional techniques is usually time consuming, uneconomic and generally revolves around the problem of protein denaturation versus the achievement of a satisfactory purification.

III. Selection of Ligand

In an attempt to establish the value of affinity chromatography as a means of protein purification we shall examine in this article the purification of lactate dehydrogenase. The conventional purification scheme for dogfish muscle lactate dehydrogenase is a multistep process taking nearly two weeks to complete; it involves several ammonium sulphate fractionations, both anion and cation ion-exchange chromatography and a gel filtration step. Clearly the advantage of affinity chromatography over this conventional procedure lies in the utilization of the specific biological interaction, and thus the limitations imposed by the lack of appreciable differences in physico-chemical properties are overcome. However, consideration of the reaction sequence of lactate dehydrogenase (Fig. 2) suggests that there are several potential ligands which could be immobilized in the synthesis of a suitable affinity adsorbent.

Two criteria determine the selection of the ligand to be immobilized: firstly, the ligand should have chemically modifiable groups which permit immobilization without impairing the interaction between the macromole-

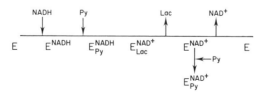

FIG. 2. The ordered reaction sequence of lactate dehydrogenase. E = enzyme, Py = pyruvate and Lac = lactate.

cule and the ligand; and secondly, the ligand should have an affinity for the macromolecule in the region of 10^{-4} to 10^{-8} M in free solution. Many macromolecules have several ligands which satisfy the above requirements and in these instances the reaction sequence should be taken into account. Some enzymes have an ordered reaction sequence where substrates are bound in relation to a coenzyme; in the case of several dehydrogenases the coenzyme binds first, thus it follows that a substrate column will require chromatography in the presence of the coenzyme. These binding characteristics assist one both in selecting the ligand and in determining the chromatographic conditions of adsorption and desorption. In the case of lactate dehydrogenase both the substrate and the coenzyme can be used as ligands, and as we shall see below, each has its own advantages in individual chromatographic separations of this enzyme.

An analogue of pyruvate, oxamate, can be immobilized to agarose (Fig. 3) and used as an affinity ligand for the purification of lactate dehydrogenase (Fig. 4). The ordered reaction mechanism of lactate dehydrogenase makes it

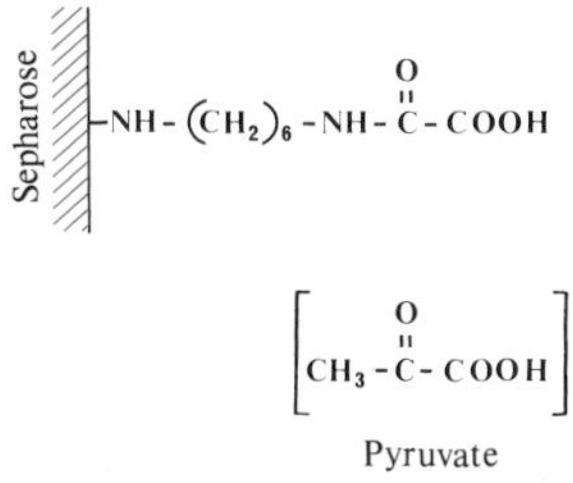

$$\text{Sepharose} \quad -NH-(CH_2)_6-NH-\overset{\overset{\displaystyle O}{\|}}{C}-COOH$$

$$\left[\;CH_3-\overset{\overset{\displaystyle O}{\|}}{C}-COOH\;\right]$$

Pyruvate

FIG. 3. The structure of (6-aminohexyl)-oxamate-Sepharose.

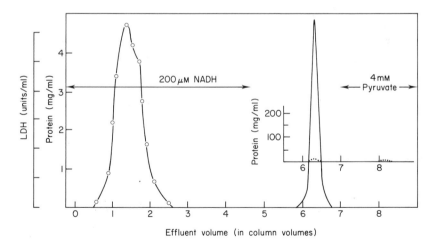

Effluent volume (in column volumes)

FIG. 4. The separation of lactate dehydrogenase from a crude placental extract on immobilized oxamate. —— LDH; —O— protein (or ....). Reproduced with permission from O'Carra and Barry (1972).

necessary to include NADH in the irrigant buffer so that the conformational change in the protein structure required for the binding to immobilized oxamate can be achieved. Elution of the enzyme is simply effected by removing NADH from the chromatographic buffer (O'Carra and Barry, 1972). This is an example of "double affinity chromatography", in that whilst all biomolecules with an affinity for oxamate will presumably bind to the adsorbent, only those requiring the presence of NADH, to effect binding, will be eluted when the nucleotide is removed. Substrate affinity adsorbents, like immobilized oxamate, have a limited potential in that they are usually only applicable to one specific protein; thus the concept of group-specific affinity adsorbents has been developed on the principle that the immobilized ligand should be common to a variety of enzyme reactions.

Group-specific affinity matrices whilst lacking in the selectivity of substrate affinity matrices have numerous advantages as listed in Table III. However, an inherent disadvantage to group-specific affinity matrices, and even more so to substrate affinity matrices, is that many of the interactions between biomolecule and its complementary ligand involve comparatively small molecules. With small molecules the chemical positions available for substitution and immobilization are limited and this can result in the biomolecule having a decreased affinity for the derived ligand. This parameter is particularly relevant to dehydrogenases since they can tolerate little or no change in the structure of their non-nucleotide substrates without a significant decrease in their affinities. In this context lactate dehydrogenase is one of the few dehydrogenases which may be purified using immobilized substrate.

Nicotinamide adenine dinucleotide (Fig. 5) has numerous chemically distinguishable groups, capable of independent substitution, through which it may be coupled to an inert matrix. The five most widely studied substitutions are: (a) N^6 on adenine (Craven et al., 1974 a, b; Lindberg et al., 1973; Barry and O'Carra, 1973); (b) C-8 on adenine (Lee et al., 1974; Trayer et al., 1974); (c) via phosphate esters (Harvey et al., 1974a; Trayer et al., 1974; Berglund and Eckstein, 1972); (d) via ribose using the methods of Lamed et al. (1973); and (e) nicotinamide substitution (Woenckhaus and Vutz, 1974). Each of the methods used for immobilization of NAD^+ is

TABLE III

Advantages of group-specific affinity matrices

Matrix applicable to the purification of more than one biomolecule.
The chemistry of ligand immobilization is restricted to one ligand as opposed to a whole range of substrates and inhibitors for each individual biomolecule.
The range of dissociation constants with respect to immobilized ligand is broad enough to achieve a separation of several like biomolecules through elution, using a gradient of competing ligand or alternative eluant.
Many proteins possess group-specific binding sites and therefore will be susceptible to purification on such matrices.

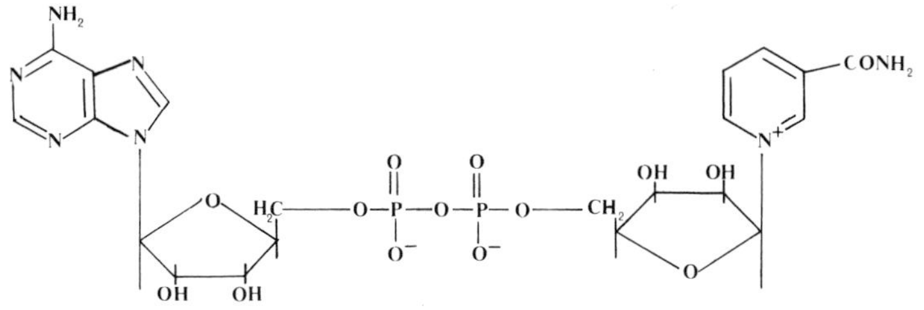

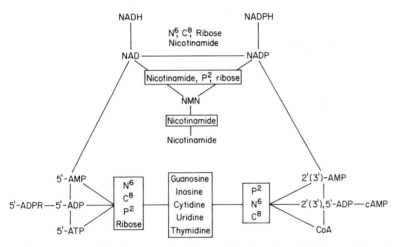

FIG. 5. The structure of nicotinamide adenine dinucleotide (NAD$^+$).

FIG. 6. The interrelationship of mononucleotides and dinucleotides and their possible modes of immobilization.

applicable to the coupling of the respective mononucleotides (Fig. 6). Similarly, the chemical substitutions employed in the preparation of adenosine mononucleotide adsorbents can be utilized in the immobilization of the whole range of purine and pyrimidine nucleotides and nucleosides (Fig. 6). In this context it should be noted that it is not essential for the whole coenzyme to be used as an affinity ligand. Thus N^6-(6-aminohexyl)5'-AMP-Sepharose has been successfully used to purify lactate dehydrogenase in a two-step purification scheme (Kaplan et al., 1974). Even greater selectivity is achieved when the ligand is immobilized to an agarose-polyacrylamide copolymer (Ultrogel AcA 34); Fig. 7 illustrates the one-step purification of lactate dehydrogenase from a crude dogfish muscle homogenate (Doley et al., 1976).

IV. SYNTHESIS OF AFFINITY ADSORBENTS

One of the principal limitations to the application of affinity chromatography is the intricate and complex chemistry required for the synthesis of

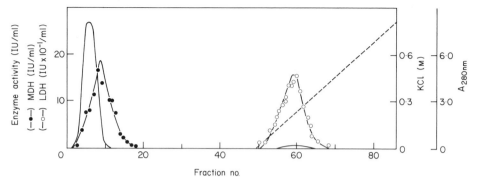

FIG. 7. The separation and purification of lactate dehydrogenase from dogfish muscle on N⁶-AMP-Ultrogel AcA-34. (———) protein; (●) malate dehydrogenase; (○) lactate dehydrogenase and (– – –) KCl concentration.

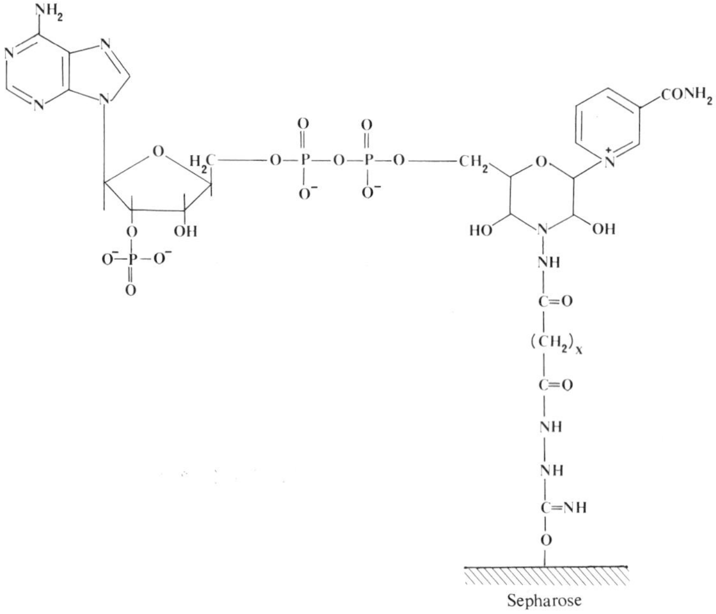

FIG. 8. The structure of ribose-immobilized $NADP^+$-Sepharose.

affinity adsorbents. Recently, a comparatively simple procedure has been devised (Lamed et al., 1973; Wilchek and Lamed, 1974) for the immobilization of nucleotides via the ribose moiety (Fig. 8). An inherent advantage of this method is the elimination of a charged group at the matrix surface, which one encounters on coupling amines to cyanogen bromide activated polymers; the resultant adsorbent is thus free of any associated "detergent" effects (see Wilchek, 1974; Dean and Harvey, 1975). Unfortunately, adsorbents prepared in this manner are liable to ligand leakage (Table IV), although it has been shown that this can be overcome by multipoint attachment (Wilchek and Miron, 1974) or periodate oxidation of the agarose

TABLE IV

Stability of ribose-immobilized NMN

Incubation time (days)	NMN lost/g gel (µMoles)	NMN bound/g gel (µMoles)	NMN bound (%)
0	—	10·1	100
4	·35	9·66	95·6
5	·53	9·57	94·7
7	·65	9·45	93·6
10	·75	9·35	92·6
13	·87	9·23	91·4

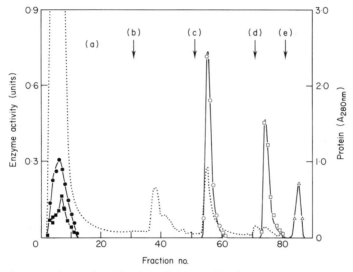

FIG. 9. The separation and purification of three NADP⁺-dependent enzymes on NADP-Sepharose. ...... protein; (●) phosphoglucose mutase (o) 6-phosphogluconate dehydrogenase. (■) phosphoglucose isomerase; (□) glutathione reductase (△) glucose 6-phosphate dehydrogenase.

Eluants: (a) 50 mM KH₂PO₄; 1 mM MgCl₂; 0·5 mM Na₄P₂O₇ — KOH pH 7·5 (350 ml).
        (b) 50 mM glycine: 1 mM MgCl₂; 0·5 mM Na₄P₂O₇ — KOH pH 9·0 (210 ml).
        (c) 50 mM Na₄P₂O₇; 1 mM MgCl₂ — HCl pH 9·0 (210 ml).
        (d) 50 mM Na₄P₂O₇; 1 mM MgCl₂; 1 M KCl-HCl pH 9·0 (110 ml).
        (e) 50 mM NADP⁺ (5 ml pulse) in buffer (d).

matrix and subsequent reduction, with sodium borohydride, of the intermediate Schiff base (Parikh *et al.*, 1974).

The ligand leakage noted in Table IV is more prevalent at room temperature; however, this does not preclude the use of these adsorbents in protein purification at 0°–4°C. Figure 9 illustrates the separation of three NADP⁺-dependent enzymes from the 45–70% $(NH_4)_2SO_4$ fraction of a sheep liver homogenate. The two marker enzymes employed in this study, phosphoglucose mutase and phosphoglucose isomerase, were located in the void volume of the column together with 97% of the applied protein. The subsequent elution profile illustrates several of the principles involved in the desorption phase of affinity chromatography (Table II). The pH of the effluent buffer was changed to effect the elution of further non-specific protein and the three NADP⁺-dependent enzymes were eluted by (i) *a competitive inhibitor*—pyrophosphate. elution of 6-phosphogluconate dehydrogenase; (ii) *salt*-KCl elution of glutathione reductase and finally (iii) *competitive ligand*-NADP⁺ elution of glucose-6-phosphate dehydrogenase. The principle enzyme under examination in this experiment was 6-phosphogluconate dehydrogenase and the above elution sequence resulted in a 52-fold increase in specific activity activity plus a 81% recovery of the protein. In this instance elution was effected by a simple change of irrigant buffer, further purification (to homogeneity) can be achieved by applying a gradient of the eluent across the same affinity adsorbent (Griffiths *et al.*, 1977).

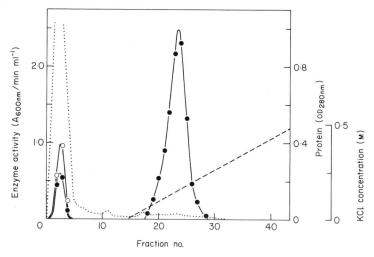

FIG. 10. Separation and purification of NAD(P)H-linked diaphorases from *Scenedesmus obliquus* on NADP⁺-Sepharose. (o) NADH- and (●) NADPH-linked diaphorases; (.....) protein; and (———) KCl concentration.

This principle is illustrated in Fig. 10 which shows the purification and separation of two diaphorase activities from the 45–75% (NH₄)₂SO₄ fraction of a *Scenedesmus obliquus* extract. The diaphorase exhibiting both NADH and NADPH activities was located in the void volume of the column, whereas the NADPH-linked diaphorase was retained by NADP⁺-Sepharose and eluted with a 150-fold purification by a linear gradient of KCl (Harvey and Dean, 1976).

V. ELUTION PROCEDURES

The range of elution procedures which can be employed in the desorption of enzymes from group-specific adsorbents compensates for the obvious decreased specificity during the adsorption stage. The relationship between the adsorbent and the specific elution technique is shown in Table V. Affinity adsorbents utilizing the biospecificity of single protein interactions, e.g. antibody–antigen; steroid receptors; and substrate–enzymes, will not require a sophisticated elution procedure and desorption can be achieved by low pH and/or chaotropic reagents. However, with such adsorbents an extremely high affinity can be observed between the macromolecule and the immobilized ligand and in these instances elution is effected by competing free ligand. In direct contrast, the interaction of a macromolecule with either hydrophobic or ion-exchange adsorbents is generally completely non-specific. Specific elution in these instances is usually unpredictable and requires, in the case of ion-exchange resins sophisticated gradients of inhibitors or substrates (Pogell, 1962; Carminatti et al., 1969).

The immobilization of a ligand common to a group of enzymes in several different ways has two inherent advantages. Firstly, each enzyme will not recognize the same portion of the ligand and this will enhance the purification of individual enzymes and, secondly, some insight into the nature of each individual enzyme-ligand interaction can be obtained. Two affinity adsorbents prepared by immobilizing 5'-AMP are illustrated in Fig. 11. When the nucleotide is coupled through the N^6 of the adenine the phosphate group is available for enzymic interaction, whereas this group is used in the immobilization procedure for the preparation of P^2-ADP-Sepharose, hence leaving the adenosine moiety of the nucleotide free for enzymic interaction. The affinity of several kinases and dehydrogenases for these two adsorbents is presented in Table VI. These results hopefully reflect the nature of the enzyme–nucleotide interaction and suggest that while the adenosine moiety is essential in the binding of myokinase and glyceraldehyde 3-phosphate dehydrogenase it has a quite different role in the interaction of 5'-AMP with yeast alcohol dehydrogenase and glycerokinase. Due to the different ligand concentrations of the two adsorbents, plus the additional phosphate group

TABLE V

The relationship between the mechanism of adsorption and the specificity of the elution procedure

Elution procedure	Adsorbent			
	Receptors immuno adsorbents substrates	Group specific	Hydrophobic or detergent	Ion exchange
Competing ligand or substrate		✓	✓	✓
Salt effects		✓	✓	
Temperature		✓		
pH	✓	✓		
Chaotrophic	✓			

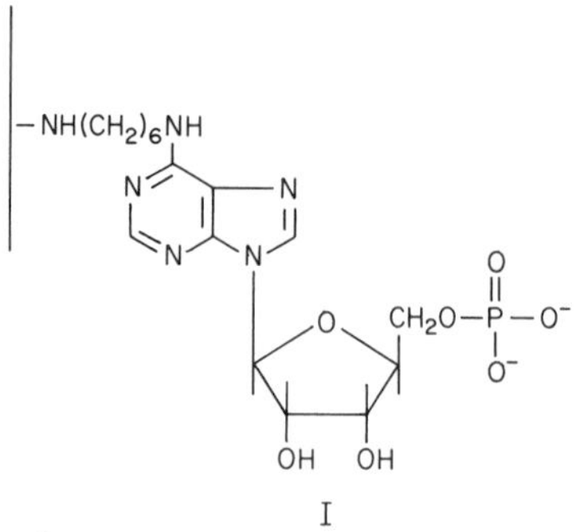

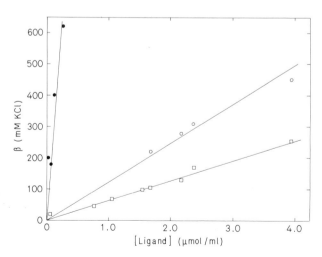

I

N^6-(6-Aminohexyl) 5'-AMP-Sepharose

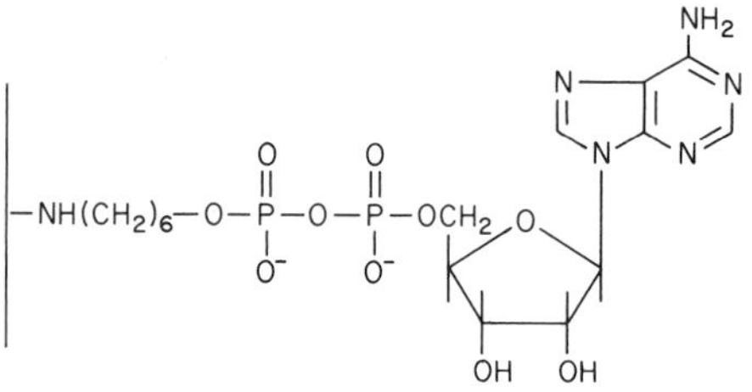

II

P^1-(6-Aminohexyl)-P^2-(5'-adenosine)-pyrophosphate-Sepharose

FIG. 11. The structure of (I). N^6-(6-aminohexyl)-5'-AMP-Sepharose and (II). P^1-(6-amino-hexyl)-P^2-(adenosine)-pyrophosphate-Sepharose.

FIG. 12. The effect of ligand concentration on the binding (β) of three enzymes to N^6-(6-aminohexyl) 5'-AMP-Sepharose (o) malate dehydrogenase; (●) lactate dehydrogenase (H_4) and (□) glycerokinase. Reproduced with permission from Harvey *et al.* (1974a).

TABLE VI

The effect of position of ligand immobilization on enzyme binding

Enzyme	Binding β N^6-AMP	P^2-AMP
Hexokinase (yeast)	0	0
Creatine kinase (rabbit muscle)	0	0
Alcohol dehydrogenase (yeast)	400	0
Glycerokinase (*Candida mycoderma*)	122	0
D-glucose 6-phosphate dehydrogenase (yeast)	0	170
D-glyceraldehyde 3-phosphate dehydrogenase (rabbit muscle)	0	1M^a
Myokinase (rabbit muscle)	0	380
Lactate dehydrogenase (pig heart)	1M^a	1M^a
Lactate dehydrogenase (rabbit muscle)	1M^a	1M^a
3-Phosphoglycerate kinase (yeast)	70	260
Pyruvate kinase (rabbit muscle)	100	110
Malate dehydrogenase (pig heart)	65	490

^{a} Elution could not be affected by 1 M KCl, enzyme eluted by a 200 μl pulse of 5×10^{-3} M NADH.

β The KCl concentration required to elute the enzyme with maximal activity.

Ligands: N^6-(6-aminohexyl) 5′-AMP-Sepharose (1·5 μmoles AMP/ml).

P^1-(6-aminohexyl)-P^2-(5′adenosine)-pyrophosphate-Sepharose (6·0 μmoles AMP/ml).

TABLE VII

Effect of ligand on the binding of rabbit muscle glyceraldehyde 3-phosphate dehydrogenase

Ligand concentration (μmoles/ml sepharose)	β(mM KCl)	Bound enzyme (%)
N^6-(6-aminohexyl)5′-AMP-Sepharose		
1·3	140	20
1·7	185	47
2·2	305	100
2·4	350	100
P′-(6-aminohexyl-P^2-(5′-adenosine)-pyrophosphate-Sepharose)		
1·2	200	54
6·0	1000^a	100
N^6-(6-aminohexyl)-NAD-Sepharose		
2·0	1000+	100

^{a} Elution could not be effected by 1 M KCl, enzyme eluted by a 200 μl pulse of 5×10^{-3} M NADH.

β The KCl concentration required to elute the enzyme with maximal activity.

in the preparation of P^2-ADP-Sepharose, interpretation of such data must be made cautiously.

It could be implied from the data presented in Table VI that a stronger reaction exists between malate dehydrogenase and P^2-ADP-Sepharose than with N^6-AMP-Sepharose. However, on increasing the concentration of

TABLE VIII

The interaction of various enzymes with group-specific affinity adsorbents

Ligand	Enzymes	References
AMP	ADH, GDH, GK, G3PDH, G6PDH, GP, βOHBDH, ISCDH, LDH, MDH, PGK, PK, TDH	1, 2, 3, 4, 5, 6, 7 8, 9, 10, 11, 12, 13 14, 15, 16, 17
ADP	G3PDH, G6PDH, LDH, MDH, MK, PGK, PK, UGDH	18, 16, 17
ATP	ATPases, AK, GAK, GK, HK, PGK	19, 20, 21
2'5'ADP	GDH, G6PDH, GR, 6PGDH	4
cAMP	HisK, PrK	22, 23
GTP	GDH, NTPS	24, 25
UDP	UGDH	17
NMNH	ADH, LDH	26
NAD⁺	ADH, GDH, G6PDH, G3PDH, GR, βOHBDH, ISCDH, LipDH, LDH, MDH, TDH	2, 5, 7, 27, 28, 9 29, 30, 10, 21, 11, 13 31
NADP⁺	AR, GDH, G6PDH, GR, ISCDH, LDH, 6PGDH	32, 33, 33a, 8, 30, 10, 21 34, 35
Blue dextran/ Cibacron Blue	ADH, G3PDH, G6PDH, LDH, MDH, PFK, PK, 6PGDH, βOHSDH	36, 37, 38

Enzyme abbreviations:
ADH alcohol dehydrogenase, AR aldehyde reductase, AK L-arabinose kinase, GAK D-galactose kinase, GDH Glutamate dehydrogenase, G6PDH D-glucose 6-phosphate dehydrogenase, G3PDH glyceraldehyde 3-phosphate dehydrogenase, GK glycerokinase, GP glycogen phosphorylase, GR glutathione reductase, βOHBDH β-hydroxybutyrate dehydrogenase, HK hexokinase, HisK histone kinase, βOHSDH 3(17)β-hydroxysteroid dehydrogenase, ISCDH isocitrate dehydrogenase, LipDH lipoamide dehydrogenase, LDH lactate dehydrogenase, MDH malate dehydrogenase, MK myokinase, NTPS dihydroneopterintriphosphate synthetase, PFK phosphofructose-kinase, PGK phosphoglycerate kinase, PK pyruvate kinase, PrK protein kinase, 6PGDH 6-phosphogluconate dehydrogenase, TDH threonine dehydrogenase, UGDH UDP-glucose dehydrogenase.

1. Anderson et al. (1974); 2. Barry and O'Carra (1973); 3. Brodelius and Mosbach (1973); 4. Brodelius et al. (1974); 5. Comer et al. (1975); 6. Craven et al. (1974a); 7. Craven et al. (1974b); 8. Kaplan et al. (1974); 9. Lee et al. (1974); 10. Lowe et al. (1972); 11. Lowe et al. (1973b); 12. Lowe et al. (1974); 13. Mosbach et al. (1972); 14. Ohlsson et al. (1972); 15. Sorensen and Wang (1975); 16. Trayer and Trayer (1974); 17. Trayer et al. (1974); 18. Harvey et al. (1974a); 19. Anderton et al. (1973); 20. Chan and Hassid (1975); 21. Lowe et al. (1973a); 22. Severin et al. (1974); 23. Wilchek et al. (1971); 24. Godinot et al. (1974); 25. Jackson et al. (1973); 26. Woenckhaus and Vutz (1974); 27. Grover and Hamms (1974); 28. Hocking and Harris (1973); 29. Lindberg et al. (1973); 30. Lowe and Dean (1971); 31. Visser and Strating (1975); 32. De Flora et al. (1973); 33. Griffiths et al. (1977); 33a. Lamed et al. (1973); 34. Lowe and Mosbach (1974); 35. Tabahoff and von Wartburg (1975); 36. Easterday and Easterday (1974); 37. Heyns and de Moor (1974); 38. Thompson et al. (1975).

TABLE IX

Parameters in affinity chromatography

Matrix
  Innocuous
  Uniform, porous structure
  Chemically adaptable
  Physically and chemically stable to experimental procedure
Extension arms
  Chemical nature
  Length
  Hydrophobic and "detergent" chromatography
Ligand
  Selection
  Position of immobilization
  Concentration
    Capacity
    Uniform distribution
    Non-uniform distribution
Column bed dimensions
  Volume
  Diameter
  Length
Batchwise versus column
Flow rate
Enzyme concentration
Protein concentration—other than enzyme under consideration
Conditions of adsorption and desorption
  pH effects
  Temperature effects
  Non-specific desorption
  Specific desorption
  Double affinity chromatography

---

immobilized N6-AMP it was noted that the strength of the enzyme–ligand interaction (β) increased linearly (Fig. 12). Similar observations have been recorded for glyceraldehyde 3-phosphate dehydrogenase (Harvey and Dean, 1976): the capacity of the adsorbent and also the strength of the enzyme–ligand interaction increase at higher ligand concentrations (Table VII). Significantly, when the complete cofactor, NAD^+, was immobilized through the N^6-position the resulting adsorbent had a greater affinity for glyceraldehyde 3-phosphate dehydrogenase than either N^6-AMP or P^2-ADP-Sepharose. Thus the investigator, by selection of either the ligand or its position of immobilization, can effect the purification of an enzyme on a narrow range of group-specific matrices. The potential of group-specific nucleotide matrices in the separation and purification of enzymes is illustrated in Table VIII, where it can be seen that the investigator has numerous ligands at his disposal to separate a wide variety of enzymes.

The previous observations on the interaction of enzymes with immobilized nucleotides with respect to position of immobilization and ligand concentration, serve to illustrate one of the basic parameters involved in the design of affinity chromatography adsorbents. Table IX lists some of the other parameters which should be taken into account both in the design of the affinity adsorbent and also in the chromatographic procedure. The application of various inert materials as support matrices and the role of the spacer arm in affinity chromatography have recently been reviewed (Scouten, 1974; May and Zaborsky, 1974; Dean and Harvey, 1975). The parameters associated with the adsorption/desorption stages of the chromatographic procedure, illustrated in this text by the isolation of 6-phosphogluconate dehydrogenase on $NADP^+$-Sepharose, have been enumerated elsewhere (Dean et al., 1974; Harvey et al., 1974b).

ACKNOWLEDGEMENTS

We would like to acknowledge the help of Dr G. Dietz and Miss S. G. Doley and thank them for allowing us to cite some of their results. The Science Research Council and the Ministry of Defence funded parts of this work.

REFERENCES

Allen, R. H. and Majerus, P. W. (1972). J. biol. Chem. **247**, 7702.

Anderson, L., Jörnvall, H., Åkeson, Å. and Mosbach, K. (1974). Biochim. biophys. Acta **364**, 1.

Anderton, B. A., Hulla, F. W., Fasold, H. and White, H. A. (1973). FEBS Letts. **37**, 333.

Astell, C. R. and Smith, M. (1972). Biochemistry **11**, 4114.

Barry, S. and O'Carra, P. (1973). FEBS Letts. **37**, 134.

Berglund, O. and Eckstein, F. (1972). Europ. J. Biochem. **28**, 492.

Brodelius, P. and Mosbach, K. (1973). FEBS Letts. **35**, 223.

Brodelius P., Larsson, P-O. and Mosbach, K. (1974). Europ. J. Biochem. **47**, 81.

Carminatti, H., Rozengurt, E. and Jimenez de Asua, L. (1969). FEBS Letts. **4**, 307.

Chan, P. H. and Hassid, W. Z. (1975). Analyt. Biochem. **64**, 372.

Chan, W. W. C. (1970). Biochem. biophys. Res. Commun. **41**, 1198.

Comer, M., Craven, D. B., Harvey, M. J., Atkinson, A. and Dean, P. D. G. (1975). Europ. J. Biochem. **51**, 201.

Craven, D. B., Harvey, M. J., Lowe, C. R. and Dean, P. D. G. (1974a). Europ. J. Biochem. **41**, 329.

Craven, D. B., Harvey, M. J. and Dean, P. D. G. (1974b). FEBS Letts. **38**, 320.

Cuatrecasas, P. (1972). Adv. Enzymol. **36**, 29.

Cuatrecasas, P. and Illiano, G. (1971). Biochem. biophys. Res. Commun. **44**, 178.

Dean, P. D. G. and Harvey, M. J. (1975). Proc. Biochem. **10** (7), 5.

Dean, P. D. G., Craven, D. B., Harvey, M. J. and Lowe, C. R. (1974). Adv. exp. Med. Biol. **42**, 99.

De Flora, A., Giuliano, F. and Morelli, A. (1973). Ital. J. Biochem. **22**, 258.

Doley, S. G., Harvey, M. J. and Dean, P. D. G. (1976). FEBS Letts. **65**, 87.

Easterday, R. L. and Easterday, I. M. (1974). Adv. exp. Med. Biol. **42**, 123.

Edelman, G. M., Rutishauser, U. and Millette, C. F. (1971). Proc. natn. Acad. Sci. U.S.A. **68**, 2153.

Feldman, K., Zeisel, H. and Helmreich, E. (1972). *Proc. natn. Acad. Sci. U.S.A.* **69**, 2278.

Fridkin, M., Cashion, P. J., Agarwal, K. L., Jay, E. and Khorana, H. G. (1974). *In* "Methods in Enzymology" (Jakoby, W. B. and Wilchek, M., eds), Vol. 34, p. 645. Academic Press, New York and London.

Gawronski, T. H. and Wold, F. (1972). *Biochemistry* **11**, 442.

Godinot, C., Julliard, J. H. and Gautheron, D. C. (1974). *Analyt. Biochem.* **61**, 264.

Griffiths, P., Houlton, A., Adams, M. J., Harvey, M. J. and Dean, P. D. G. (1977). *Biochem. J.* **161**, 561.

Grover, A. K. and Hamms, G. G. (1974). *Biochim. biophys. Acta* **356**, 309.

Guilbault, G. G. (1970). *In* "Enzymatic methods of Analysis" Pergamon Press, New York.

Harvey, M. J. and Dean, P. D. G. (1976). (unpublished observations.)

Harvey, M. J., Lowe, C. R., Craven, D. B. and Dean, P. D. G. (1974a). *Europ. J. Biochem.* **41**, 335.

Harvey, M. J., Craven, D. B., Lowe, C. R. and Dean, P. D. G. (1974b). *In* "Methods in Enzymology" (Wilchek, M. and Jakoby, W. B., eds), Vol. 34, p. 242. Academic Press, New York and London.

Heyns, W. and de Moor, P. (1974). *Biochim. biophys. Acta* **358**, 1.

Hocking, J. D. and Harris, J. I. (1973). *FEBS Letts.* **34**, 280.

Hofmann, K., Smithers, M. J. and Finn, F. M. (1966). *J. Am. chem. Soc.* **88**, 4107.

Horvath, C., Sardi, A. and Woods, J. S. (1973). *J. appl. Physiol.* **34**, 181.

Inman, D. J. and Hornby, W. E. (1972). *Biochem. J.* **129**, 255.

Inouye, H., Fuchs, S., Sela, M. and Littauer, U. Z. (1973). *J. biol. Chem.* **248**, 8125.

Jackson, R. J., Wolcott, R. M. and Shiota, T. (1973). *Biochem. biophys. Res. Commun.* **51**, 428.

Jovin, T. M. and Kornberg, A. (1968). *J. biol. Chem.* **243**, 250.

Kaplan, N. O., Everse, J., Dixon, J. E., Stolzenbach, F. E., Lee, C-Y., Lee, C-L. T., Taylor, S. S. and Mosbach, K. (1974). *Proc. natn. Acad. Sci. U.S.A.* **71**, 3450.

Katzen, H. M. and Soderman, D. D. (1972). *In* "The Role of Membranes in Metabolic Regulation" (Mehlman, M. A. and Hanson, R. W., eds), p. 205. Academic Press, New York and London.

Kenyon, A. J., Gander, J. E., Lopez, C. and Good, R. A. (1973). *Science* **179**, 187.

Krug, F., Desbuquois, B. and Cuatrecasas, P. (1971). *Nature New Biol.* **234**, 268.

Lamed, R., Levin, Y. and Wilchek, M. (1973). *Biochim. biophys. Acta* **304**, 231.

Lee, C-Y., Lappi, D. A., Wermuth, B., Everse, J. and Kaplan, N. O. (1974). *Arch Biochem. Biophys.* **163**, 561.

Lindberg, M., Larsson, P-O. and Mosbach, K. (1973). *Europ. J. Biochem.* **40**, 187.

Lowe, C. R. and Dean, P. D. G. (1971). *FEBS Letts.* **14**, 313.

Lowe, C. R. and Mosbach, K. (1974). *Europ. J. Biochem.* **49**, 511.

Lowe, C. R., Mosbach, K. and Dean, P. D. G. (1972). *Biochem. biophys. Res. Commun.* **48**, 1004.

Lowe, C. R., Harvey, M. J., Craven, D. B. and Dean, P. D. G. (1973a). *Biochem. J.* **133**, 499.

Lowe, C. R., Harvey, M. J., Craven, D. B., Kerfoot, M. A., Hollows, M. E. and Dean, P. D. G. (1973b). *Biochem. J.* **133**, 507.

Lowe, C. R., Harvey, M. J. and Dean, P. D. G. (1974). *Europ. J. Biochem.,* **41**, 347.

Marshall, D. L. (1974). *Adv. exp. Med. Biol.* **42**, 345.

Massey, V., Komani, H., Palmer, G. and Elion, G. B. (1970). *J. biol. Chem.* **245**, 2837.

Mawal, R., Morrison, J. F. and Ebner, K. E. (1971). *J. biol. Chem.* **246**, 7106.

May, S. W. and Zaborsky, O. R. (1974). *Sep. and Purif. Methods* **3**, 1.

Messing, R. A. ed. (1975). *In* "Immobilized Enzymes for Industrial Reactors" Academic Press, New York and London.

Mosbach, K., Guilford, H., Ohlsson, R. and Scott, M. (1972). *Biochem. J.* **127**, 625.

Neurath, A. R., Prince, A. M. and Lippin, A. (1973). *J. gen. Virol.* **19**, 391.

O'Carra, P. and Barry, S. (1972). *FEBS Letts.* **21**, 281.

Olsen, R. W., Meunier, J. C. and Changeux, J. P. (1972). *FEBS Letts.* **28**, 96.

Ohlsson, P., Brodelius, P. and Mosbach, K. (1972). *FEBS Letts.* **25**, 234.

Parikh, I., March, S. and Cuatrecasas, P. (1974). *In* "Methods in Enzymology" (Wilchêk, M. and Jakoby, W. B., eds), Vol. 34, p. 77. Academic Press, New York and London.

Pogell, B. M. (1962). *Biochem. biophys. Res. Commun.* **7**, 225.

Sampson, D., Han, T., Hersh, L. S. and Murphy, G. P. (1974). *J. surg. Oncol.* **6**, 39.

Schmidt, J. and Raftery, M. A. (1972). *Biochem. biophys. Res. Commun.* **49**, 570.

Scouten, W. H. (1974). *International Lab.* **Dec**, p. 13.

Severin, E. S., Kochetkov, S. N., Nesterova, M. V. and Gulyaev, N. N. (1974). *FEBS Letts.* **49**, 61.

Shih, H. Y. and Martin, M. A. (1973). *Proc. natn. Acad. Sci. U.S.A.* **70**, 1697.

Sica, V., Nola, E., Puca, G. A., Parikh, I. and Cuatrecasas, P. (1973). *Fedn. Proc.* **32**, 1297.

Sluyterman, L. A. E. and Wijdenes, J. (1970). *Biochim. biophys. Acta.* **200**, 593.

Soderman, D. D., Germershausen, J. and Katzen, H. M. (1973). *Proc. natn. Acad. Sci. U.S.A.* **70**, 792.

Sorensen, D. B. and Wang, P. (1975). *Biochem. biophys. Res. Commun.* **67**, 883.

Starkenstein, E. (1910). *Biochem. Z.* **24**, 210.

Sundberg, L. and Hoglund, S. (1973). *FEBS Letts.* **37**, 70.

Tabahoff, B. and von Wartburg, J. P. (1975). *Biophys. biochem. Res. Commun.* **63**, 957.

Tomino, S. and Paigen, K. (1970). *In* "The Lac Operon" (Zipser, D. and Beckwith, J., eds), p. 223. Cold Spring Harbour Laboratory for Quantitative Biology.

Thompson, S. T., Cass, K. H., and Stellwagen, E. (1975). *Proc. natn. Acad. Sci. U.S.A.* **72**, 699.

Trayer, I. P. and Trayer, H. R. (1974). *Biochem. J.* **141**, 775.

Trayer, I. P., Trayer, H. R. Small, D. A. P. and Bottomley, R. C. (1974). *Biochem. J.* **139**, 609.

Truffa-Bachi, P. and Wofsy, L. (1970). *Proc. Natn. Acad. Sci. U.S.A.* **66**, 685.

Visser, J. and Strating, M. (1975). *Biochim. Biophys. Acta* **384**, 69.

Wallach, D. F. H., Kranz, B., Ferber, E. and Fischer, H. (1972). *FEBS Letts.* **21**, 29.

Weintraub, B. (1970). *Biochem. Biophys. Res. Commun.* **39**, 83.

Weith, H. L., Wiebers, J. L. and Gilham, P. T. (1970). *Biochemistry* **9**, 4396.

Wilchek, M. (1970). *FEBS Letts.* **7**, 161.

Wilchek, M. (1974). *Adv. exp. Med. Biol.* **42**, 15.

Wilchek, M. and Givoi, D. (1973). *In* "Peptides" (Nesvadba, H., ed.), p. 203. North Holland Publishing Co., Amsterdam.

Wilchek, M. and Jakoby, W. B. (eds) (1974). *In* "Methods in Enzymology" Vol. 34, p. 3. Academic Press, New York and London.

Wilchek, M. and Lamed, R. (1974). *In* "Methods in Enzymology" (Wilchek, M. and Jakoby, W. B., eds.), Vol. 34, p. 475, Academic Press, New York and London.

Wilchek, M. and Miron, T. (1974). *In* "Methods in Enzymology" (Wilchek, M. and Jakoby, W. B., eds.), Vol. 34, p. 72. Academic Press, New York and London.

Wilchek, M., Salomon, Y., Lowe, M. and Selniger, Z. (1971). *Biochem. biophys. Res. Commun.* **45**, 1177.

Woenckhaus, C. and Vutz, H. (1974). *H.S.Z. Physiol. Chem.* **355**, 1271.

Worthington, C. C. (1974). *Adv. exp. Med. Biol.* **42**, 235.

CHAPTER 14

Subcellular Fractionation Techniques in Enzyme Distribution Studies

R. M. LEECH

Department of Biology, University of York, York, England

I. Isolation of Subcellular Components .	. . . . .	290
A. Choice of Plant Material	. . . . . .	292
B. Tissue Homogenization Methods .	. . . . .	295
C. Cellular Fractionation Procedures	. . . . .	296
II. Non-aqueous Methods of Subcellular Fractionation	. . .	310
III. Other Methods of Cellular Fractionation .	. . . .	311
A. Thin Layer Counter-current Distribution	. . . .	311
B. The Use of Sephadex Columns for Particle Separation	. .	312
C. Electrophoretic Methods of Separation .	. . . .	312
IV. Characterization of Subcellular Components	. . . .	313
A. Assessment of Contamination and Organelle Damage	. .	313
B. Assessment of Organelle Heterogeneity in Enriched Fractions of One Organelle .	. . . .	320
C. Assessment of Enzyme Activity .	. . . . .	322
References .	. . . .	323

The first attempts to isolate and characterize subcellular components were made almost 100 years ago. One particularly notable achievement in plant studies at that time was by Engelman in 1881 who isolated green fragments from algal cells and demonstrated that they were able to produce oxygen when illuminated; this work was brilliantly extended by Hill in his classical quantitative studies (1937, 1939). The great impetus to analytical subcellular studies came more recently with the development of differential centrifugation procedures innovated notably, by Albert Claude and his associates in the 1940s and the parallel development of analytical electron microscopical techniques introduced by Palade and Sjöstrand in the 1950s. By the dual application of these techniques to animal cells it became possible to isolate populations of subcellular organelles and examine their structural and metabolic characteristics in situations uncomplicated by the presence of other cellular components. As a result, great progress has been made in the study of subcellular enzyme localization, and a few standard tissues, notably rat liver cells, can be described in considerable detail.

Much of our understanding of the mechanisms underlying the function of plant cells is also based on studies of isolated subcellular fractions, although the heterogeneity of the populations of isolated organelles, and damage suffered to them during isolation are still major problems which remain only partly resolved for many plant tissues. It is now abundantly clear that the naive application of a standard isolation procedure to a new tissue, or even to the same tissue at a different stage of development, can frequently lead to erroneous conclusions about enzyme location, since no two biological tissues respond identically to homogenization and cell fractionation procedures. With plant cells these problems can be magnified since the two major difficulties in the successful isolation of plant cell organelles concern (a) the effective breakage of the cell wall and (b) the minimization of the effects of disruption of the vacuole and the mixing of its contents with the components of the cytoplasm. Both the physical properties of plant cell walls and the composition of cellular vacuoles vary dramatically from one plant tissue to another and in different states of development, so it is necessary to consider the method best suited to individual tissues and to each problem of enzyme location.

At the present time, not only biochemical cytologists but also plant ecologists, plant chemists and plant physiologists are becoming increasingly interested in carrying out enzyme distribution studies as they attempt to explain ecological adaptations and physiological responses at the cellular level. It therefore seems appropriate to review the problems and difficulties attendant on cell fractionation studies and to attempt to evaluate the extent to which such procedures have now become standardized. The present article discusses the parameters which should be taken into account in the selection and modification of isolation procedures and the criteria which should be used to assess the success of a method in relation to the biological problem it is used to examine. No attempt will be made to catalogue the variety of procedures available, particularly as these far outnumber the workers in the field. Since my own practical experience is with the problems of isolation of various plastid components, the salient principles will be mainly illustrated with examples from procedures for plastid isolation, in particular the chloroplasts of higher plants, but recent innovations for the isolation of other organelles of higher plants will also be reviewed.

I. Isolation of Subcellular Components

In the investigation of the location of subcellular enzyme activities the ultimate goal is the preparation of homogeneous suspensions containing only one species of organelle or membrane system and the assay of these suspensions under controlled conditions facilitating maximum activation of the enzyme(s) of interest. The organelles must be changed as little as possible both structurally and metabolically from their intracellular state and no other cellular material should be present in the suspensions. Provided the organelles, such as chloroplasts, are isolated, and can be shown to be clean and morphologically and functionally intact, subsequent osmotic rupture during enzymic assay will allow entry of substrates (Lilley and Walker, 1973, 1974; Lilley *et al.*, 1973, 1974) and enzyme digestions can also be carried out at this stage to remove adsorbed proteins or particles, such as ribosomes, in a controlled manner (Ellis, 1976). If a "balance sheet" for the whole cell is needed, then clean suspensions of each cell organelle or membrane system need to be assayed and corrections made for the yield of each organelle fraction (compared with its concentration in the cell). In practice, this situation is rarely achieved and a variety of compromise methods are commonly adopted in which corrections are made for the presence of contaminating organelles by using "marker enzymes" and additional corrections made for organelle breakage as assessed by quantitative light or electron microscopical examination.

Several steps are involved in the procedure adopted to establish a subcellular enzyme location. They may be considered in sequence as follows:

A. the choice of a suitable plant tissue and assessment of its physiological function;

B. the choice of a method of tissue homogenization;

C. the choice of a cellular fractionation procedure;

D. methods for the assessment of (i) the contamination and (ii) the structural and functional damage to the isolated organelles;

E. the choice of an enzyme assay procedure;

F. the choice of a parameter by which to express the enzyme activity measurements;

G. the assessment of the relationship between the enzyme activity measurements and cellular function.

The method adopted early in the procedure will modify the choices available during later stages. For example, the final assessment of the *in vitro* activity in terms of intracellular function depends on methods having been used to measure the levels of enzyme activity in the isolated organelles, which enable their activity to be referred back to the physiological activity of the tissue from which the organelles were removed. The choice of a particular cellular fractionation procedure will also influence the enzyme assay procedures adopted, and the validity of the criteria which can be used to quantify the results. Many ingredients of the isolation medium can affect the subsequent enzyme assays and in particular the increasingly common practice of including bovine serum albumin in the isolation medium negates any possibility of expressing enzyme results in terms of mg of protein. Such inter-relationships will be referred to where necessary, but for convenience each stage in the experimental procedure will be considered in turn.

1. *In Enzyme Activity Studies*

Examples abound which demonstrate clearly that the successful solution of a problem of enzyme localization depended directly on the suitability of the tissue chosen for investigation. When the problem of interest concerns the biochemical characteristics of a synthetic mechanism, then clearly a tissue of choice is one showing high physiological activity of the phenomena under investigation, but it is also most important that the tissue should be one from which enzymically active organelles can be quickly and easily isolated. At the start of a new investigation, an extended survey of potentially suitable tissues may well be rewarding. In studies of the C4-photosynthesis the soft leaved grasses such as *Digitaria* (Black, 1973) and dicotyledons such as *Amaranthus* and plants with low phenol content (Coombs *et al.*, 1973a, b, c; 1974), has proved far more amenable to detailed biochemical investigations than the fibrous mature leaves of maize and sugar cane in which the C4 photosynthetic pathway was originally discovered (Hatch and Slack, 1966; Kortschak *et al.*, 1965; Black, 1973; Laetsch, 1974). Another example is the choice of young castor bean tissues by Beevers (1975) for the investigation of the conversion of fat to glucose; a choice which led to the successful isolation and characterization of the microbodies named glyoxysomes. Young roots of corn and onion are rich in dictyosomes and plasma membranes and an excellent source of preparations of these membranes (Van der Woude *et al.*, 1974; Leonard and Van der Woude, 1976). Mitochondria are traditionally isolated from germinating shoots (Hallaway, 1965; Millard *et al.*, 1965; Opik, 1968) and nuclei from such sources as pear (Tautvydas, 1971) and soybean hypocotyl (Chen *et al.*, 1975). Significantly, the few plant species which yield chloroplasts capable of active prolonged CO_2 photoreduction in isolation are almost without exception food crops, notably the pea (*Pisum-sativum*) and spinach (*Spinacia oleracea*) in which the vacuolar pH of the leaf cells is near neutrality and so cell rupture causes least damage to the organelles (Walker, 1967). The variety of plant may be very important, for example the cv. True Hybrid 102 Arthur Yates variety of spinach gives routinely higher rates of CO_2 fixation than other spinach varieties (Lilley and Walker, 1975; Lilley *et al.*, 1975).

2. *In Ecological Studies*

In other situations the aim of the investigation may be concerned with the biochemical basis for specific physiological responses to environmental treatments and here the obvious choice for experimental tissue may be the wrong one, i.e. to use for biochemical investigations the plant in which the physiological response was initially demonstrated. A far better procedure in the long term is to establish whether the physiological response can also be elicited in a species which has been proved to be amenable to sophisticated biochemical investigations. Many attempts to relate ecological findings to biochemical changes have foundered because the plants used have yielded either totally inactive cellular fractions or cell homogenates with such high phenolic content, acidity or viscosity as to be unsuitable for further investigation. In contrast, the outstanding achievement of Wildman's group, in successfully relating their genetical and ecological observations to the biochemical observations on the structure and functions of Fraction I protein, was in part, a result of his choice of the tobacco plant as the experimental material, initially chosen because it was a very good biochemical source of Fraction I protein and subsequently examined genetically (Kung, 1976).

3. *In Developmental Studies*

Where temporal changes are of great importance, such as in the study of development, additional considerations need to be borne in mind in the choice of tissue for investigation. If the comparative biochemistry of cells at different stages of development is attempted, then the characteristics of the young cells act as controls for the changes manifested at later stages in the development. It is most important that the age of the tissue and the state of its development are not confused. An example may illustrate this point. In the study of plastid development, etiolated leaves from plants grown for several days in the dark are subsequently illuminated to induce chloroplast development, and the biochemical changes in the leaves followed for periods up to 92 hours in the light. At the end of the experiment the etiolated illuminated plants are therefore four days older than at the beginning, yet frequently the control tissue is harvested from the dark grown plants at the start of the illumination phase and not, as should be correctly done, at its end. In developmental studies, it is particularly useful if the samples at different developmental stages can be taken from the same organ or tissue at the same time, rather than from different batches of plants grown and harvested at different time intervals. In roots and also in some monocotyledonous leaves (Leech et al., 1973; Robertson and Laetsch, 1974; Stocking, 1975), the cells are produced largely from an intercalary meristem so that in both organs they are in linear array with the youngest cells nearest the meristem and the oldest furthest away. Samples of the cells taken from tissue slices from successively more acropetal sections will therefore give a sequence of developing cells. Such a natural system comes nearest to approaching the advantages of an artificial continuous cell culture undergoing more or less synchronous development and from which successive examples can be taken and analysed. Cultures from which uniform populations of cells of differing ages can be sampled and analysed clearly have considerable advantages as the starting material for studies of organelle development. Moore and Beevers (1974) fractionated cells taken from such suspension cultures of soy bean at different developmental stages. On linear sucrose gradients, clear peaks of enzyme activities were found corresponding to endoplasmic reticulum, dictyosomes, mitochondria and microbodies and the sedimentation behaviour of the isolated subcellular particles changed as the culture aged.

In comparative studies, the choice of parameters for the expression of the enzyme activities is very important (see later). In the changing developmental (or senescing) situation all the usual bases for comparison are changing themselves, e.g. fresh weight, dry weight and protein content, so the only sound basis for comparison is either the cell or the organelle itself. Enzyme activities expressed on a per cell basis or a per plastid basis for example, are then directly comparable from one stage of development to another.

Another factor which becomes particularly accentuated in developmental studies is the problem of the heterogeneity of the cells in the starting material. In earlier work, attention was only infrequently paid to this problem. For example, the mitochondria in a suspension isolated from a tissue containing many cell types are of multiple origin and may well not be identical. The differences in the chloroplasts from palisade and from mesophyll cells of the same leaf has also been only infrequently considered (McClendon, 1952, 1953). Recently, elegant procedures have been published which involve initial dissection of the tissue into areas of more uniform cell type; the separated samples of tissue are then used for subsequent organelle preparation. Such techniques have been particularly valuable in the comparative studies of the enzyme complements of bundle sheath and mesophyll cells (Kanai and Edwards, 1973a, 1973b) and also chloroplasts (Anderson *et al.*, 1971) from plants showing C4 metabolism. Prior mechanical separation of the two types of cell or of the protoplasts derived from them, gives a relatively homogeneous cell population from which cellular components can be subsequently isolated.

Isolated protoplasts and also suspension cultures of single cells of higher plants are becoming increasingly used as sources of subcellular fractions. Protoplasts can be rapidly prepared with relative ease from leaf tissue such as tobacco in sufficient quantity for enzymic analysis and the organelles can be removed with minimum mechanical breakage. Soy bean cell suspension cultures (Moore and Beevers, 1974) have also already been used for the successful isolation of subcellular organelles and it is to be expected that considerably more attention will be paid to the biochemistry of the development of these synchronized uniform cell populations in the future.

An alternative solution to the problems encountered when examining the organelles from tissue of several cell types is to preferentially destroy one type of cell. For example, the mesophyll cells of maize and sorghum leaves can be prevented from developing by a brief intense cold treatment early in their development (Slack *et al.*, 1974). After such treatment, in the pale green band of treated tissue seen at the later stages of leaf development, only the bundle sheath cells are normal so their enzyme complement can be compared with the enzyme complement of similar tissue of the same age in which both bundle sheath and mesophyll cells are still functional.

Plants which are to be used for biochemical investigations should be grown under carefully controlled, standardized conditions. Environment chambers in which the illumination, humidity and temperature can be controlled are often preferable to greenhouses: uniform nutrient status should also be maintained. The growth of plants in individual water culture is particularly valuable for the subsequent isolation of active chloroplasts from the leaves (Jensen and Bassham, 1966; Lilley *et al.*, 1975) and nuclei from peas (Tautvydas, 1971). Large diurnal variations in enzyme activities have now been reported from several laboratories (Naguchi and Tamaki, 1962; Steer, 1973), so the time of harvesting of the plants in relationship to the beginning of the light period also needs to be standardized. A factor often overlooked is the genetic make-up of the seeds from which the plants are grown. Genetically homogeneous seed of a pure line should be used wherever this is possible and the biochemical cytology of the plants grown in standard conditions first established: variations in the plants grown under environmental conditions which are more relevant to a specific ecological or physiological situation can then be examined.

B. TISSUE HOMOGENIZATION METHODS

The successful isolation of undamaged cell components depends to a large extent on the method used for the initial homogenization of the tissue from which they are derived. In general, less attention has been paid to the standardization of methods for tissue breakage than to the subsequent subfractionation procedures. (Many methods are an art best learnt from personal demonstration by the originator.) A multiplicity of methods for homogenization exist and have been previously reviewed (Hallaway, 1965; Opik, 1968; Walker, 1971). Many of the early methods originally used for animal tissues or for the extraction of a particular type of macromolecule have in practice proved unsuitable for organelle isolation from plants. Methods involving extensive grinding, pulverizing, osmotic shock, ultrasonic disruption or rapid pressure change are generally too violent and destructive for higher plant tissue and have now been superseded. The method of tissue breakage should be as gentle and as rapid as possible. Experience has shown that mechanical methods based on rapid cutting, chopping or slicing of the tissue are the most likely to lead to the successful isolation of active, undamaged subcellular fractions from plant tissue. Such methods have been particularly valuable in the isolation of glyoxysomes from castor bean endosperm (Kagawa *et al.*, 1973), plasma membrane from stems (Van der Woude *et al.*, 1974) and the isolation of chloroplasts still showing the amoeboid movements so clearly seen *in vivo* (Honda *et al.*, 1962; Wildman *et al.*, 1962).

The objective of tissue breakage is to cut open a proportion of the cells so their contents flow rapidly into the isolation medium. Homogenization for two periods of 15 seconds is usually adequate to release sufficient organelles from soft tissues such as leaves, buds and young roots: extension of the homogenization period is often counter-productive, as the organelles suffer increasing damage with time. The yield of organelles is readily increased by prior rapid chopping of the material before homogenization (Leonard and Van der Woude, 1976). A modification of this approach in which the tissue is pressed through the mesh of a fine sieve has proved successful with particularly soft tissue, such as avocado pear mesocarp (Weiare and Kekwick, 1975) and in the isolation of nuclei (Lyndon, 1963) and plasma membranes (Leonard and Van der Woude, 1976). Homogenization of several batches of tissue is much better than bulk isolation. For particularly small quantities, hand chopping of the tissue with razor blades is often the method of choice and the blades can be used singly or several mounted together on a metal holder.

Two mechanized chopping methods incorporating sharp razor blades have recently proved most useful for the isolation of plastids. The first is a modification of the electric carving knife from which the blade is removed and replaced by a holder set at right-angles containing two one-sided razor blades. Operation of the modified knife gives a rapid chopping action of each blade independently. A more sophisticated mechanical version incorporating the moving blade principle is the Yeda Press recently marketed by Yeda, Research and Development Co. Ltd, Rehovet, Israel, which is reported to be very successful in operation. To date, the most commonly used homogenizers have been modified liquidizers fitted with basal rotating blade assemblies, such as the Waring Blendor and the MSE Atomix. Provided particular care is taken to avoid any heating of the sample, and the blades are frequently replaced or sharpened (about every two to four weeks if the blade is in constant use), this method is good for routine preparations. The utensils should always be pre-cooled before use and the partial freezing of the isolation medium to a frozen slurry of the consistency of "melting snow" is also recommended. The Polytron has recently been favoured in several laboratories for plastid isolation (Schwenn et al., 1973; Lilley et al., 1975; Ellis, 1976).

C. CELLULAR FRACTIONATION PROCEDURES

Cell fractionation procedures are now available for most plant subcellular components which yield biochemically active suspensions enriched in one specific organelle. Recent review articles or methodological papers giving some of the best available techniques for the isolation of each subcellular particle or membrane system are collected together in Table I. In most modern procedures, rapid gentle homogenization and filtration is followed by brief centrifugation to concentrate the organelle of interest. Further enrichment is usually achieved by isopycnic centrifugation through a density gradient. All these procedures (with the notable exception of the method of James and Das (1957) for chloroplast thylakoid isolation) yield considerably enriched fractions but all still contain varying amounts of other cellular material and the organelles themselves are damaged to different degrees. The components of the medium in which the tissue is first homogenized greatly affect the organelles as they are released from the cells so their selection is critical for the successful isolation of active, undamaged organelles.

The characteristics of the cells being examined, the methods chosen for fractionation and the methods chosen for enzyme assay will all influence the choice of an isolation procedure.

TABLE I

Equilibrium densities in sucrose gradients

	g/cm^2	References
Endoplasmic reticulum	1·10–1·12	1, 2, 3
Dictyosomes	1·12–1·15	2, 4
Plasma membranes	1·16–1·18	2, 4, 5
Broken plastids	1·15–1·18	6, 7
Mitochondria	1·17–1·20	8, 6, 7
Intact plastids	1·21–1·24	9, 6, 4, 7
Microbodies	1·19–1·24	8, 9, 6, 7

1. Hill (1939); 2. Leonard and Van der Woude (1976); 3. Lord *et al.* (1973); 4. Moore and Beevers (1974); 5. Van der Woude *et al.* (1974); 6. Miflin and Beevers (1974); 7. Rocha and Ting (1970); 8. Huang and Beevers (1971); 9. Miflin (1974).

1. Isolation Media

In tackling a new problem of enzyme localization, the usual procedure is to select an isolation medium previously used with success in a similar investigation: this is generally an aqueous isolation medium. At first sight, the list of ingredients may appear excessively long and complicated but many of the ingredients will have been selected with extreme care for special reasons. Other ingredients may have been added much more empirically and the guiding principle should always be to use the simplest medium which yields active preparations of a suitable degree of purity for the investigation of the current problem. Osmolarity, ionic strength and pH are frequently the most critical factors in the maintenance of structural and functional integrity of the organelles. Other compounds may have been added because they are apparently beneficial to structural or functional preservation, although their exact mode of action may be unknown, and for a few compounds the reasons for their addition may even have been forgotten over the course of time.

TABLE II

Cellular fraction	Recent isolation methods	Recent reviews	Possible markers	Comments	Problems of isolation
Chloroplasts	Lilley and Walker (1974)	Hall (1972)	Triose-phosphate isomerase chl. RNA (rRNA) Fraction I protein	Intact and active	Mixed populations with different degrees of damage
Chl. thylakoids	James and Das (1957)	—	Δ3t-hexadecenoic chlorophyll	No contamination No stroma	—
Etioplasts	Wellburn and Wellburn (1971) Leese et al. (1972)	Leech (1975)	Protochlorophyllide; carotenoids	Intact	Should be isolated in the dark
Proplastids	Thomson et al. (1972) Miflin and Beevers (1974)	—	Acetolactate synthetase	Little contamination	Probably variably sized particles
Nuclei	Chen et al. (1975)	—	Specific nucleic acids	Can be observed visually	Bounding membrane continuous with endoplasmic reticulum
Mitochondria	Kollöffel (1975)	Leaver and Harmey (1973)	Cytochrome oxidase monoamine oxidase	No satisfactory method for leaves	Bacterial contamination main problem

Peroxisomes	Tolbert (1971a)	Tolbert (1971b)	Catalase	Marker enzyme diagnostic	Very delicate organelles with a single bounding membrane
Glyoxysomes	Huang and Beevers (1971)	Beevers (1975)	Glyoxylate cycle	Marker enzymes diagnostic	
Dictyosomes	Van der Woude et al. (1974)	—	Latent IDPase?	Marker not unequivocal	Enriched fraction only: other membranes present
Plasma membrane	Leonard et al. (1976)	—	Latent IDPase? K$^+$-ATPase, pH 6·5	Neither marker unequivocal	As above
Endoplasmic reticulum	Shore and Maclachan (1975) Leonard et al. (1976)	—	NADH-cyt c reductase insensitive to 1 μM antimycin A		Periodate–phosphotungstic chromic stain diagnostic in vivo and in vitro
Tonoplast	No method	—	—	No marker	Contaminant in other membrane fractions
Cytoplasm (non-aqueous)	Bird et al. (1973)	—	cyt RNA	—	Present as dehydrated fragments

Clearly, if preliminary experiments show that a particular isolation medium is adequate for the investigation in hand, then the next step should be to consider each ingredient in turn to discover the quantitative significance of its addition and the rationale behind its inclusion in the medium. The medium can then be suitably modified taking into account the nature of the new investigation.

Two recent innovations have proved particularly valuable. The introduction of zwitterionic buffers, pioneered by Good has enabled much more precise qualitative and quantitative control of the ionic component of the medium. These buffer compounds are either N-substituted taurines or N-substituted glycines with pKa ranging from 6·15–8·35 (Good et al. 1966). Another frequent addition to modern isolation procedure is bovine serum albumin in concentrations of 0·1% and 5% (w/v) which binds fatty acids and basic proteins (Leese et al., 1972; Miflin and Beevers, 1974): only the purest available commercial fractions should be used, e.g. Cohn Fraction V. Polyvinylpyrrolidone (PVP) may also be added as it acts as an absorbant of phenolics and tannins released on cell rupture (Hanson et al., 1965; Hulme et al., 1965; O'Neal et al., 1972), and also may act as a general membrane preservative. The addition of only one extra component to a medium may make the difference between the isolation of inactive and active organelles. For example, the addition of polyvinylpyrrolidone (PVP) led to the first successful isolation of active mitochondria from apple fruit tissue (Jacobson, 1968), and the addition of dithiothreitol (DTT) made possible the demonstration of the presence of NADPH-dependent malic dehydrogenase in mesophyll chloroplasts of C4 plants (Johnson and Hatch, 1970). Sulphydryl protectants are in general valuable additions to isolation media.

Several ingredients added to isolation media as potential membrane preservatives have the effect of sticking organelles together so they coprecipitate and cannot readily be reseparated, even by gradient centrifugation. Bovine serum albumin, Ficoll (Miflin and Beevers, 1974; Leonard and Van der Woude, 1976) and polyvinyl pyrolidone all have this effect if added to media. Their addition to isolation media should always be carefully considered and avoided if possible.

Ideally the effect of additional ingredients to isolation media should be tested at several different concentrations, alone and in combination with other components. This would generally involve an unacceptable period of preliminary experimentation and in most instances only a few critical components are rigorously tested. It is worth remembering that the effects of isolation media on gross structural changes in the larger organelles such as nuclei and plastids and on the cells from which they are derived can be rapidly monitored by light microscopy (Albertsson and Baltscheffsky, 1963; Leech, 1968; Hall, 1972; Chen et al., 1975). Tests on functional preservation are of course a great deal lengthier.

The history of the experiments designed to determine the rates of CO_2 reduction by isolated chloroplasts from Calvin-cycle plants will serve as an example to illustrate that attention to the details of isolation procedure can be rewarding. Intact leaves of temperate plants such as spinach or pea, grown under standard growth room conditions, reduce CO_2 to sugar at rates of c. 100 μmoles/mgchl/h (Lilley and Walker, 1975). In early experiments the chloroplasts from similar leaves isolated in 0·35 MNaCl would reduce CO_2 at rates no greater than 1–5 μmoles/mgchl/h. Little improvement in the rate of CO_2 reduction *in vitro* was found until Walker in 1964 (Walker, 1964), using pea chloroplasts, obtained rates of 25 μmoles CO_2/mgchl/h. He used suspensions isolated in 0·3 M sorbitol (instead of sodium chloride or sucrose) in which a higher proportion of the plastids had intact outer membranes (Walker, 1965). These chloroplasts apparently still retained many enzymes of the Calvin-cycle (Walker, 1967), in particular, the soluble carboxylating enzyme ribulose bis phosphate carboxylase and also, presumably, the translocating systems recently shown to be located in the chloroplast envelope membrane (Werdan and Heldt, 1972; Heber, 1974). The rapid removal of the chloroplasts from the cytoplasmic and vacuolar fluids was critical in the success of Walker's experiments and the brief homogenization and rapid isolation procedure he used was designed to recover a small number of whole chloroplasts in a short time. Chloroplasts were isolated at pH 6·7 and assayed at pH 7·5. Further increases in the rates of photosynthetic CO_2 reduction were later obtained by the addition of Calvin-cycle intermediates during fixation (Walker *et al.*, 1967; Walker, 1973) and the final demonstration that isolated chloroplasts could assimilate CO_2 at rates equivalent to the intact tissue came from the work of Jensen and Bassham (Jensen and Bassham, 1966; Bassham and Jensen, 1967). The exclusion of all except catalytic concentrations of phosphate from their media, the use of a zwitterionic buffer and sodium pyrophosphate and the hydroponic culture of the plants were critical factors contributing to the demonstration of high rates of CO_2 reduction recorded in Jensen and Bassham's experiments. The role of phosphate in the activation and inhibition of the carboxylating reaction is particularly critical and the sensitivity of CO_2 fixation to very small changes in phosphate concentrations has been made clear by the careful studies of Walker and his co-workers who, have studied the role of phosphate and have found that the CO_2 fixation is sensitive to very small changes in phosphate concentration (Cockburn *et al.*, 1967, 1968). The close similarity of the CO_2 rates in intact cells and isolated chloroplasts established firmly that the chloroplast is the site of the complete photosynthetic process in green plants (Walker, 1970; Walker and Crofts, 1970).

2. *Fractionation Procedures*

(a) *Differential centrifugation procedures.* Differential centrifugation is the most frequently used method for the separation of cellular components from a tissue homogenate. This technique separates particles according to their size, shape and density but primarily according to their size. In the classical standard procedures an initial short run at low speed (e.g. 50–$200 \times g$) removes whole cells and cell walls. Nuclei are usually collected at $c.\ 500 \times g$. Successive supernatants are then centrifuged at 1000–$5000 \times g$ to sediment populations enriched in plastids, at $10\ 000$ to $25\ 000 \times g$ for fractions enriched with mitochondria, between $25\ 000 \times g$ and $50\ 000 \times g$ for membrane fractions and upwards of $100\ 000 \times g$ for pellets rich in ribosomes (Hallaway, 1965). These are only general divisions and the choice of centrifugal force varies considerably from one method to another. Each fraction collected as a pellet is a heterogeneous population containing several types of organelle, membranes and particles, but enriched in one type of component. The fraction is also a mixed population with respect to the organelle present in highest concentration since while some of the particles may remain undamaged, many will be fragmented to varying degrees. Further purification of the "mitochondrial fraction" yields microbodies (glyoxysomes and peroxisomes) and further fractionation of the "membrane pellet" yields plasma membrane fractions. As previously mentioned, it is now clear that speed of removal of the organelle from the discharged cytoplasmic and vacuolar material is critical in the preservation of membrane function. Many recently published methods eliminate the initial low speed spin and substitute a filtration procedure in which filtration through a thick pad of fibrous cloth (Walker, 1971), nylon bolting silk (Ridley and Leech, 1968; Leese et al., 1972), miracloth (Hawke et al., 1974) or cheese cloth (Leonard and Van der Woude, 1976) is used to remove the cell debris. Rapid sedimentation of a fraction enriched in the organelle of interest then follows. Procedures of this type are used for plastid and mitochondrial isolation. In general the choice is between a rapid procedure yielding a highly active suspension of organelles of limited purity and a longer procedure in which contaminating organelles are removed but with some sacrifice in yield or in activity of the organelles.

The degree of purification required in an organelle fraction depends on the function of the organelle which is being experimentally examined. For example, if the phosphorylation mechanism of isolated mitochondria is investigated, suspensions with high rates of oxidation of citric acid cycle intermediates and good P:2e ratios are required and some contamination of the suspensions with cell membrane fragments and ribosomes is acceptable. When the protein synthesizing capacity of mitochondria is examined, however, purification procedures need to be much more rigorous. Since the protein synthetic capacity of the mitochondrial suspensions accounts for only 1% of the tissue synthesis, the possibility that the mitochondrial activity reflects the presence of cytoplasmic ribosomes or even bacteria in the suspensions must be seriously considered. In such circumstances a longer isolation procedure involving density gradient purification to yield a preparation completely free of extraneous non-mitochondrial nucleic acid species would be mandatory.

(b) *Gradient centrifugation.* Working procedures which involve resuspension of the pellet in medium and resedimentation are found to be relatively ineffective either for particle enrichment or for the removal of cellular contaminants (Leech and Ellis, 1961; Miflin and Beevers, 1974). Pellets prepared by differential centrifugation can be most effectively further concentrated in the particle of interest, and a higher proportion of the extraneous cellular material removed, by some form of gradient centrifugation. Equilibrium centrifugation in a density gradient is known as isopycnic centrifugation. Organelles accumulate in the centrifuge tube at a position corresponding to their solvated equilibrium density at a unique position in the density gradient where all effective centrifugal forces on them vanish. The gradient may be set up before centrifugation or may be formed on application of a centrifugal field. Such methods have been used with success for the isolation of glyoxysomes (Huang and Beevers, 1971; Moore and Beevers, 1974), chloroplasts (Leech, 1964; Rocha and Ting, 1970; Miflin and Beevers, 1974), dictyosomes (Morré and Mollenhauer, 1964; Van der Woude *et al.*, 1974), mitochondria (Moore and Beevers, 1974; Kollöffel *et al.*, 1975), and plasma membranes (Leonard and Van der Woude, 1976).

In present day studies the requirement is for a suspension of functional organelles and the degree of contamination or morphological preservation which is acceptable will vary from one investigation to another. Different methods of plastid isolation will afford examples of the types of procedure which are available and their value for different investigations. Photosynthetically active chloroplasts for example must be prepared extremely rapidly and loss of activity inevitable on purification is unacceptable in kinetic experiments. When examining the photoreactions of isolated thylakoids it is regarded as academic to consider as relevant any activity of mitochondrial membranes which are still present. It is regarded as more important to ensure that the thylakoids are free of stroma and envelope membranes and that the thylakoids themselves are in a uniform state of preservation. However, in other studies where similar enzyme activities are suspected to be located in several different cellular organelles, the exact organelle complement of a suspension is critical. For more rigorous purification continuous and or discontinuous density gradient centrifugation is used. In practice, discontinuous gradients appear to be far more effective for the separation and concentration of plant organelles than continuous gradients, spun to equilibrium. The best discontinuous gradient for a particular purpose can only be discovered from the results of preliminary experiments in which the density and the depth of the layers, the centrifugal force and the time of centrifugation are varied. Again it should be emphasized that organelles from different species of plant or different tissues at different developmental stages often differ greatly in size, weight and density—the criteria used for selective sedimentation in gradient centrifugation. Each tissue and each fractionation problem needs to be considered individually. Recent procedures for the isolation of chloroplasts and other cell fractions will be considered in turn.

(c) *Organelle purification*

(i) *Procedures for the removal of contaminants from chloroplast suspensions.*
For many enzyme location studies plastid preparations free from cell
contamination are required. Mitochondria are the most prevalent contaminating particle in chloroplast fractions prepared by differential centrifugation
(Leech and Ellis, 1961). An early method which effectively removed mitochondria from chloroplast suspensions involved centrifugation of an enriched
plastid fraction through a density gradient made up of glycerol and sucrose
(James and Das, 1957). Electron microscopical examination of the cleaned
up plastid fraction revealed the complete absence of mitochondria and
membranes but the plastids themselves were devoid of their binding
envelopes and stroma and were merely naked and stripped thylakoid systems
(Leech, 1968). These clean, naked thylakoids have been very useful in analysis
of enzymes of the thylakoids such as NADP-glutamate dehydrogenase
(Kirk and Leech, 1972) and photophosphorylation (Leech, 1966) but are of
no value in studies of the biochemistry of the complete chloroplast. For these
latter investigations, chloroplasts with their bounding envelope still intact
are required.

One simple, quick method of cleaning up plastid preparations in use
routinely in our laboratory is to spin the resuspended pellet from a fast
differential centrifugation through a 10 ml, 10 cm band of 0·6 M sucrose
(Leese *et al.*, 1972). This removes 95% of the broken chloroplasts and virtually all the mitochondria and smaller membranes, and the pellet can be
rapidly resuspended and used for enzyme assay. Isopycnic centrifugation of
leaf homogenates and root homogenates yields enriched plastid fractions
which appear to be contaminated with both microbodies and mitochondria.
This is almost certainly because the equilibrium densities of plastids, microbodies and mitochondria are very similar. The apparent equilibrium densities in sucrose density gradients of a variety of subcellular fractions are given
in Table I. Separation of the larger particles, such as plastids, can be achieved,
if advantage is taken of their higher sedimentation coefficients by using short
centrifugation times and discontinuous gradients. None of the particles are
centrifuged to equilibrium by these methods but they are particularly useful
for the separation of plastids which approach their equilibrium density
relatively quickly. Suspensions consisting of virtually 100% intact chloroplasts with few broken chloroplasts and less than 5% cellular contamination
can be collected at the interface of a 46%/50% (w/v) discontinuous sucrose
density gradient in which depth of the sucrose band is approximately 6 cm
(Leech, 1964). The chloroplast layer to be purified is placed on the top of the
gradient and centrifugation at $1000 \times g$ for ten minutes is sufficient to give
good separation. This method has been successfully scaled up for use in
large batch preparations using the zonal centrifuge (Price and Hirvonen,
1966). Photosynthetic CO_2 fixation rates have been reported for these
chloroplasts (maximum reported is 40 micromoles/mgchl/h), and the

chloroplasts have been used extensively in enzyme location studies (Kirk and Leech, 1972; Rathnam and Das, 1974) in photosynthetic control investigations (West and Wiskich, 1968) and as a starting material for the isolation of chloroplast envelope membranes. There are indications that spinning the plastids through the sucrose interface causes considerable disruption and this may account for the rather low yields of intact plastids obtained by this method. Isopycnic centrifugation using a gradient of the silica sol Ludox AM and polyethylene glycol yields very much more active preparations of chloroplasts capable of rates of photosynthetic O_2 evolution and CO_2 fixation of the order 80 µmoles/mgchl/h (Morgenthaler *et al.*, 1974). The selection of this particular silica sol is critical since previous materials used have proved to be toxic.

Two more recent methods also take advantage of the fact that despite similar equilibrium densities, the sedimentation coefficients of plastids and microbodies or mitochondria vary considerably. In a method devised by Rocha and Ting (1970) sedimentation velocity and isopycnic density gradient centrifugation are combined in sequence, but up to 13% contamination with mitochondria and microbodies still remained in the plastid suspension. Miflin and Beevers (1974) have published a more sophisticated method of plastid purification in which the tissue homogenate is applied directly to a semi-linear sucrose gradient and centrifuged only briefly. After 15 minutes centrifugation, only the plastids have reached their equilibrium density: the other organelles band preferentially higher up the gradient. The criteria used to assess the efficiency of this technique were maximum recovery of plastid marker enzymes (triosephosphate isomerase and NADP-triosephosphate dehydrogenase) and minimum contamination with other organelles as shown by their marker enzymes. Catalase and cytochrome c oxidase were used as marker enzymes for microbodies and mitochondria respectively, and the plastid fractions showed 2% contamination with these organelles. The method has been used successfully to confirm findings on the location of nitrate reductase and other enzymes of amino acid biosynthesis in the plastids of roots and leaves (Miflin, 1974), and to confirm the location of NADPH-isocitric dehydrogenase in chloroplasts. No electron microscopical description of the Miflin and Beevers' preparations is so far available.

An ever-present problem in organelle studies is the continuing progressive change taking place in the organelle subsequent to isolation from the cell and at all stages during fractionation procedure and plastids are no exception. The extremely brief centrifugation times currently in vogue are an attempt to reduce this complication to a minimum. Even during short centrifugal spins starch grains are dragged through the chloroplast envelope so the starch content of leaves should always be reduced to a minimum before plastid isolation. Subjection of organelles responsive to osmotic changes to sucrose density gradient centrifugation also affects the characteristics of the isolated particles and there is some evidence that passage across the interface between two discontinuous bands of differing density can also cause membrane rupture. A less commonly appreciated hazard is the problem of electrostatic charging of centrifuge tubes, particularly those made of polycarbonate. Plastid envelopes are dragged away from those chloroplasts which come into contact with the surfaces of such tubes and therefore cellulose nitrate tubes should always be used for density gradient centrifugation when good yields of clean plastids are required.

(ii) *The isolation of etioplasts.* Morphologically well-preserved etioplasts were first isolated using differential centrifugation procedures in the early 1960s by Klein and Poljakoff (1960). More recent procedures rely on the additions of bovine serum albumin to the isolation medium (Jacobson, 1968) and the time for isolation can be considerably shortened if the filtration and brief rapid centrifugation used successfully for chloroplast isolation is adopted. The Sephadex filtration method of Wellburn and Wellburn (1971) yields excellent preparations of intact etioplasts (from *Arena*) with virtually no cytoplasmic contamination. Once isolated, etioplasts seem to be remarkably robust and are readily freed from mitochondrial and membrane contamination. Prolamellar body dispersion and chlorophyll formation can be followed on illumination of isolated etioplasts (Wellburn and Wellburn, 1973; Horton and Leech, 1975a, 1975b), and phytochrome activity has recently been located in the envelopes of etioplasts isolated by differential centrifugation and by the Sephadex method (Evans and Smith, 1976).

(iii) *The isolation of etiochloroplasts.* On the illumination of etiolated leaves, prolamellar body dispersion and chlorophyll synthesis immediately occur in the plastids which are now referred to as etiochloroplasts (Leech, 1976). The Sephadex separation method has been applied to the isolation of these plastids and fractions isolated by the same procedure as used for etioplasts differ quantitatively and qualitatively in their protein complement (Cobb and Wellburn, 1973). No electron microscopical investigation of these fractions has been published.

(iv) *The isolation of proplastids.* In 1969 it was stated that as proplastids had not been isolated, nothing was known of their enzymic complement. Since then procedures for their isolation have been published (Thomson *et al.*, 1972; Miflin and Beevers, 1974) and limited examinations have been made of their enzymic activity. Nitrite reductase and acetolactate synthetase have been shown to be located in the purified proplastids of pea roots (Miflin, 1974). They also possess glutamine synthetase and glutamic dehydrogenase.

(v) *The purification of suspensions of mitochondria.* The first active suspensions of plant mitochondria were prepared from the hypocotyls of mung bean (Millard *et al.*, 1965) and this tissue, the inflorescence of the cauliflower and corn roots are the most common source for plant mitochondria.

The methods developed to 1967 have been ably reviewed by Opik (1968). The use of non-green tissue means that proplastids and their membranes are the primary source of contamination in most mitochondria suspensions, but because they are so difficult to remove the presence of plastids is generally ignored when oxidative reactions of the citric acid cycle and oxidative phosphorylation are being investigated. But the presence of contaminants including proplastids, ribosomes and bacteria cannot be ignored when reactions suspected of being located in more than one cellular site are investigated. In particular, scrupulously purified suspensions are needed for studying protein synthesis. The best recent method which employs a discontinuous sucrose gradient was developed by Leaver (unpublished) who used the presence of non-mitochondrial DNA and RNA as indices of contamination. The mitochondrial band was collected at the 1·2 M/1·35 M sucrose interface and shown to contain virtually no bacteria, cytoplasmic or plastid ribosomes or DNA. The fraction had high respiratory activity and high P:2e ratio. A similar gradient procedure has been used successfully to study subcellular arginase distribution by Kollöffel et al. (1975). The methods of Rocha and Ting (1970) and Miflin and Beevers (1974) also utilizing discontinuous density gradients, and already referred to in connection with plastid isolation, apparently yield mitochondrial preparations in which microbody and plastid contamination is considerably reduced. The isolation of mitochondria from leaf tissue freed from thylakoids still remains a problem and the most reliable methods available are probably still those of Pierpoint (see Chapter 10).

(vi) *The isolation of nuclei.* In the first methods for the isolation of plant nuclei (Johnston et al., 1957), non-aqueous techniques were used but these fractions were undoubtedly heavily contaminated with cytoplasm and the nuclei were damaged. Subsequently, aqueous methods have been developed for the isolation of nuclei from cereal embryos and pea epicotyls but early procedures were often lengthy and the yield of nuclei too low to allow extensive biochemical investigations (Rho and Chipchase, 1962; Lyndon, 1963). Two recent methods employing rapid homogenization and filtration, through a series of nylon screens and further purification of the nuclei on discontinuous sucrose gradients appear to have led to a substantial increase in the yield of nuclei from pea stems (Tautvydas, 1971) and soy bean epicotyls (Chen et al., 1975) respectively. The media employ non-ionic buffers, and contain sulphydryl protectants and the methods should be consulted for the most up-to-date published information. For non-enzymic work nuclei from ungerminated wheat embryos can be further purified on Metrizamide gradients (Trewavas, unpublished).

(vii) *The isolation of ribosomes.* Suspensions of cytoplasmic (80S) ribosomes isolated from plant tissue are always contaminated with (70S) ribosomes from mitochondria and plastids of the same tissue (Lyttleton, 1962). Pure fractions of chloroplast ribosomes can however be prepared by treatment of plastid suspensions with ribonuclease. All the RNA and ribosomes external to the intact plastids will be destroyed and after thorough washing, on osmotic rupture of the plastid, ribosomes are released. An up-to-date account of the problems and techniques available for plant ribosomal preparations and assay can be found in Ellis (1976).

(viii) *Microbody isolation (peroxisomes and glyoxysomes)*. The name microbody was originally used to describe a particle in mammalian liver cells distinguished by a boundary of a single membrane: it was not until the late 1960s that microbodies were recognized as a major class of organelles in many plant cells (Tolbert, 1971b). The presence of microbodies in several plant tissues was reported intermittently by electron microscopists but they were not recognized in subcellular fractions since they were ruptured by harsh grinding and their enzymes lost to the cytosol fractions. Microbodies are generally circular in profile, 0·2–1·5 microns in diameter, and have a single bounding membrane, a matrix, and no ribosomes (de Duve *et al.*, 1966; Douglass *et al.*, 1973). They coprecipitate with chloroplasts or mitochondria (Miflin and Beevers, 1974). Two types of plant microbody have been recognized and are distinguished by their enzymic complement. Beevers (1969) discovered microbodies in germinating castor beans and demonstrated their functional role in the glyoxylate cycle and named them glyoxysomes. The microbodies from leaves were given the name peroxisome by Tolbert since their enzyme complement resembled that of the peroxisomes of liver and kidney.

Peroxisomes have been isolated from a number of leaves of both C_3 and C_4 plants and their appearance in isolation resembles very closely the particles in the leaf cells shown by Newcomb and his collaborators (Frederick and Newcomb, 1971). An excellent summary of work in this field is provided by Tolbert (1971b). There are three steps in the isolation of microbodies from plant tissue. Firstly the plant material is gently chopped, preferably with razor blades and the homogenates rapidly centrifuged to pellet a fraction enriched in microbodies. The particles are then layered onto a discontinuous sucrose density gradient in which the microbodies exchange the water from their matrix with the sucrose of the gradient across the highly permeable microbody membrane. The result is that the particles behave as if they had specific densities of $1·24–1·26 \times cm^{-3}$, characteristic of their high protein to lipid content (Huang, 1975). In the gradient, the microbodies become concentrated in a narrow band at the 1·9 M/2·0 M interface on the density gradients. Mitochondria, plastid fragments and membranes are recovered from other parts of the gradient. Tolbert's publications (1971a) give details of peroxisomal isolation and those of Beevers' and co-workers describe the methods of the preparation of glyoxysomes (Cooper and Beevers, 1969a, b; Beevers, 1975).

(ix) *The isolation and characterization of membrane fractions*. One of the current central problems of great interest in membrane biology is determination of the relationship between the physico-chemical properties of membranes and their biochemical functions. Isolation, purification and characterization of each membrane component of the plant cell is a prerequisite for the more detailed studies of these membranes. Supernatants remaining after cell homogenates have been spun at 10 000 × g contain a mixture of fragments of membranes of multiple origin. In addition to fragments derived from cell membranes, those from dictyosomes, mitochondria, nuclei and plastids are also present. Fractions enriched in plasma membranes (Leonard and Van der Woude, 1976) in dictyosomes and in endoplasmic reticulum (Van der Woude *et al.*, 1974), can be isolated from suitable young plant tissue, particularly from stems and roots. All the fractions are heterogeneous and while enriched in one species of membrane, also contain a proportion of other membranes. Diagnostic identification of the isolated membranes depends on specific markers, either specific stains or enzymes which can be shown to be unequivocally restricted to one species of membrane only. Very few reliable marker systems are available at the present time, but one stain which appears to be specific for plasma membranes both *in vivo* and *in vitro* is the periodate phosphotungstic acid-chromic acid stain developed by Roland *et al.* (1972). Thylakoid fragments can always be readily recognized by their chlorophyll content and corrections made for the presence of chlorophyll–containing fragments in all membrane fractions isolated from leaves. A valuable enzyme marker for chloroplast envelopes is their ability to incorporate UDP-galactose into galactolipids (Douce *et al.*, 1973a; Douce, 1974). NAD-cytochrome c reductase activity, insensitive to Antimycin A, at present seems to be a reasonably reliable indication of the presence of endoplasmic reticulum (Shore and Maclachan, 1975; Vigil, 1970; Lord *et al.*, 1972; Lord *et al.*, 1973; McKersie and Thompson, 1975; Leonard and Van der Woude, 1976), but a similar enzyme, sensitive to Antimycin A is present in mitochondria (Leonard and Van der Woude, 1976). K^+-stimulated ATPase activity at pH 6·5 is clearly a plasma membrane enzyme but the activity is not an unequivocal marker since it is also found in other (unidentified) membranes. Cytochrome c oxidase is a reliable marker for mitochondria inner membranes and the outer membrane may be recognized by its monoamine oxidase activity (Opik, 1968). Latent IDPase, originally suggested as a marker of dictyosomes is not specific, since it is found in several different membrane fractions. No enzymic or microscopical method is at present available to enable the tonoplast membrane to be recognized in isolated fractions.

Membrane systems can now be examined in isolation but in general, the present methods only allow a semi-quantitative examination of their enzymic properties and indicate rather than prove specific enzyme locations in the cell. Immunological techniques and the improved fractionation procedures should resolve some of the uncertainties in the future.

(x) *The isolation of Golgi apparatus and dictyosomes* (*individual stacks of cisternae*). Morré *et al.* (1965) first isolated a dictyosome fraction from onion stem. The isolated membranes were still intact and recognizable (electron microscopically) as Golgi cisternae (Morré *et al.*, 1965). Purification of the fraction by density gradient centrifugation resulted in the loss of recognizable stacks of cisternae but addition of 1% (w/v) glutaraldehyde to the homogenization medium stabilized the membranes and after this treatment, highly purified fractions were obtained (Morré, 1970). The fractions were used successfully to monitor enzymic transformations *in vivo* following feeding of radioactively-labelled metabolites to the intact issue (Morré, 1970). For example the purified dictyosomes were shown to have become preferentially labelled *in vivo* with choline $1\text{-}2\text{-}^{14}C$ and acetate-$2\text{-}^{14}C$ and also to incorporate glucose and leucine. After glutaraldehyde fixation these membranes could not be used for *in vitro* enzyme studies but recently Van der Woude *et al.* (1974) have used a new method to isolate a fraction enriched in dictyosome membranes from sucrose density gradients. No glutaraldehyde was used in the preparation and $\beta,1\text{-}4$ glucose synthetase was shown to be associated with the dictyosome fraction, confirming previous findings of Ray *et al.* (1969). This enzymic activity cannot however be used as a marker for dictyosomes since it is also associated with plasma membrane fractions.

II. NON-AQUEOUS METHODS OF SUBCELLULAR FRACTIONATION

Non-aqueous isolation procedures were originally introduced by Behrens and Thalacker (1957) to prepare subcellular components by a method which prevented redistribution of their water soluble components during isolation. Although originally used for the isolation of nuclei, non-aqueous procedures have recently been more frequently used for leaf fractionation. The tissue is rapidly frozen and dried without thawing, ground in non-polar solvents, particularly mixtures of cyclohexane and carbon tetrachloride, and separated on density gradients made up of the same organic solvents in different proportions in the complete absence of water. Independently Heber (1957, 1960) and Stocking (1959), published similar procedures for plastid isolation and subsequently examined their fractions for cytoplasmic contamination (Stocking and Ongun, 1962; Ongun and Stocking, 1965). In most modern modifications (Heber and Willenbrink, 1964; Keys, 1968; Roberts *et al.*, 1970; Stocking, 1971 and Bird *et al.*, 1973), the tissue homogenate is applied directly in the middle or the bottom of a gradient made up of layers differing in density from 1·24–1·36 g/cm. The least contaminated band of chloroplasts from tobacco leaves is bouyant between densities 1·24 and 1·28; for tomato leaf chloroplasts the equivalent location is 1·28–1·32 (Bird *et al.*, 1973). A further small reduction in cytoplasmic contamination can be achieved by removing the "chloroplast" band and spinning it to the top of several subsequent gradients. The most useful marker enzyme originally used was pyruvic kinase, as it was thought to be restricted exclusively to the cytoplasm, the main contaminant of non-aqueous fractions. However, if cytoplasmic ribosomal RNA is used as an alternative index of cytoplasmic contamination in the chloroplast fractions the suspensions are shown to be never less than 15% contaminated. This contamination has proved virtually impossible to remove and can only be reduced to c. 10% by extensive centrifugation through several successive density gradients. In non-aqueous solvents, dehydrated pieces of cytoplasm remain together as fragments and the components do not disperse as in aqueous media (Stocking *et al.*, 1968). This means that all the cytoplasmic organelles and their enzymes are retained together as contaminants of the chloroplast fraction. Non-aqueously isolated chloroplasts have lost many membrane lipids and pigments and are not able to carry out any photoreactions. Many of their soluble enzymes still remain active but because of the high level of cytoplasmic contamination it would be unwise to assign an enzyme to a particulate location on the basis of fractionation using non-aqueous solvents alone. Corroborative evidence, using alternative isolation procedures is always needed.

Notwithstanding the problem of cytoplasmic contamination, the non-aqueous method was used with great success to establish the subcellular distribution of both ribulose bis phosphate carboxylase and also NADP-triosephosphate dehydrogenase. In a carefully designed series of experiments Heber *et al.* (1963) showed conclusively that virtually 100% of the activity of both of these enzymes was located in the chloroplast stroma. At the time when these experiments were carried out this conclusion could not have been reached by using any other method currently available (Smillie, 1963), although the restriction of these enzymes to the chloroplast has subsequently been confirmed using improved aqueous techniques (Jensen and Bassham, 1966). Non-aqueous methods have been used to examine the location of the enzymes of sucrose synthesis (Bird *et al.*, 1965), and nitrogen metabolism (Ritenour *et al.*, 1967; Santarius and Stocking, 1969), and have provided an important alternative method of looking at enzyme locations. In some instances the high levels of cytoplasmic contamination, previously unsuspected, have led to conclusions about enzyme location which have not been supported by other studies using alternative methods of organelle isolation. Since sucrose synthesis has not so far been unequivocally demonstrated in intact suspensions of isolated chloroplasts freed from whole cells, it is possible the small activity of sucrose synthetase and sucrose phosphate synthetase (Bird *et al.*, 1965) demonstrated in non-aqueously isolated chloroplasts may be accounted for as cytoplasmic contaminations.

III. OTHER METHODS OF CELLULAR FRACTIONATION

A. THIN LAYER COUNTER-CURRENT DISTRIBUTION

The adaptation of the analytical technique of counter-current distribution in polymer two phase systems for the separation of cells, organelles and membranes is largely the work of Albertsson's group in Umea, Sweden

(Karlstam and Albertsson, 1969; Larsson and Albertsson, 1974). Albertsson and Baltscheffsky (1963) introduced an adaptation of the technique using thin layers of the two phases, so the time required for complete separation of cell particles could be reduced to a few hours. Using this technique, Larsson *et al.* (1971) were able to separate intact and broken chloroplasts (i.e. Class A and Class C chloroplasts) in two phase systems of aqueous solutions of polyethylene glycol and dextran. The clean separation achieved is largely due to the sensitivity of the procedure to differences in the surface properties of particles and reflects the differences between the thylakoid membrane surface and the external face of the outer chloroplast envelope membrane. (The third class of separated "plastids" were vesicles with plastid, mitochondria and microbody inclusions. Their origin is unknown but it is possible they are produced during the isolation procedure). Treffry (1975) has recently used counter-current distribution to examine the changes in surface properties of the internal membranes of etiochloroplasts as they green in the light. Two main classes of membrane differing in affinity properties in the two phase system can be distinguished and the proportion of these two classes of membrane, changes as the plastid develops. The phase systems (6·3% (w/v) polyethyleneglycol and 6·3% (w/v) dextrose) are made up in 0·2 M sucrose buffered at pH 7·8 to reduce damage to the plastids. No enzyme activity studies have so far been made on organelles separated by counter-current distribution but the length of time required for the separation suggests the technique may be more profitably applied to problems concerned with less labile organelle components.

B. THE USE OF SEPHADEX COLUMNS FOR PARTICLE SEPARATION

Excellent separation of intact etioplasts from oat leaves has been achieved using Sephadex G25 as a macromolecular sieve by Wellburn and Wellburn (1971). The sephadex is used as a slurry and some skill is needed in pouring the leaf homogenate onto the column at the correct rate but the method yields suspensions of well-preserved etioplasts free from contaminants. Its successful use with etiolated leaves of other species has not been reported but it may be possible to adapt the method to separate different morphological types of plastid or even plastids of differing size and shape. It is possible nuclei could also be separated by this method.

C. ELECTROPHORETIC METHODS OF SEPARATION

Although cooled controlled systems are now available for the separation of organelles electrophoretically, these do not seem to have been adapted in studies of plant organelles.

IV. Characterization of Subcellular Components

A. Assessment of Contamination and Organelle Damage

1. *Examination of Isolated Fractions by Light Microscopy*

Larger cell organelles such as nuclei and chloroplasts can be examined relatively easily under the light microscope. This means that the morphological heterogeneity of the particles in a suspension can be quickly and easily assessed. Isolated chloroplasts have been extensively and carefully examined visually and the best method is to view the suspensions by phase contrast microscopy when the chloroplasts still bounded by the double envelope appear spherical and opaque with a shining appearance (Kahn and Wettstein, 1961; Spencer and Ont, 1965; Leech, 1966; Walker, 1967 and Lilley *et al.*, 1975). In contrast ruptured chloroplasts are larger, darker and their margin is rather fuzzy and frequently their membranes can be seen. However, many chloroplasts are always seen with an appearance intermediate between these two and indeed different methods of preparation of the slide for viewing leads to different degrees of breakage. The intact and broken chloroplasts have been given the name Class A and Class C chloroplasts respectively by Hall (1972). In a recent attempt to increase the accuracy of counting proportions of different types of chloroplasts the Quantimet has been used (Lilley *et al.*, 1975), and the suggestion made that chloroplasts may break and reseal during isolation.

Whilst it is certain that intact plastids can exhibit amoeboid characteristics *in vitro* as well as in the cell, it is still not clear whether some additional manifestations of the so-called mobile phase may not reflect the presence of a skin of cytoplasm adhering to and surrounding the isolated plastids. Certainly, the presence of groups of organelles encased in cytoplasm and often surrounded by a membrane are frequently encountered in material aqueously isolated from some coenocytic algae (e.g. *Acetabularia*) with particularly viscous cytoplasm rich in mucopolysaccharides (Leech, 1972 for review) Larsson and Albertsson (1974), using a special blending technique isolated membrane bound vesicles containing mitochondria and microbodies in addition to chloroplasts from spinach leaves. It is probable these assemblies were formed during the fractionation procedure, which included counter-current distribution separations, since only very few similar aggregations were encountered when full-grown spinach leaves were extracted more conventionally.

Gross contamination of fractions with whole cells, can be seen visually and mitochondrial and bacterial contamination is evidenced by the presence of particles in rapid Brownian movement but these cannot be identified by light microscopy. Smaller fragments are beyond the resolution of the light microscope.

2. *Examination of Subcellular Organelle Fractions by Electron Microscopy*

Frequently, the only unequivocal means of characterizing a subcellular fraction morphologically is by viewing thin sections of fixed embedded and pelleted material in the electron microscope. There are few laboratories without access to such facilities these days and it has been proposed that every publication providing evidence for a subcellular enzyme location should include a cytological investigation of the fractions which have been analysed biochemically, and some quantitative assessment of the degree of heterogeneity of the preparation. Such assessments can now be carried out with relative ease, since rapid techniques for sampling, fixation and dehydration without further damage to the tissues and methods for rapid assessment of electron microscopical profiles are now available. One recently published procedure (Bain and Gore, 1971) allows sectioning of blocks of material 24 hours after the samples have been taken. Published electron micrographs should be chosen carefully so as to be representative of the several samples examined and a large enough field shown at low magnification to indicate fairly the pellet composition. Electron micrographs showing only a few organelles at high magnification can never be representative and may give a false impression of the population of material. Each stage in the preparation of the final electron micrographs should be as carefully monitored as possible to ensure representative non-artefactual sampling of the material.

(a) *Sampling and fixation techniques for electron microscopy.* The salient point of this part of the procedure is that uniform samples should be taken from the same organelle suspension used in enzyme assay and the material must be fixed at zero time of the assay procedure. If any additional centrifugations are carried out or there is a delay in sampling, the cytological investigations will clearly not correspond to the biochemical ones. Progressive deterioration of organelles occurs immediately following cell rupture and samples taken on completion of the biochemical assays will give a false assessment of the degree of morphological damage to the preparations. Several samples should be taken, each from a separate isolation performed on a separate occasion and the heterogeneity of the different samples compared. Pipettes with wide orifices are valuable since further rupture of the organelles is avoided by their use. After fixation the suspension is usually pelleted and the pellet then treated as a tissue sample for subsequent dehydration and embedding. A recently published microtechnique for rapid sampling of cell and organelle suspensions should have many applications in analytical plant studies (Bullock and Christian, 1976). In this method, a suspension of organelles is mixed with bovine serum albumin, gelled in a few minutes by the addition of a suitable fixative and then embedded. The gelled mixture contains evenly dispersed particles and is treated as a piece of tissue in the further preparative procedures for electron microscopy. The new method supersedes less satisfactory techniques which involved prior fixation before gel embedding or the use of heat.

A second method of great value in ensuring adequate sampling of the material in the suspension is the adjustment of the size of the aliquot taken from the organelle suspension so that when it is pelleted in a small plastic centrifuge tube (the type used for cytocrit analysis) a verticle section is of suitable size so that it can be laid over a grid slit and its complete profile recorded in a single photographic plate. In the case of chloroplasts, the chlorophyll concentration of the pellet is a good index of the concentration of chloroplasts in the suspension and the procedure is outlined below whereby this calibration is used to calculate the size of the sample analysed.

(b) *Fixation, dehydration and embedding for electron microscopy.* Samples are rapidly mixed with an equal volume of double strength fixative made up in the same medium in which the organelle is suspended. A change in the sugar used as an osmoticum, e.g. from sorbitol to sucrose should be avoided at this stage, since it leads to graininess of the membranes in the final electron micrographs.

Reference is often made to the possible further damage to isolated organelles which may occur during preparation for electron microscopy. Such artefacts can now almost certainly be eliminated by careful attention to the components of the solution used in fixation and dehydration. There is now substantial evidence that the osmolarity and pH of the fixative are critical for organelle preservation and several procedures now include sucrose or sorbitol in the fixative. Violent step-wise changes in osmolarity during fixation should, of course, be avoided. It is less widely appreciated that the osmolarity of the solutions used for dehydration is equally critical (Leech, 1964). The osmolarity of the osmoticum should be progressively reduced as the solvent concentration is increased. In our experience, rapid progressive dehydration is preferable to a slow step-wise process. In the case of chloro-phyll-containing membranes dehydrated in acetone, the 70%/80%/90% transitions are critical. If any green coloration appears in the dehydration solutions, more acetone should immediately be added to prevent further leakage. Quantitative analysis of the proportions and the species of the lipid molecules lost from membranes during several dehydration procedures have been published and should be consulted before a particular protocol is adopted (Ongun et al., 1968a, 1968b; Swanson et al., 1973). Experience shows that etioplasts should be fixed and embedded in complete darkness as there is some evidence that dispersion of the prolamellar body can still occur up to the stage of resin polymerization.

A variety of embedding media have been used for isolated cellular material. We have found Spurr's resin very satisfactory for several different types of suspension and the lack of contrast sometimes encountered when using this resin can be improved by judicious staining of the sections in potassium permanganate.

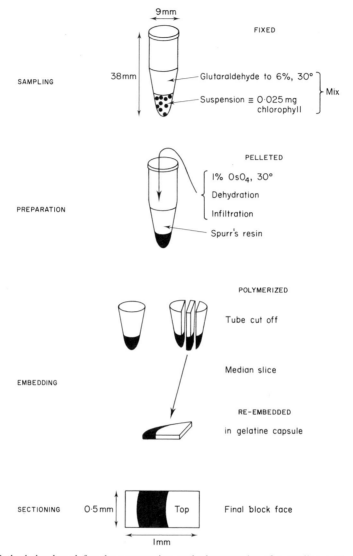

FIG. 1. Method developed for the preparation and photography of a median section of a complete pellet of isolated chloroplasts. Fixation was at room temperature; dehydration in ethanol, embedding in Spurr's resin (Bain and Gove, 1971). The blocks were ready for sectioning, 10 h after the samples were taken.

FIG. 2. Electron micrographs of a median section of a whole pellet (A) of isolated maize chloroplasts and fractions from the pellet (B–D). The chloroplasts were prepared as described by Leese *et al.* (1972) and a suspension containing 25 μg chlorophyll pelleted. The pellet was prepared as shown in Fig. 1: fixation, dehydration and embedding were according to Bain and Gove (1971). The sections on a formvar film on R150 grids were viewed in a HV 12A Hitachi electron microscope. A was photographed using the lower magnification stage at an intermediate aperture position. (H. M. Nice, unpublished photograph.)

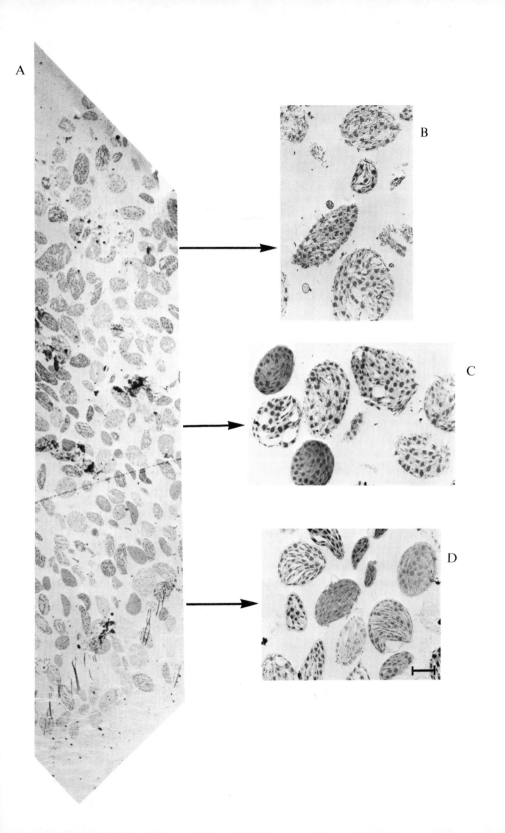

(c) *Observation and assessment of thin sections of pelleted material by electron microscopy*

(i) *Pellets of subcellular particles.* The nature of the sampling technique used in viewing the sections of embedded material is just as critical as the method of sampling used before fixation. Ideally only one or two sections should be taken from each of several different pellets prepared on different occasions and about 20 sections compared. Since objectivity in observation is very important (a remarkably difficult thing to achieve) a good safeguard is to conceal the rationale for the experiment from the electron microscopist until after the pellets have been observed and assessed. Photographs of vertical sections of whole pellets (Fig. 2) are, of course, the easiest to observe for the assessment of the composition of suspensions.

The skill and time needed to obtain representative electron micrographs of pellets isolated organelles may have previously been a deterrent in the completion of these necessary electron micrograph studies. Rapid, reliable fixation and sectioning procedures now exist and it is quite possible to view sections of pellets less than 24 hours after the samples have been fixed. We have recently developed a quick, simple method for assessing the composition of chloroplast suspensions which can be modified for the examination of other cellular fractions. A suitable volume of the suspension in a small plastic (2 ml) cytocrit centrifuge tube, with a tapered base is fixed, pelleted, dehydrated and infiltrated with resin. The tube is then cut away and the whole pellet re-embedded. The volume of sample is chosen so that a vertical section of the whole pellet is a suitable one to be aligned along the slit of an EM grid. A single, low power photograph ($\times 500$) of this section will illustrate the spectrum of the contents of the whole pellet from top to bottom (Fig. 2A). Higher resolution pictures can be taken at specific positions to assess the heterogeneity quantitatively. An example of such a photograph is shown in Fig. 2(B–D); the details of the method are given in Fig. 1. In the case of chloroplasts, the chlorophyll concentration of the suspension is a reliable guide to the pellet size. For other organelles the activity of the marker enzymes, protein content or the concentration of a contaminant such as starch or an inorganic component such as oxalate may be useful parameters for assessing the concentration of organelles. When it is not possible to section a whole pellet, care should be taken that the top, middle and bottom of the pellets are all sampled, heterogeneity assessed at each level and then the quantitative contribution of each portion of pellet to the whole suspension accounted for. When observing thin sections it is often less ambiguous to count the total number of *profiles* of each organelle or membrane type from perhaps 20 or 30 sections rather than attempt a semi-quantitative assessment of the contribution of each type of organelle to a suspension. If a quantitative assessment is needed, section thickness and the shape and average dimensions of all the particles present need to be known. For example, if equal numbers of profiles of mitochondria and swollen (spherical chloroplasts) are represented in electron micrographs then, since the average chloroplast diameter is six times the average mitochondrial diameter, six chloroplast profiles will be present for every single mitochondrial profile, the true mitochondria to chloroplast ratio 6:1, i.e. there are six times as many mitochondria as plastids in the "chloroplast suspension". At the present time adequate statistical techniques are not available to describe the heterogeneity of mixed suspensions of organelles, and one hopeful approach would seem to be to apply to populations of organelles the techniques used successfully to describe biological populations.

(ii) *Pellets containing only membranes.* Fragments of tonoplast, plasma membranes, mitochondria and chloroplast membranes are virtually in-distinguishable morphologically and in most cases enzyme markers must be used to indicate the presence of specific membrane types, since few specific stains are available. As mentioned above, a stain which has found con-siderable use recently is periodate phosphotungstic acid-chromic acid, a specific stain for plasma membranes introduced by Morré's group (Roland *et al.*, 1972). This stain is highly specific, for plasma membranes *in vivo* and isolated membranes which become electron opaque on treatment with the stain can be specifically identified as plasma membranes, although oc-casionally a small proportion of the plasma membranes may remain unstained (Leonard and Van der Woude, 1976). Recently it has been found that if the temperature is raised to 38° then greater intensity of staining is achieved and this modification of the technique should be used in future investigations (Leonard and Van der Woude, 1976). Since isolated membrane fragments originating from many different locations in the cell have approxi-mately the same dimensions and appearance after isolation, quantification of the composition of mixed membrane fractions is less complex than quantifying the contents of particulate suspensions. However, identification of the species of membranes present is still a major problem. Membrane associations which are still intact after isolation such as assembled dictyo-somes or chloroplast granal stacks can be readily identified but smaller fragments of membranes most frequently reseal as vesicles of various shapes and sizes and those of different origin are not easily distinguishable. The specific electron microscopical stain for plasma membranes can however be used to quantify the enrichment of fractions with plasma membranes. A transparent overlay of parallel lines is placed on the micrograph and the point of line-membrane intersect quantified for each recognizable membrane component. The spacing of parallel lines is adjusted suitably for the magni-fication and pellet density. In preparations from onion roots examined successfully in this way, a one cm spacing was ideal for micrographs magni-fied × 20 000 (Van der Woude *et al.*, 1974).

B. ASSESSMENT OF ORGANELLE HETEROGENEITY IN ENRICHED FRACTIONS OF ONE
ORGANELLE

Even when it is possible to reduce the level of contamination of a suspension of a specific organelle to an acceptable level, the heterogeneity of structure and apparent function of organelles themselves is an additional complicating factor which needs to be considered. This aspect of biochemical cytology has been most thoroughly investigated in chloroplast suspensions and the effects of plastid breakage and membrane permeability on the manifestation of enzyme activity are receiving increasing attention. Work using chloroplasts in which attempts have been made to relate morphological integrity to biochemical function will be reviewed to illustrate the parameters which are important in these assessments.

The examination of electron micrographs of suspensions of chloroplasts capable of rapid rates of sustained photosynthesis reveals a mixed population of plastids in which the majority (70%) have entire double bounding envelopes, others have ruptured envelopes and many intermediate forms are also present showing varying degrees of structural damage, including partial loss of the envelope and stroma contents. The two extremes of chloroplast preservation, intact chloroplasts and ruptured chloroplasts, can be readily distinguished by phase contrast microscopy (Kahn and Wettstein, 1961; Leech, 1964; Spencer and Ont, 1965 and Walker, 1967). The intact chloroplasts appear highly refractive, opaque with a shining appearance and a distinct halo: the broken chloroplasts larger ($\times 2$–3) and have a dull, granular appearance and often their margins are indistinct. It has been recognized for some time that the functional capabilities of intact and broken chloroplasts, and certainly the intermediate forms, differ greatly. Suspensions containing a high proportion of intact chloroplasts can sustain CO_2 fixation and PGA-dependent O_2 evolution, but are relatively impermeable to NADP and adenine nucleotides (Walker, 1964; Heber and Santarius, 1965; Walker, 1965; Jensen and Bassham, 1966; Robinson and Stocking, 1968; Stocking et al., 1969 and Heber and Santarius, 1970) and ferricyanide. Broken chloroplasts are inactive in CO_2 fixation, CO_2-dependent O_2 evolution (Walker and Hill, 1967; and Walker and Lilley, 1974) and phosphoglycerate-dependent O_2 evolution, but are able to evolve O_2 and photophosphorylate in the presence of additional cofactors. In an attempt to reduce the confusion associated with the nomenclature used to describe isolated plastids with differing appearances and biochemical functions, Hall (1972) proposed that the term Type or Class A be adopted to describe preparations with large numbers of intact chloroplasts capable of high rates of CO_2 fixation (50–250 μmol/mg chl/h) without additional substrates, and Type or Class C to describe suspensions of broken chloroplasts inactive in CO_2 fixation but able to phosphorylate and reduce ferricyanide.

Suspensions containing high proportions of chloroplasts which have low rates of CO_2 fixation (less than 5 µmoles CO_2/mgchl/h) but still appear intact microscopically are described in Hall's terminology as Class B. Those chloroplasts have clearly sustained some damage during extraction since they are now permeable to NADP, ferricyanide and ADP yet ferredoxin is still retained and need not be added back to such chloroplasts to sustain NADP reduction. Type B chloroplasts exhibit good photosynthetic control (West and Wiskich, 1968). Using suspensions containing a high proportion of Class B chloroplasts, clear concentration effects of NADP and ADP on fatty acid biosynthesis can be demonstrated (Hawke et al., 1974) in contrast to the extremely slow rate of penetration of pyridine nucleotides and adenine nucleotides recorded for Class A chloroplasts (Walker, 1965; Robinson and Stocking, 1968; Heber, 1974). In every suspension of freshly isolated chloroplasts prepared by modern techniques, chloroplasts of Type A, Type B and Type C will all be present in different proportions so the biochemical attributes of the suspension will reflect the mixed nature of the plastid population. The categories proposed by Hall are useful ones and his nomenclature, particularly for Class A and Class C chloroplasts is now generally adopted. He also proposed terms to describe chloroplasts which have been purposefully further damaged by ultrasonic or shock treatment following isolation as Class A or Class C plastids. After controlled breakage, substrates can penetrate more readily and the biochemical capabilities of subchloroplast fractions can also be examined.

The probability that chloroplasts can break and their envelopes reseal again after isolation is suggested by the results of some recent experiments by Lilley et al. (1975). They measured the rate of ferricyanide reduction in numerous chloroplast suspensions after osmotic shock and since ferricyanide does not penetrate Class A chloroplasts assumed an inverse linear relationship between states of ferricyanide reduction and degree of intactness of the chloroplasts. But they also found that the apparent percentage intactness, as determined by ferricyanide, does not relate linearly to the protein loss from the plastids which occurs on chloroplast rupture nor to the rates of CO_2-dependent O_2 evolution in the same chloroplast suspensions. A second order regression curve relates both protein loss and the rate of CO_2-dependent O_2 evolution to the apparent intactness as measured by ferricyanide reduction, i.e. the suspension with low % intact chloroplasts have proportionately less protein and a lower rate of O_2 evolution. Lilley et al. (1975) explain their findings by suggesting the possibility that plastids may break and reseal during isolation procedure so that soluble protein, including the carboxylating enzymes are lost from the stroma but because the resealing of the chloroplast envelope reseals the barrier to ferricyanide, the chloroplast still has the appearance of morphological integrity under phase microscopy. This is a plausible suggestion and a more rigorous demonstration that resealing occurs for example by measuring losses of chloroplast RNA would be of great interest. It is possible the "intermediate" class chloroplasts described by Ridley and Leech, recognizable by their intermediate size and dull appearance in the phase microscope may also be resealed plastids since they were prepared by very brief osmotic rupture of intact plastids which were then rapidly resuspended in isotonic sucrose (Ridley and Leech, 1968). The mechanics involved in the resealing of double membranes such as the chloroplast envelope would be complex although it is possible that the chloroplasts able to reseal may be only bounded by a single membrane in isolation. It is frequently difficult to convincingly demonstrate the doubleness of a membrane bounding isolated chloroplasts.

C. ASSESSMENT OF ENZYME ACTIVITY

Subcellular fractionation studies are carried out for a variety of reasons. The aim of the most straightforward type of investigation is to establish unequivocally the subcellular location of a specific enzyme or a multienzyme complex. Many preliminary experiments are designed on the assumption that there is a single cellular location for the enzyme under study, and although rather infrequently encountered this is the simplest situation to investigate. Fractionation procedures for most plant cellular components are now available which yield suspensions considerably enriched in one specific cellular organelle or membrane system and it is relatively easy to establish if over 90% of the cellular activity is associated with one particular fraction. Confirmatory experiments should always be carried out using fractions with higher degrees of purification, preferably ones virtually completely freed of cellular contamination. The only plant cell organelles which can be currently isolated in this way are plastids.

Similar enzyme activities of multiple origin in cellular homogenates are much more difficult to establish than a single enzyme location. The problems are compounded when low enzyme activities are associated with fractions enriched in one organelle but which still contain a proportion of other cellular components. The problem is to determine whether the activity is a reflection of low activity of the enzyme located in the major component or a high activity associated with a minor component or components. The specific activities of different isoenzymes may differ widely causing further problems in quantitative analysis. A useful approach is to prepare fractions enriched in the organelles of interest by the best methods currently available, characterize the fractions as far as possible by enzyme markers and electron microscopical examination and compare the enzyme activities in fractions showing different degrees of enrichment. Similar analyses can be carried out on other cellular components of interest. The next approach is to further separate and characterize the enzymes themselves using electrophoresis and standard protein separation techniques to separate isozymes, and immunological techniques (Huang et al., 1974) to identify the different proteins at different subcellular sites. If it is technically possible to perform parallel immunological assays on slices of the plant tissue, the intracellular site(s) of activity can be adequately identified. Demonstration of the mere presence of an enzyme in association with a particular organelle is of course insufficient to establish its contribution to the physiology of the tissue from which it is derived. A quantitative assessment relating one aspect of the physiology of a tissue to the biochemical attributes of its organelles involves a major research effort over many years. An investigation of this kind is described by Walker (1973) who investigated the relationship between photosynthetic induction phenomena in whole plants and intact isolated chloroplasts. In his review paper (Walker, 1973), the quantitative characteristics of induction phenomena *in vivo* are compared and related to the accumulation of metabolites brought about by the autocatalytic action of the Benson–Calvin cycle. The experimental evidence is derived from both measurements of whole plants and detailed analyses of isolated chloroplast and reconstituted chloroplast systems and the paper is a model presentation of this type of approach.

The clarification of physiological phenomena in biochemical terms remains one of the most challenging and most difficult tasks which a plant biologist can set himself and cannot be completed quickly or by taking methodological short cuts. Enzyme location studies are one facet of attempts to understand the complex pattern of cell metabolism and it is now clear that in many of the integrated metabolic pathways of the plant cell, several organelles collaborate functionally. Detailed knowledge of these interactions will depend to a large extent on careful subfractionation studies over the next era, and research in this field should see exciting developments in the reconstitution of isolated characterized subcellular components into multiple organelle systems *in vitro*, capable of carrying out integrated cellular reactions.

REFERENCES

Albertsson, P. A. and Baltscheffsky, H. (1963). *Biochem. biophys. Res. Commun.* **12**, 14.

Anderson, J. M., Boardman, N. K. and Spencer, D. (1971). *Biochim. biophys. Acta* **245**, 253.

Bain, J. and Gove, D. (1971). *J. Microsc.* **93**, 159.

Bassham, J. A. and Jensen, R. G. (1967). *In* "Harvesting the Sun" (San Pietro, A., Green, F. A. and Army, T. J., eds), pp. 79–100. Academic Press, New York and London.

Beevers, H. (1969). *Ann. N.Y. Acad. Sci.* **168**, 313.

Beevers, H. (1975). *In* "Recent Advances in the Chemistry and Biochemistry of Plant Lipids" Galliard, T. and Mercer, E. I., eds), pp. 287–299. Academic Press, New York and London.

Behrens, M. and Thalacker, R. (1957). *Naturwissenschaften* **44**, 621.

Bird, I. F., Porter, H. K. and Stocking, C. R. (1965). *Biochim. biophys. Acta* **100**, 366.

Bird, I. F., Cornelius, M. J., Dyer, T. A. and Keys, A. J. (1973). *J. exp. Biol.* **24**, 211.

Black, C. C. (1973). *A. Rev. Pl. Physiol.* **24**, 253.

Bullock, G. and Christian, R. A. (1976). *Histochem. J.* (in press).

Chen, Y-M., Lin, C-Y., Chang, H., Guilpoyle, T. and Key, J. L. (1975). *Pl. Physiol., Lancaster* **56**, 78.

Cobb, A. H. and Wellburn, A. R. (1973). *Planta* **114**, 131.

Cockburn, W., Baldry, C. W. and Walker, D. A. (1967). *Biochim. biophys. Acta.* **143**, 614.

Cockburn, W., Walker, D. A. and Baldry, C. W. (1968). *Pl. Physiol., Lancaster* **43**, 1415.

Cooper, T. G. and Beevers, H. (1969a). *J. biol. Chem.* **244**, 3507.

Cooper, T. G. and Beevers, H. (1969b). *J. biol. Chem.* **244**, 3514.

Coombs, J., Baldry, C. W. and Bucke, C. (1973a). *Planta* **110**, 95.

Coombs, J., Baldry, C. W. and Bucke, C. (1973b). *Planta* **110**, 109.

Coombs, J., Baldry, C. W. and Brown, J. E. (1973c). *Planta* **110**, 121.

Coombs, J., Maw, S. L. and Baldry, C. W. (1974). *Planta* **117**, 279.

Douce, R. (1974). *Science N.Y.* **183**, 852.

Douce, R., Holtz, R. B. and Robertson, R. N. (1973a). *J. biol. Chem.* **248**, 7215.

Douce, R., Mannella, C. A. and Bonner, W. D. (1973b). *Biochim. biophys. Acta.* **292**, 105.

Douglass, S. A., Criddle, R. A. and Breidenbach, R. W. (1973). *Pl. Physiol., Lancaster* **51**, 902.

de Duve, C. and Bauohuin, P. (1966). *Physiol. Rev.* **46**, 323.

Ellis, R. J. (1976). *In* "Perspectives in Experimental Biology" (Sutherland, N., ed.), Vol. 2, pp. 283–298. Pergamon Press, Oxford.

Engelmann, T. W. (1881). *Bot. Z.* **39**, 441.

Evans, A. and Smith, H. (1976). *Nature, Lond.* **259**, 323.

Frederick, S. E. and Newcomb, E. H. (1971). *Planta* **96**, 152.

Good, N. E., Winget, G. D., Winter, W., Connolly, T. N., Izaway, S. and Singh, R. M. M. (1966). *Biochemistry* **5** (2) 467.

Hall, D. O. (1972). *Nature New Biol.* **235**, 125.

Hallaway, M. (1965). *Biol. Rev.* **40**, 188.

Hanson, J. B., Wilson, C. M., Chrispeels, M. J., Krueger, W. A. and Swanson, H. R. (1965). *J. exp. Bot.* **16**, 282.

Hatch, M. D. and Slack, C. R. (1966). *Biochem. J.* **101**, 103.

Hawke, J. C., Rumsby, M. G. and Leech, R. M. (1974). *Phytochemistry* **13**, 403.

Heber, U. (1957). *Ber. dt. bot. Ges.* **70**, 371.

Heber, U. (1960). *Z. Naturf.* **15**, 100.

Heber, U. (1974). *A. Rev. Pl. Physiol.* **25**, 393.

Heber, U. and Willenbrink, J. (1964). *Biochim. biophys. Acta* **82**, 313.

Heber, U. and Santarius, K. A. (1965). *Biochim. biophys. Acta* **109**, 390.

Heber, U. and Santarius, K. A. (1970). *Z. Naturf.* **256**, 718–728.

Heber, U., Pon, N. G. and Heber, M. (1963). *Pl. Physiol., Lancaster* **38**, 355.

Hill, R. (1937). *Nature* **139**, 881.

Hill, R. (1939). *Proc. R. Soc.* B **127**, 192.

Honda, S. I., Hongladoram, T. and Wildman, S. G. (1962). *Pl. Physiol.* **37**, 41.

Horton, P. and Leech, R. M. (1975a). *Pl. Physiol., Lancaster* **55**, 393.

Horton, P. and Leech, R. M. (1975b). *Pl. Physiol., Lancaster* **56**, 113.

Huang, A. H. (1975). *Pl. Physiol., Lancaster* **55**, 870.

Huang, A. H. C. and Beevers, H. (1971). *Pl. Physiol., Lancaster* **48**, 637.

Huang, A. H., Bowman, P. D. and Beevers, H. (1974). *Pl. Physiol., Lancaster* **54**, 364.

Hulme, A. C., Jones, J. D. and Wooltorton, L. S. C. (1965). *New Phytol.* **64**, 152.

Jacobson, A. B. (1968). *J. Cell Biol.* **38**, 238.

James, W. O. and Das, U. S. R. (1957). *New Phytol.* **56**, 325.
Jensen, R. G. and Bassham, J. A. (1966). *Proc. natn. Acad. Sci. U.S.A.* **56**, 1095.
Johnson, H. S. and Hatch, M. D. (1970). *Biochem. J.* **119**, 273.
Johnston, F. B., Nasatin, M. and Stern, H. (1957). *J. biochem. biophys. Cytol.* **6**, 53.
Kagawa, T., Lord, J. M. and Beevers, H. (1973). *Pl. Physiol., Lancaster* **51**, 61.
Kahn, A. and Wettstein, D. (1961). *J. Ultrastruct Res.* **5**, 557.
Kanai, R. and Edwards, G. E. (1973a). *Pl. Physiol., Lancaster* **52**, 484.
Kanai, R. and Edwards, G. E. (1973b). *Naturwissenschaften* **60**, 157.
Karlstam, B. and Albertsson, P. A. (1969). *FEBS Letts* **5**, 360.
Keys, A. J. (1968). *Biochem. J.* **108**, 1.
Kirk, P. A. and Leech, R. M. (1972), *Pl. Physiol., Lancaster* **50**, 228.
Klein, S. and Poljakoff-Mayber, A. (1960). *Expl. Cell Res.* **24**, 143.
Kollöffel, C., Henk, D. and Dijke, U. (1975). *Pl. Physiol. Lancaster* **55**, 507.
Kortschak, H. P., Hart, C. E. and Burr, G. O. (1965). *Pl. Physiol., Lancaster* **40**, 209.
Kung, S. (1976). *Science N.Y.* **191**, 429.
Laetsch, W. M. (1974). *A. Rev. Pl. Physiol.* **25**, 27.
Larsson, C. and Albertsson, P-A. (1974). *Biochim. biophys. Acta* **357**, 412.
Larsson, C., Collin, C. and Albertsson, P. A. (1971). *Biochim. biophys. Acta* **245**, 425.
Leaver, C. J. and Harmey, M. A. (1973). *Biochem. Soc. Symp.* **38**, 175.
Leech, R. M. (1964). *Biochim. biophys. Acta* **79**, 637.
Leech, R. M. (1966). *In* "NATO Symposium, Biochemistry of Chloroplasts" (Goodwin, T. W., ed.), Vol. 1, p. 65. Academic Press, New York and London.
Leech, R. M. (1968). *In* "Plant Cell Organelles" (Pridham, J. B., ed.), p. 137. Academic Press, New York and London.
Leech, R. M. (1972). *In* "Biology and Radiobiology of Anucleate Systems, II Plant Cells". (Bunotto, S., ed.), pp. 27–49. Academic Press, New York and London.
Leech, R. M. (1976). *In* "Perspectives in Experimental Biology" (Sutherland, N., ed.), Vol. 2, pp. 145–162. Pergamon Press, Oxford and New York.
Leech, R. M. and Ellis, R. J. (1961). *Nature, Lond.* **190**, 790.
Leech, R. M., Rumsby, M. G. and Thomson, W. W. (1973). *Pl. Physiol., Lancaster* **52**, 240.
Leese, B. M., Leech, R. M. and Thomson, W. W. (1972). *In* "Second International Congress of Photosynthesis" (Forti, G., Avron, M. and Melandri, A., eds), Vol. 3, pp. 1485–1494. Junk, The Hague.
Leonard, R. T. and Van der Woude, W. J. (1976). *Pl. Physiol., Lancaster* **57**, 105.
Lilley, R. McC. and Walker, D. A. (1973). *Biochim. biophys. Acta* **314**, 354.
Lilley, R. McC. and Walker, D. A. (1974). *Biochim. biophys. Acta* **368**, 296.
Lilley, R. McC. and Walker, D. A. (1975). *Pl. Physiol., Lancaster* **55**, 1087.
Lilley, R. McC., Schwenn, J. D. and Walker, D. A. (1973). *Biochim. biophys. Acta* **325**, 596.
Lilley, R. McC., Holborrow, K. and Walker, D. A. (1974). *New Phytol.* **73**, 657.
Lilley, R. McC., Fitzgerald, M. P., Rienits, K. G. and Walker, D. A. (1975). *New Phytol.* **75**, 1.
Lord, J. M., Kagawa, T. and Beevers, H. (1972). *Proc. natn. Acad. Sci. U.S.A.* **69**, 2469.
Lord, J. M., Kagawa, T., Moore, T. S. and Beevers, H. (1973). *J. Cell Biol.* **57**, 659.
Lyndon, R. F. (1963). *J. exp. Bot.* **14**, 419.
Lyttleton, J. W. (1962). *Expl. Cell Res.* **26**, 312.
McClendon, J. H. (1952). *Am. J. Bot.* **39**, 275.
McClendon, J. H. (1953). *Am. J. Bot.* **4**, 260.
McKersie, B. and Thompson, J. E. (1975). *Pl. Physiol., Lancaster* **56**, 518.
Miflin, B. J. (1974). *Pl. Physiol., Lancaster* **54**, 550.
Miflin, B. J. and Beevers, H. (1974). *Pl. Physiol., Lancaster* **53**, 870.

Millard, D. L., Niskich, J. T. and Robertson, R. N. (1965). *Pl. Physiol., Lancaster* **40**, 1129.
Moore, T. S. and Beevers, H. (1974). *Pl. Physiol., Lancaster* **53**, 261.
Morgenthaler, J. J., Price, C. A., Robinson, J. M. and Gibbs, M. (1974). *Pl. Physiol., Lancaster* **54**, 532.
Morré, D. J. (1970). *Pl. Physiol., Lancaster* **45**, 791.
Morré, H. and Mollenhauer, H. H. (1964). *J. Cell Biol.* **23**, 295.
Morré, D. J., Mollenhauer, H. H. and Chambers, J. F. (1965). *Expl. Cell Research* **38**, 672.
Naguchi, M. and Tamaki, E. (1962). *Archs Biochem. Biophys.* **98**, 197.
O'Neal, D., Hew, C. S., Latzko, E. and Gibbs, M. (1972). *Pl. Physiol., Lancaster* **49**, 607.
Ongun, A. and Stocking, C. R. (1965). *Pl. Physiol.* **40**, 825.
Ongun, A., Thomson, W. W. and Mudd, J. B. (1968a). *J. Lipid Res.* **9**, 409.
Ongun, A., Thomson, W. W. and Mudd, J. B. (1968b). *J. Lipid Res.* **9**, 416.
Opik, H. (1968). *In* "Plant Cell Organelles" (Pridham, J. B. ed.), pp. 47–89. Academic Press, New York and London.
Price, C. A. and Hirvonen, A. P. (1966). *Pl. Physiol.* **9**, 41.
Rathnam, C. K. M. and Das, V. S. R. (1974). *Can. J. Bot.* **52**, 2599.
Ray, P. M., Shininger, T. L. and Ray, M. M. (1969). *Biochemistry N.Y.* **64**, 606.
Rho, J. H. and Chipchase, M. I. (1962). *J. Cell Biol.* **14**, 183.
Ridley, S. M. and Leech, R. M. (1968). *Planta* **83**, 20.
Ritenour, G. L., Joy, K. W., Bunning, J. and Hageman, R. H. (1967). *Pl. Physiol., Lancaster* **42**, 233.
Roberts, G. R., Keys, A. J. and Whittingham, C. P. (1970). *J. exp. Bot.* **21**, 683.
Robertson, D. and Laetsch, W. H. (1974). *Pl. Physiol., Lancaster* **54**, 148.
Robinson, J. M. and Stocking, C. R. (1968). *Pl. Physiol., Lancaster* **43**, 1597.
Rocha, V. and Ting, I. P. (1970). *Archs Biochim. Biophys.* **140**, 398.
Roland, J. C., Lembi, C. A. and Morré, D. J. (1972). *Stain Technol.* **47**, 195.
Santarius, K. A. and Stocking, C. R. (1969). *Z. Naturf.* **24b**, 1170.
Schwenn, J. D., Lilley, R. McC. and Walker, D. A. (1973). *Biochim. biophys. Acta* **325**, 586.
Shore, G. and Maclachlan, G. A. (1975). *J. Cell Biol.* **64**, 557.
Slack, C. R., Roughan, P. G. and Barrett, H. C. M. (1974). *Planta* **118**, 57.
Smillie, R. M. (1963). *Can. J. Bot.* **41**, 123.
Spencer, D. and Unt, H. (1965). *Aust. J. biol. Sci.* **18**, 197.
Spurr, A. R. (1969). *J. Ultrastruct. Res.* **26**, 31.
Steer, B. T. (1973). *Pl. Physiol., Lancaster* **51**, 744.
Stocking, C. R. (1959). *Pl. Physiol., Lancaster* **34**, 56.
Stocking, C. R. (1971). *Meth. Enzym.* **13**, 221.
Stocking, C. R. (1975). *Pl. Physiol., Lancaster* **55**, 626.
Stocking, C. R. and Ongun, A. (1962). *Am. J. Bot.* **49**, 284.
Stocking, C. R., Williams, G. R. and Ongun, A. (1963). *Biochem. biophys. Res. Commun.* **10**, 416.
Stocking, C. R., Shumway, L. K., Weier, T. E. and Greenwood, D. (1968). *J. Cell Biol.* **36** (1), 270.
Stocking, C. R., Robinson, J. M. and Weier, T. E. (1969). *In* "First International Congress of Photosynthesis" (Metzner, H., ed.), pp. 258–266. International Union of Biological Sciences.
Swanson, E. S., Thomson, W. W. and Mudd, J. B. (1973). *Can. J. Bot.* **51**, 1221.
Tautvydas, K. J. (1971). *Pl. Physiol., Lancaster* **47**, 499.
Thomson, W. W., Foster, P. and Leech, R. M. (1972). *Pl. Physiol., Lancaster* **49**, 270.
Tolbert, N. E. (1971a). *Meth. Enzym.* **23**, 665.

Tolbert, N. E. (1971b). *A. Rev. Pl. Physiol.* **22**, 45.

Treffry, T. (1975). *Planta* **126**, 11.

Van der Woude, W. J. Lembi, C. A. and Morré, D. J. (1974). *Pl. Physiol., Lancaster* **54**, 333.

Vigil, E. L. (1970). *J. Cell Biol.* **46**, 439.

Walker, D. A. (1964). *Biochem. J.* **92**, 226.

Walker, D. A. (1965). *Pl. Physiol., Lancaster* **40**, 1157.

Walker, D. A. (1967). *In* "NATO Symposium, Biochemistry of Chloroplasts" (Goodwin, T. W., ed.), Vol. 2, p. 53. Academic Press, New York and London.

Walker, D. A. (1970). *Nature, Lond.* **226**, 1204.

Walker, D. A. (1971). *Meth. Enzym.* **23**, 211.

Walker, D. A. (1973). *New Phytol.* **72**, 209.

Walker, D. A. and Hill, R. (1967). *Biochim. biophys. Acta.* **131**, 330.

Walker, D. A. and Crofts, A. R. (1970). *A. Rev. Biochem.* **39**, 389.

Walker, D. A. and Lilley, R. McC. (1974). *Pl. Physiol., Lancaster* **54**, 950.

Walker, D. A., Cockburn, W. and Baldry, C. W. (1967). *Nature, Lond.* **216**, 597.

Weiare, P. J. and Keckwick, R. G. O. (1975). *Biochem. J.* **146**, 425.

Wellburn, A. R. and Wellburn, F. A. M. (1971). *J. exp. Bot.* **22**, 972.

Wellburn, A. R. and Wellburn, F. A. M. (1973). *Ann. Bot.* **37**, 11.

Werdan, K. and Heldt, H. W. (1972). *Biochim. biophys. Acta,* **283**, 430.

West, K. R. and Wiskich, J. T. (1968). *Biochem. J.* **109**, 527.

Wildman, S. G., Hongladoram, T. and Honda, S. I. (1962). *Science* **138**, 434.

CHAPTER 15

Microscopic Cytochemistry in Enzyme Localization and Development

J. L. HALL, M. J. AL-AZZAWI AND J. L. FIELDING

School of Biological Sciences, University of Sussex, Brighton, England

I. Introduction 329
II. Range of Methods 332
   A. Simultaneous Capture Mechanisms 334
   B. The Substrate Film Methods 335
   C. Autoradiographic Procedures 335
   D. Immunochemical Methods 336
III. Factors Affecting Localization 337
   A. Preservation of Enzyme Activity and Cell Structure . . . 337
   B. Effects of Aldehyde Fixation on Enzyme Activity 340
   C. Precision of the Localization Procedure 343
   D. Specificity of the Reaction 345
   E. Controls 346
IV. Applications 346
   A. Peroxidase Localization and Differentiation in Roots . . . 347
   B. Enzymic Changes During Leaf Abscission 349
   C. Enzymic Changes in Washed Storage Tissue Discs 353
References 361

I. Introduction

   The aim of enzyme microscopic histochemistry is the identification and localization of specific enzyme activities in cells and tissues. Although the various methods designed to achieve this localization have certain basic limitations, they may be extremely valuable in the study of certain problems, particularly when used as a complementary technique to cell fractionation and other biochemical procedures. The major advantage of these microscopic techniques is that intact tissue sections are used which, if prepared carefully, preserve the normal morphological structure of cells and tissues. Thus enzyme activity may be related to a specific cell or cell type or organelle, a localization which may not be possible using biochemical techniques. For example, the analysis of enzyme activity of tissue homogenates assumes a homogeneous mass of cells, when clearly this is rarely so, although it may not be possible to separate the various cell types for specific analysis. In this case,

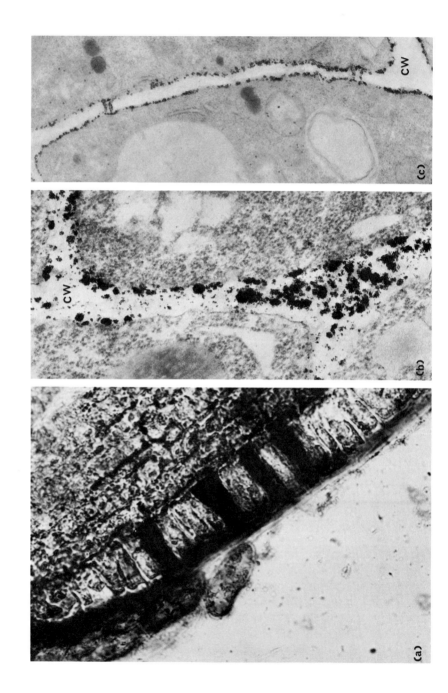

(a)

(b)

CW

(c)

CW

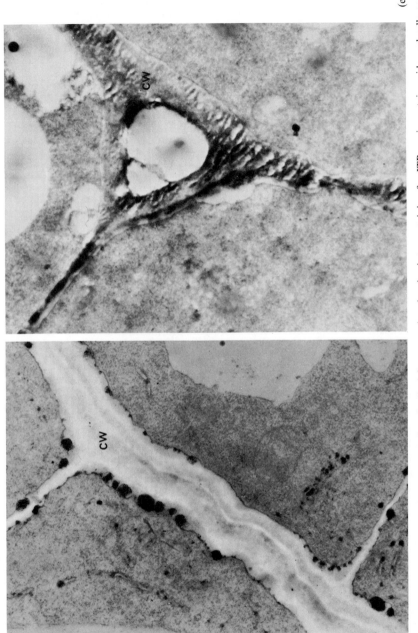

(e)

(d)

Fig. 1. Localization of enzymes in maize roots. (a) Longitudinal frozen section showing intense staining for ITP-ase in certain epidermal cells ×440. (From Hall, 1969a.) (b–e) Electron micrographs showing the different sites of activity of various cell surface enzymes in maize roots. (b) β-glycerophosphatase ×42 000. (c) ATP-ase ×16 500. (d) Adenyl cyclase ×26 250. (e) Peroxidase ×18 500. β-glycerophosphatase and peroxidase appear to be essentially cell wall enzymes while ATP-ase and adenyl cyclase are associated with the plasma membrane. The methods used are in the following: (b, c) Hall (1971); (d) Al-Azzawi and Hall (1976); (e) Hall and Sexton (1972) CW (cell wall.)

microscopic histochemistry may be used with advantage to demonstrate the relative activities of an enzyme in the various cell types within an organ such as the root (see Jensen, 1955; Hall, 1969a, b; Sutcliffe and Sexton, 1969; Goff, 1975; and Fig. 6). Avers (1958), studying the root epidermis of grasses by microscopic histochemistry, was able to show that the root hair initials possessed higher activity for certain enzymes than other epidermal cells (Fig. 1a). Such resolution between tissues and individual cells would not be possible using biochemical techniques. At the intracellular level, enormous advances have been made with the development of cell fractionation procedures, although these techniques still suffer from a number of limitations (see Leech, Chapter 14); these may be improved by the use of microscopic cytochemistry. For example, loss or gain of an enzyme by an organelle during isolation may be more readily assessed if its localization can be demonstrated in the intact cell. Again, there are still certain cell structures which cannot be reliably isolated from plants free from contamination; these include the plasma membrane, tonoplast and cell wall. Cell wall fractions, which possess a variety of enzyme activities, may also contain fragments of plasma membrane and other organelles; microscopic cytochemistry can provide valuable information as to the site of localization of these enzymes at the cell surface (Fig. 1b-e). The development of microscopic techniques may also help in the improvement of procedures for the isolation of problematic cell fractions; a thorough microscopic study of the plasma membrane and tonoplast using enzymic and other staining procedures could reveal reliable markers for subsequent use in cell fractionation.

The major limitations of these microscopic techniques when compared to biochemical procedures are that they are largely qualitative rather than quantitative, and that the range of enzymes that may be localized is relatively restricted. Thus it is not possible to make a detailed biochemical analysis of, say, a certain organelle in intact sections as may be carried out on an isolated cell fraction. Other general problems, such as the denaturation or diffusion of the enzyme during the preparation or assay procedure, are also associated with biochemical and cell fractionation studies. The extent of the problem may vary with different enzymes and different procedures and the use of microscopic and biochemical techniques together may help to clear up many uncertainties. The specific problems associated with microscopic histochemistry will be discussed in later sections of this chapter.

II. Range of Methods

The range of microscopic techniques which may provide information relating to the localization, function and development of enzymes is large. For example, any localization of a metabolite or macromolecule may provide indications as to the nature of the enzyme activity associated with its synthesis or utilization. This chapter, however, will be concerned with methods which aim to directly localize a specific enzyme activity.

Modern enzymic methods began with the development independently by Gomori and Takamatsu in 1939, of a procedure for the localization of alkaline phosphatase. Since this time the number of procedures available has increased rapidly although this is still well below that for biochemical assays. A list of enzymes for which there are published microscopical methods is shown in Table I, although the reliability of these procedures may vary widely. There may be several procedures available for the localization of some enzymes. Not all of these procedures can be applied to electron microscopy although there are examples in most of these groups which may be studied at the ultrastructure level. The range is fairly restricted being largely confined to the hydrolases and oxidoreductases; no procedures are available for the localization of isomerases and only one for a ligase. However,

TABLE I

List of enzymes for which microscopic localization techniques have been published

Phosphatases
   acid and alkaline phosphatase
   adenosine triphosphatase
   nucleoside diphosphatase
   adenyl cyclase
   DNA-ase
Esterases
   cholinesterase
   lipase
Sulphatases
Glycosidases
   α-amylase
   β-glucuronidase
   β-glucosidase
   thioglucosidase (myrosinase)
Proteases
   aminopeptidase
   endopeptidases
Dehydrogenases
   succinate dehydrogenase
   NAD- and NADP-dependent dehydrogenases
Oxidases and peroxidases
   cytochrome oxidase
   phenol oxidase
   peroxidases
   catalase
Transferases
   RNA polymerase
   phosphorylase
   aspartate aminotransferase
Malate synthase

this number will undoubtedly increase and the development of immunochemical techniques, in particular, should greatly expand the range.

The methods used for the microscopic localization of enzymes are quite
varied but may be classified into a number of broad groups.

A. SIMULTANEOUS CAPTURE MECHANISMS

These are the oldest and most common methods of enzyme localization
and involve the capture of the end product of the enzyme reaction as a
coloured or electron dense precipitate. The method is illustrated diagram-
matically as follows:

$$\text{Substrate} \xrightarrow[\text{reaction}]{\text{enzymic}} \substack{\text{Primary} \\ \text{reaction} \\ \text{product}} \xrightarrow[\text{reaction}]{\text{capture}} \substack{\text{Final} \\ \text{reaction} \\ \text{product}}$$

Examples of simultaneous capture methods for light microscopy
include:

a. the azo dye methods for phosphatases and esterases in which naphthols
   released from synthetic substrates are trapped by coupling with a
   diazonium salt;
b. the lead phosphate method for phosphatases in which released inorganic
   phosphate is trapped by lead ions;
c. the method for dehydrogenases in which tetrazolium salts act as
   electron acceptors and are converted to insoluble, coloured formazans;
d. the indoxyl acetate method for esterase in which the indoxyl group
   released is oxidized by a ferricyanide–ferrocyanide reagent to form an
   indigo precipitate.

With post-coupling procedures, the primary reaction product is assumed
to be sufficiently insoluble to remain *in situ* during the incubation period;
its site of formation is subsequently revealed by transfer to a coupling
reaction medium. Post-coupling procedures have been extensively criticized
and are not widely used (see Pearse, 1968).

For enzyme localization with the electron microscope, the final reaction
product must be electron dense; there are a number of methods which
produce this result and most may be divided into three groups (see Shnitka
and Seligman, 1971). They are:

a. metal-salt precipitation procedures such as the lead phosphate procedure for phosphatases which is listed above;
b. ferricyanide reduction methods for oxidoreductases in which ferrocyanide formed by the enzymic reduction of ferricyanide is captured
   by cupric ions to yield insoluble electron-dense copper ferrocyanide.

A technique has been developed for the demonstration of malate synthase in glyoxysomes using this principle (Trelease, *et al.*, 1974);

c. methods based on the osmiophilic principle in which the final product of the enzyme reaction is osmiophilic and so osmium staining may be used to localize this product. The most successful methods are those employing 3,3'-diaminobenzidine (DAB) for certain oxidases and peroxidases; oxidation of DAB results in the formation of a water- and lipid-insoluble osmiophilic polymer.

It is beyond the scope of this chapter to give details of the staining media used in these various procedures. Full details of these may be found in various texts, the most comprehensive being Burstone (1962), Chayen *et al.* (1973), Jensen (1962) and Pearse (1968), while the applications of these techniques to electron microscopy are reviewed by Shnitka and Seligman (1971) and details given by Hayat (1973). Most of these techniques have been developed for animal tissues and it is important to stress that the procedures should be modified to optimize conditions for the particular enzyme and tissue under study. Factors such as pH, cofactor and inhibitor concentrations, and temperature should be carefully checked, preferably by biochemical methods.

B. THE SUBSTRATE FILM METHODS

In this method, a film of substrate (e.g. RNA in gelatin) spread on a slide is placed in contact with a tissue section on another slide. After incubation the slides are separated and the unaffected substrate in the film stained. Comparison of the unstained areas in the film with the corresponding tissue sections show the sites of enzyme activity. This method has been used for the localization of ribonucleases, deoxyribonucleases, proteases, amylase and hyaluronidase; the technical problems and limitations have been thoroughly reviewed by Daoust (1965). Substrate film methods have limited use but do allow the demonstration of certain enzymes which are difficult to localize by other procedures. In addition, they avoid certain problems, such as product diffusion, associated with capture methods. However, the resolution achieved by these procedures is low and usually allows localization only at the tissue level.

C. AUTORADIOGRAPHIC PROCEDURES

There are a few procedures which make use of autoradiography for the localization of enzyme activity. For example, RNA polymerase has been localized by following the incorporation of a radioactive precursor, uridine-H^3 triphosphate, into RNA by autoradiography using frozen sections (Fisher, 1968). This procedure may be applied to both light and electron microscopy.

D. IMMUNOCHEMICAL METHODS

These methods make use of the specificity of antibodies to detect specific enzymes. A purified enzyme is injected into a suitable animal which makes a highly specific antibody to itself. This is purified from the globulin fraction of the animal's blood. When tissue sections are treated with the antibody it should react only with the enzyme for which it was made. A detector reagent is necessary to visualize the antigen–antibody reaction. In early experiments for light microscopy, a fluorescent label was usually conjugated directly to the antibody. For electron microscopy, ferritin has frequently been used although resolution may be improved by the use of enzyme markers. For example, peroxidase-labelled antibodies can be detected by the DAB staining procedure described earlier and this appears to be very successful when applied to animal tissues (see Shnitka and Seligman, 1971); it may be more difficult to apply to plants due to the high levels of peroxidase present in most tissues.

The specificity of this procedure may be improved if the marker is not conjugated to the antibody directly but through an immunological bridge which avoids possible loss of specificity of the antibody by conjugation with the label. A number of steps are involved in the process:

    a. a specific antiserum to the antigen to be localized is prepared and the tissue treated with this antiserum;

    b. after washing, the tissue is treated with antiserum prepared in another animal species against the gamma globulin of the species used in the first step. The antibody of the first step acts as antigen in this second reaction;

    c. the tissue is then treated with antibody prepared against the label (e.g. peroxidase) in the same species as used in (a). It reacts with the free binding site of the second antibody; this is non-specific and so a high concentration is used;

    d. the tissue is finally treated with peroxidase which reacts as antigen with the specific binding sites of peroxidase antibody. The peroxidase may then be stained with the DAB procedure to localize the sites of the original antigen.

This technique will not be discussed further here since so far there have been very few examples of its application to plant tissues. However, immuno-cytochemistry clearly has great potential since it should be possible to localize any enzyme if it can be isolated and purified biochemically. Details of these methods are described by Sternberger (1973) and interesting applications to the study of (Na^+, K^+) ATP-ase and NADPH-cytochrome c reductase in animal cells are reported by Kyte (1976) and Morimoto et al. (1976) respectively. Immunofluorescence has been used for the study of α-amylase in barley (Jacobsen and Knox, 1973) and the principles and limitations of this technique have been reviewed by Mayersbach (1967). Immunochemical localization of cellulose at the ultrastructural level has been described by Bal *et al.* (1976) using pea epicotyls.

III. Factors Affecting Localization

There are a number of general factors associated with the reliable localization of enzymes by microscopic methods in addition to the specific problems associated with any one technique. According to Jensen (1962), the general problems that have to be considered are:

a. the maintenance of enzyme activity;
b. the maintenance of cytological sites;
c. the precision of the localization procedure;
d. the specificity of the reaction.

A. PRESERVATION OF ENZYME ACTIVITY AND CELL STRUCTURE

The first two requirements listed above are clearly closely related to each other and are involved with the process of specimen preparation. It is obviously desirable to maintain as much enzyme activity as possible and at the same time preserve structural detail. If one of these aims fails appreciably then the value of enzyme histochemistry is greatly reduced.

Since chemical fixation may partially or wholly denature many enzymes, fresh and frozen sections have often been used for enzyme localization with the light microscope. These methods have, in turn, a number of drawbacks. For example, with fresh tissue, thin sections are more difficult to obtain and, even with frozen sections, morphological preservation may not be as good as prefixed material (Sexton *et al.*, 1971; and Fig. 2a, b). Even more critical may be the problems of penetration of the reaction medium into fresh or frozen sections and loss of materials during the incubation period. Presumably the normal permeability properties of membranes will impede the movement of the components of the staining medium into the cell and organelles. Thus sections of tissues fixed with glutaraldehyde required shorter incubation times for acid phosphatase localization than sections of fresh tissue (Sexton *et al.*, 1971 and Fig. 2a, b). Thus the maintenance of full membrane integrity may have disadvantages for microscopic cytochemistry. A compromise is needed in which membrane permeability is increased without appreciable loss of enzyme activity. In some systems this may be achieved by mild aldehyde fixation which at the same time may give adequate structural preservation.

A further drawback to the use of fresh or frozen tissues is that considerable disruption of structure and loss of materials may occur during incubation in aqueous staining media. This is illustrated by the measurement of loss of peroxidase activity from sections cut from fixed and fresh pea roots (Table

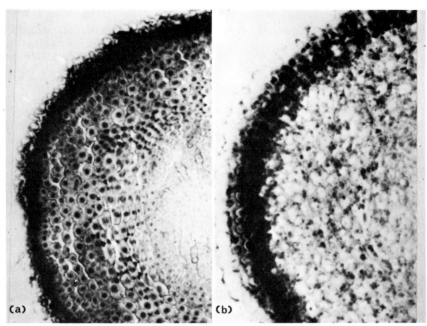

FIG. 2a, b. Transverse frozen sections of maize roots cut about 1 mm behind the tip and stained for β-glycerophosphatase activity. (a) From a prefixed root, stained for 8 min × 125. (b) From an unfixed root, stained for 12 min × 115. Note the better morphological preservation with the fixed roots. (From Sexton *et al.*, 1971.)

TABLE II

Loss of peroxidase activity from sections of fixed and unfixed pea roots

Section thickness Fixation	250 μm —	250 μm 3% Glutaraldehyde	125 μm 3% Glutaraldehyde
Wash periods			
0– 5 min	67	18	39
5– 6 min	4	3	3
6–10 min	2	6	2
10–20 min	2	5	3
20–50 min	1	5	4
Remaining in tissue	24	63	49

125 μm and 250 μm sections were transferred through a series of wash solutions for the times shown, and the activity in the wash solutions, and that remaining in the tissue determined. The results are expressed as a percentage of the total activity. From Hall and Sexton (1972).

II). It is clear that prior fixation of tissue in glutaraldehyde before sectioning greatly reduced the loss of activity from sections; part of this loss was probably from cut cells as shown by the increased leakage from thinner sections. Glutaraldehyde fixation was even more effective in the retention of β-glycerophosphatase activity of pea roots. When roots were cut into 125 μm

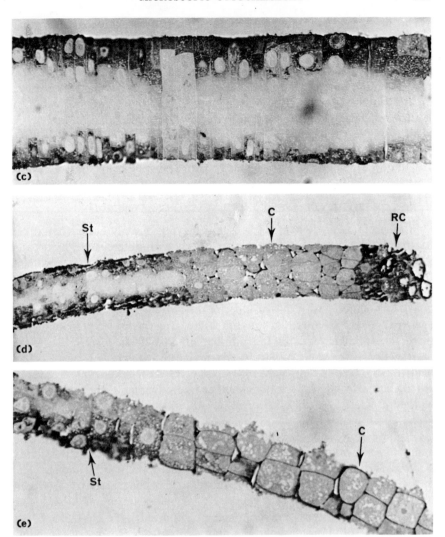

FIG. 2c–e. Penetration of the DAB staining medium for peroxidase into sections of pea roots of varying thicknesses. (c) 2 μm section of a 90 μm Vibratome-cut section of a glutaraldehyde-fixed root, which was incubated in the DAB medium for 15 min, before post-fixation and embedding in epon × 319. (d) 60 μm section treated as in (c) × 250. (e) 30 μm section treated as in (c) × 540. Note the poor penetration, particularly into the 90 μm section, and the differential penetration into root cap (RC), cortex (C) and stele (St). (From Hall and Sexton, 1972.)

sections as described in Table II, 47% of activity was lost from fresh sections but there was no detectable loss if roots had been prefixed in 3% glutaraldehyde (Sexton *et al.*, 1971). For light microscopy, pretreatment of the tissue with an inert polymer, polyvinyl alcohol (PVA), may improve cytological detail and help to reduce loss of cell contents during incubation of frozen sections of plant tissues (Gahan *et al.*, 1967).

The problems encountered with the use of fresh or frozen sections, particularly the poor preservation of structure and the diffusion and loss of soluble enzymes, are magnified when considered in relation to ultra-structural localization. Consequently, Shnitka and Seligman (1971) have concluded that for animal cells aldehyde fixation offers the best compromise for the preservation of both enzymic activity and ultrastructural detail. A similar conclusion was reached in a study of acid phosphatase activity in plant cells (Sexton et al., 1971). It is therefore important to consider the effects of aldehyde fixation on enzyme activity in more detail since it is essential to the interpretation of many histo- and cytochemical studies.

B. EFFECTS OF ALDEHYDE FIXATION ON ENZYME ACTIVITY

In general, aldehyde fixatives preserve enzymic activity better than other fixatives although the results may vary widely with different aldehydes and enzymes. Aldehydes preserve cellular fine structure by cross-linking proteins through active hydrogen, amino and imino groups. With glutaraldehyde fixation, the loss of activity which occurs with some enzymes is believed to be due to cross-linking of lysine residues of the enzyme proteins (Quiocho and Richards, 1966; Schejter and Bar-Eli, 1970). The quality of morphological preservation increases with the number of cross-links formed, although preservation of enzyme activity varies inversely with this property of the fixative (Shnitka and Seligman, 1971; Hayat, 1973). Thus enzyme recoveries are highest with inferior fixatives such as hydroxyadipaldehyde and lowest with good fixatives such as acrolein. Glutaraldehyde is probably the most useful compromise and is now the most widely used fixative for enzyme localization studies; it gives good preservation of fine structure and, with many enzymes, preserves a high proportion of the activity. Another possibility is to use mixtures of aldehydes to achieve satisfactory structural and enzymic preservation. A mixture of paraformaldehyde and glutaraldehyde has been used for the localization of a number of enzymes in animal tissues (see Hayat, 1973). The purity of commercial glutaraldehyde can also be an important factor. For example, the high concentration of inorganic phos-phates in commercial glutaraldehydes may cause inhibition of phosphatase activity (Fahimi and Drochmans, 1968); phosphates should be removed by thorough washing of the tissue after fixation. Commercial preparations also contain a number of impurities and may have a low pH value. In general, better preservation of enzyme activity is obtained if commercial glutaralde-hyde is purified; distillation appears to be the most effective method (Fahimi and Drochmans, 1968; Chayen et al., 1973).

The degree of preservation of enzyme activity after fixation is dependent on a number of factors. In a detailed study of the effects of formaldehyde fixation on acid phosphatase activity of hamster kidney, Christie and Stoward (1974) showed that the most important factors were the time and temperature of fixation, the buffer used with the fixative, and the length of the post-fixation wash. Again, in a study of nucleoside diphosphatase of onion roots, Goff and Klohs (1974) demonstrated that both the time of fixation and concentration of fixative could affect the degree of inhibition of this enzyme; glutaraldehyde was more inhibitory than formaldehyde. Another interesting observation was reported by Herzog and Fahimi (1974) in a study of mammalian catalase activity; glutaraldehyde fixation inhibited the catalytic activity of catalase but enhanced its peroxidatic activity towards certain hydrogen donors including DAB. Glutaraldehyde fixation appears to be an essential prerequisite for staining of peroxisomes with DAB.

Thus, although there have been few detailed studies of the effect of fixation on plant enzymes, it is clear that the type and concentration of the fixative and the conditions of fixation may have marked effects on the preservation of activity and that different enzymes may respond in different ways. Information relating to plants is sparse and so it is difficult to generalize. The response to fixation of any enzyme to be localized cytochemically should be carefully studied. This can be illustrated by examining the effects of fixation on peroxidase and ATP-ase activity of maize roots. When the activity of total root homogenates was determined, peroxidase showed little reduction after either glutaraldehyde or formaldehyde fixation whereas ATP-ase was appreciably reduced with increasing fixative concentration, particularly with glutaraldehyde (Fig. 3). The effect of fixation has usually been determined in total tissue homogenates but this procedure may mask differential fixation effects if activity occurs at various intracellular sites. Both peroxidase and ATP-ase occur in various organelles in plant cells and have a group character consisting of a number of different isoenzymes. There is evidence for ATP-ase that these isoenzymes may show specific intracellular localizations (Edwards and Hall, 1973) and so may respond differently to fixation. This was tested by studying the effect of fixation on peroxidase and ATP-ase activity in three subcellular fractions separated by differential centrifugation (Fig. 4). With ATP-ase activity, the cell wall fraction was relatively unaffected by fixation whereas the microsomal activity was markedly reduced, particularly by glutaraldehyde. Glutaraldehyde fixation showed more inhibition than formaldehyde in all three fractions. Further differences between cell fractions were observed in relation to peroxidase activity. Considerable microsomal activity remained, particularly after formaldehyde fixation, whereas the cell wall activity was best preserved by fixation with glutaraldehyde. The mitochondrial fraction showed a similar response to the microsomes although slightly more inhibition was observed with both fixatives. Thus it is clear that enzyme activity in different cell fractions may show very different responses to fixation. If the activity in the intact cell behaves in the same way as in isolated fractions, this is an important factor to consider in the interpretation of cytochemical observations.

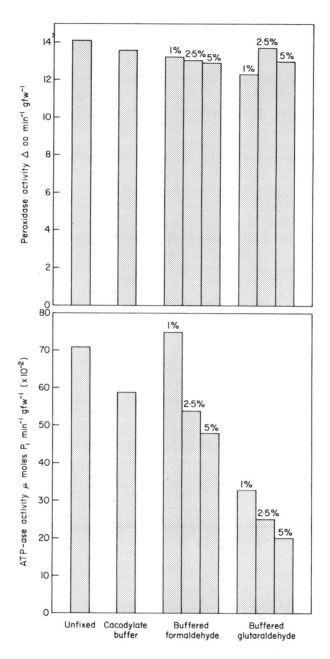

FIG. 3. Effects of fixation on peroxidase and ATP-ase activity in total homogenates of maize roots. Roots were fixed for 2 h in ice-cold fixative, washed thoroughly for 6 h in buffer, and homogenized. Peroxidase was assayed using guaiacol as hydrogen donor at pH 6·1 and measuring the increase in OD at 470 nm. ATP-ase was assayed by the release of inorganic phosphate at pH 7·0.

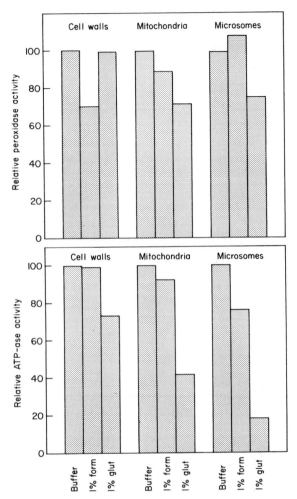

FIG. 4 Effects of fixation on peroxidase and ATP-ase activity in subcellular fractions isolated from maize roots. The fractions were isolated by techniques similar to those used by Edwards and Hall (1973) and resuspended in buffer or fixative for 2 h. After washing three times by recentrifugation and resuspension, the fractions were assayed as described for Fig. 3.

C. PRECISION OF THE LOCALIZATION PROCEDURE

The precision of an enzymic localization procedure depends on the response of the enzyme during and after fixation and on the nature of the localization technique. The major problems that may arise are due to the following:

1. diffusion of the enzyme during or after fixation;

2. diffusion of the components of the reaction medium to the enzymatic sites;
3. diffusion of the primary or final reaction products.

1. *Diffusion of the Enzyme*

The extent of enzyme diffusion will depend on the quality of fixation and on the nature of the enzyme itself. Clearly an enzyme which is tightly bound to a membrane will present fewer diffusion problems than a "soluble" enzyme. If diffusion problems are indicated by diffuse staining, this should be checked by suitable control experiments including the use of a range of fixation procedures and the overlapping section technique (see Danielli, 1953). Enzyme diffusion is probably not a major problem with well fixed tissues; there was no evidence of diffusion of acid phosphatase in fixed pea roots (Sexton *et al.*, 1971). However, significant diffusion may occur if fixed material is stored for some time before staining. Fahimi (1973) has shown that catalase may diffuse from rat liver peroxisomes to give apparent staining of nearby membranes and ribosomes if fixed tissues were stored for 16 h or more before staining with the DAB technique. There was no evidence of diffusion of the reaction product even after nine months storage.

2. *Diffusion of the Reaction Medium*

The distribution and intensity of the staining reaction may be affected by the different rates of diffusion of the components of the incubation medium (substrate, capture reagent, cofactors, buffer) to the sites of enzymatic activity. Inconsistent staining resulting from poor penetration of the reaction medium may occur with fixed as well as unfixed tissues. This problem has been investigated in relation to the localization of acid phosphatase, ATP-ase and peroxidase activity in root tissues (Hall, 1969a, b; Sexton *et al.*, 1971; Hall and Sexton, 1972). Problems arise particularly when tissue blocks are used for staining and, unfortunately, such blocks have been extensively used in cytochemical studies with the electron microscope. The patchy staining that may occur is illustrated in Fig. 2c-e, which shows the penetration of the DAB staining medium for peroxidase into pea root sections of various thicknesses. With the thicker sections, staining is confined to the outermost cells while large regions of the block remain unstained. Poux (1970) has suggested that this problem might be solved by preincubation of tissue blocks in the staining medium for several hours at 0°C. However, when this was tested, penetration was still limited to the outer layers of cells (Sexton *et al.*, 1971). This problem can be largely overcome if thin sections of tissue are incubated in the staining medium before embedding for electron micro-scopy; 20–30 μm sections cut in a cryostat or using a vibratome (Oxford Instrument Co.) have proved to be very suitable for this purpose. The latter method is particularly useful since ice crystal damage may occur when frozen sections are employed in ultrastructural studies (Hall and Sexton, 1972). However, even with this method, a concentration gradient will exist across the section.

3. *Diffusion of Reaction Products*

Diffusion of the primary reaction product is probably the major pitfall in the microscopic localization of enzyme activity. The majority of localization procedures are of the simultaneous capture type described earlier. In this case, diffusion of the reaction product will be dependent on a number of factors including the size of the enzymic site, the rate of the enzymic reaction, the diffusion coefficient of the primary reaction product and the rate of the capture reaction. The interactions of these various factors and the possible ways of estimating such diffusion will not be detailed here since they have been discussed elsewhere (Chayen *et al.*, 1973; Cornelisse and Van Duijn, 1973a and b; Holt and O'Sullivan, 1958). However, this is clearly a very complex problem. For example, fixation may reduce the rate of the enzymic reaction and so maintain a low concentration of the primary reaction product. This will reduce the possibility of precipitation and so increase the diffusion of the primary reaction product. This source of error may be reduced by increasing the concentration of the capture reagent although this, in turn, may have undesirable effects. The use of lead as a capture reagent in the localization of ATP-ase and other phosphatases introduces a number of potential pitfalls (see Essner, 1973) and increasing the concentration of lead ions should be treated with considerable caution. Adequate controls to test for possible diffusion artifacts should always be run, and these include the use of a range of concentrations of substrate and capture reagent, a range of incubation times and the overlapping section technique (see Danielli, 1953).

D. SPECIFICITY OF THE REACTION

The specificity of a localization procedure is best studied by using a range of substrates and inhibitors and varying the concentration of substrate and cofactors and the pH of the incubation medium. The effects of these variations are more difficult to assess than in biochemical assays because it is not usually possible to measure the reaction quantitatively. This is particularly the case when several enzymes are present with overlapping substrate specificities. However, variation of the substrate concentration, pH and incubation time may show clear differences in the localization sites of such enzymes. An example of this is given by the localization of surface phosphatase activity discussed earlier (Fig. 1). Artificial substrates are, of course, widely used in histochemical studies since the specificity of many enzyme reactions depends on a particular bond in the molecule rather than on the complete molecule. The kinetic characteristics of the reactions using such substrates should always be compared with those of the natural substrate by biochemical assay whenever possible.

E. CONTROLS

Most of the potential pitfalls associated with microscopic localization procedures can be identified by the use of suitable controls. Although the problems may vary greatly with different methods, there are a number of general controls that should always be run. The enzymic nature of the reaction may be tested by incubation of sections in media lacking substrate or cofactors, or containing inhibitors, and the incubation of heat- or chemically-inactivated sections in the complete medium. Diffusion artifacts may be assessed by varying the section thickness and the times of fixation, storage and incubation, and by the use of the overlapping section procedure. Artifacts due to diffusion and adsorption of enzyme and reaction product may be investigated by incubation of sections in suitable preparations of the enzyme and product. Loss of enzyme activity from sections and the effects of fixation and components of the reaction medium should be measured biochemically. The use of these controls are described in detail in relation to the localization of acid phosphatase and peroxidase activities by Sexton et al. (1971) and Hall and Sexton (1972).

IV. APPLICATIONS

Although the number of enzymes that can be localized microscopically is relatively small, these techniques have made important contributions in a number of areas of physiology and cell biology. With animal cells, the localization procedure for acid phosphatase has demonstrated the wide morphological diversity of lysosomes and the relationship between primary lysosomes and Golgi bodies (see de Duve and Wattiaux, 1966; Beck et al., 1972). Methods for the localization of thiamine pyrophosphatase and other phosphatases have been shown to be valuable in monitoring the fractionation of Golgi bodies and other membranes (Farquhar et al., 1974), while cytochemical techniques for transport ATP-ases have been successfully applied to problems associated with the amphibian epidermis (Farquhar and Palade, 1966), red cell ghosts (Marchesi and Palade, 1967) and the avian salt gland (Ernst, 1972a, b). The localization procedure for horseradish peroxidase has been shown to be a very useful tracer in the study of protein absorption and breakdown by mammalian cells (Strauss, 1967; Graham and Karnovsky, 1966) and, as discussed earlier, the DAB stain for peroxidase is probably the most useful label at present available for immunochemical studies. With plants, the DAB method for catalase has been widely used in the identification of microbodies (Silverberg, 1975; Vigil, 1973) and in studies of microbody development (Vigil, 1970; Gruber *et al.*, 1973). Localization procedures for acid phosphatases and other hydrolases have been important in establishing the autophagic properties of vacuoles (Hall and Davie, 1971; Pitt, 1975), and in the study of hydrolases in various developmental processes (for example, Gahan and Maple, 1966; Sutcliffe and Sexton, 1969; Villiers, 1971). Cytochemical methods have been applied in investigations of cell surface enzymes, including the role of ATP-ases in transport processes (see Hall, 1973) and the function of peroxidase in cell wall synthesis (see De Jong, 1967; Hepler *et al.*, 1972). The application of microscopic procedures to the study of enzyme changes in developmental processes in plants will now be illustrated in more detail with three examples. These are concerned with root differentiation, leaf abscission and the washing or ageing process in storage tissue discs.

A. PEROXIDASE LOCALIZATION AND DIFFERENTIATION IN ROOTS

The developing root has often been used as a system for studying the changes that occur in enzymes and metabolism in relation to differentiation (Brown and Broadbent, 1950; Sutcliffe and Sexton, 1969; Fowler, 1975). Its great advantage is that it may be cut into serial segments along its length for biochemical analysis, these sections representing different stages of differentiation. However, each segment represents a heterogeneous cell population and differences between segments may be related largely to changes in the dominant cell type. The application of microscopic techniques may allow a more detailed analysis of changes occurring in specific cell types. The use of biochemical analysis of serial sections in combination with microscopic histochemistry was used by Jensen (1955) in an analysis of peroxidase activity in *Vicia faba* roots. A correlation was reported between peroxidase activity and lignification, although only the first 3 mm of the root was studied and the localization achieved was limited by the section thickness and solubility of the reaction product. Later, histochemical studies of onion roots found no correlation between peroxidase and ligni- fication (De Jong, 1967; Goff, 1975), whereas peroxidase localization in sections of trees (Harkin and Obst, 1973) and wound vessel members (Hepler *et al.*, 1972) indicated a strong linkage between the two activities. These later reports did not include a biochemical analysis.

In the present study this question has been re-examined. Peroxidase activity in pea roots has been localized by light and electron microscopy and measured biochemically in serial segments cut along the root. The biochemical analysis showed that there was an overall increase in the specific activity of peroxidase in total root homogenates as root develop- ment and differentiation proceeds (Fig. 5). When activity was separated into a 500 g pellet and supernatant fraction, there was a marked change in

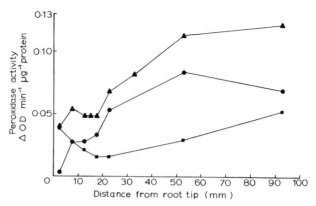

FIG. 5. Distribution of peroxidase along the axis of pea roots. Pea roots were cut into 5 mm segments and assayed for peroxidase as described in Fig. 3. Activity was measured in the total homogenate (▲), the 500 g supernatant (■) and the 500 g pellet (●).

the distribution of activity, particularly in the region up to 20 mm behind the tip. Activity in the supernatant fraction declined while that in the pellet increased, perhaps indicating an increase in cell wall activity. Activity was expressed on a total protein basis which is probably the most meaningful basis (Fowler, 1975). Since there is little change in the number of cells per segment in pea roots from about 5 mm onwards (Sutcliffe and Sexton, 1969), expression of the results on a cellular basis would make little difference to this overall pattern.

This change in peroxidase activity was then examined microscopically. Roots were fixed in glutaraldehyde, sectioned with a vibratome, and stained for peroxidase before examining by light microscopy or processing further for electron microscopy. Both guaiacol and DAB were used as hydrogen donors for light microscopy and the latter for electron microscopy. Although the solubility of the reaction product with guaiacol makes it less desirable as a histochemical reagent, the overall pattern of staining was similar for guaiacol and DAB.

In the root cap and meristematic regions, highest activity was observed in the outer root cap cells and epidermis while activity was low in the meristem (Fig. 6a); the staining in root cap and meristematic cells appeared to be largely cytoplasmic with little staining in the walls. In the region from 600 μm to 2000 μm from the tip, heavy staining was found in the cells of the root cap, epidermis, inner cortex, endodermis and areas of the phloem, while there was little activity in the outer cortex and developing xylem (Fig. 6b, d). A similar distribution of peroxidase activity in the root tip was reported by Hall and Sexton (1972). The activity in these cells was localized largely in the cell walls, particularly at the corners of the intercellular spaces, and in the membranes of the vacuoles (Fig. 6c, 7a). In older regions of the root (20–25 mm and 50–55 mm behind the tip), peroxidase activity was associated with most cell types and found in both the cell wall and cytoplasm (Fig. 6e, 7b). Staining in the protoxylem, however, was low although activity was found in the cell walls bordering the intercellular spaces (Fig. 6e), and a strong reaction was observed in the walls of the metaxylem (Figs. 6e, f). Much of the peroxidase activity associated with the cell walls of the various cell types may be ionically bound, since treatment of isolated cell wall fractions with 0·5 M-NaCl removed approximately 71% of the activity from the walls. Staining of cell wall fractions with the DAB procedure both before and after NaCl treatment indicated that much of the peroxidase washed from the walls came from the sites of intense activity bordering the intercellular spaces (Fig. 7c, d).

The xylem cells showed a strong positive reaction when stained with phloroglucinol, which is widely used as a test for lignin, although such staining was only observed in the xylem in regions beyond 17 mm from the root tip. Phloroglucinol staining was also occasionally observed in the cell walls of cortical cells bordering the intercellular spaces, a site where high peroxidase activity was found. A possible explanation of the differences reported in the literature concerning the association between lignification and peroxidase activity is that different authors have studied different regions of the root and so different stages of the process. For example, the level of differentiation of pea roots, including lignification, in terms of distance from the tip can vary with such factors as root aeration, and between varieties (Popham, 1955). In addition, it must be added that a negative response to a staining reaction for lignin does not necessarily mean that lignin is absent (Jensen, 1962). Thus the present study does not rule out a possible role for peroxidase in lignification, although the widespread occurrence of the enzyme in all tissues of the root suggests that it may also have other functions in plant metabolism.

B. ENZYMIC CHANGES DURING LEAF ABSCISSION

During the abscisson of leaves, flowers and fruits, marked changes occur in the levels of certain enzymes in the abscission zone (see Poovaiah et al., 1973; Henry et al., 1974). The precise localization of these enzymes can only be established by microscopic histochemistry since the cells of the abscission zone and surrounding tissues cannot be fully separated for biochemical analysis. It is also clearly impracticable to isolate the organelles of the abscission zone cells to establish the subcellular basis of these enzymic changes. There have been a number of histochemical studies which have demonstrated changes in enzymic activity in the abscission zone but few fine structural and cytochemical studies of leaf abscission. Although it is clear that dissolution of the cell walls is an essential part of the abscission process (see Sexton and Hall, 1974), microscopic methods are not available for the localization of wall-degrading enzymes, such as cellulase. However, some indications as to the nature of the subcellular changes that occur have

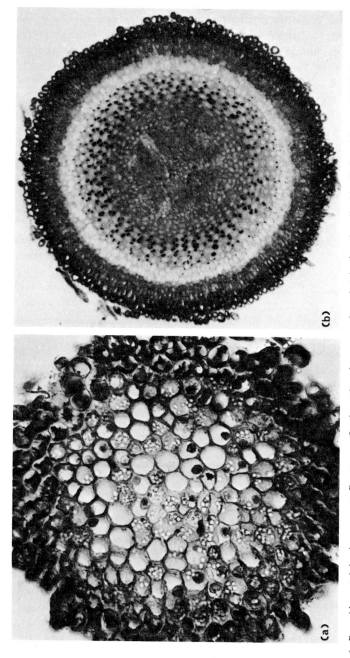

FIG. 6. Peroxidase staining in pea roots. Roots were fixed, sectioned transversely and stained as described by Hall and Sexton (1972). Px (protoxylem) and Mx (metaxylem).

(a) 250 μm behind the tip ×296;
(b) 600 μm behind the tip ×150.

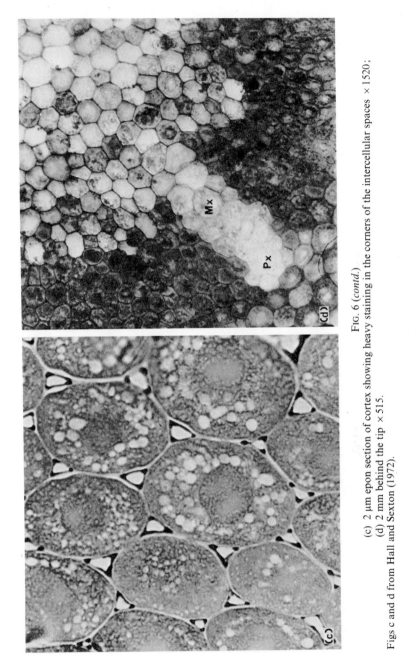

FIG. 6 (contd.)

(c) 2 μm epon section of cortex showing heavy staining in the corners of the intercellular spaces ×1520;

(d) 2 mm behind the tip ×515.

Figs c and d from Hall and Sexton (1972).

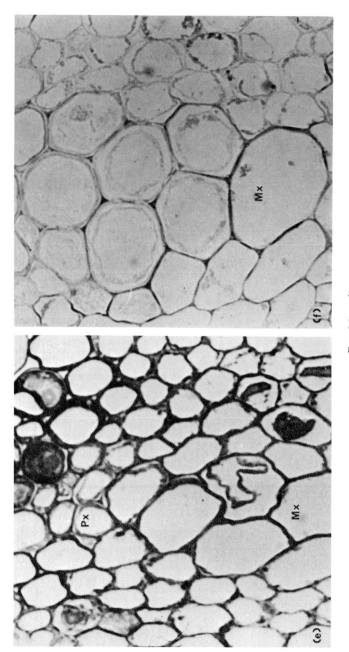

Fig. 6 (*contd.*)
(e) 20–25 mm behind the tip × 1400.
(f) 50–55 mm behind the tip × 1400.

been obtained from a study of peroxidase and acid phosphatase activity in the abscission of bean leaves (Hall and Sexton, 1974). Peroxidase has been implicated in lignification, auxin oxidation and disease resistance; all of these activities may be important at various stages of the abscission process. Acid phosphatase is of interest as a marker for cellular autolysis since there is some dispute as to the degree of cellular breakdown occurring during separation.

When the fine structure of the separation zone cells of bean leaves was examined, the major difference found between these cells and those in a similar region of control plants with no abscission zone was a large increase in the number of Golgi bodies and in the amount of endoplasmic reticulum (ER) (Sexton and Hall, 1974). When examined by light microscopy, heavy staining for acid phosphatase was observed in the abscission zone cells and in the vascular tissue (Fig. 8a). The staining in the abscission zone appeared to be largely associated with the cell walls, and this was confirmed by electron microscopy which showed reaction product chiefly in the middle lamellar region (Fig. 8b). This resembled the localization of acid phosphatase in cultured carrot cells in regions of the cell wall which eventually disintegrate (Halperin, 1969). In addition, intense staining for acid phosphatase was found in the Golgi bodies and vesicles (Fig. 8c). Staining for peroxidase activity showed a similar localization to that for acid phosphatase, being found mainly in the cell walls, Golgi bodies and ER (Fig. 8d); staining was found throughout the wall rather than in the middle lamella. One interpretation of the increase in volume of Golgi bodies and endoplasmic reticulum, and of the high enzymic activity associated with these organelles, is that they are involved in the secretion of enzymes into the cell wall. Morre (1968) has postulated that abscission must involve the secretion of abscission-specific enzymes into the walls of the separation zone cells. These results would appear to support this idea, although methods are really needed for the localization of wall-degrading enzyme, such as cellulase.

C. ENZYMIC CHANGES IN WASHED STORAGE TISSUE DISCS

The washing of storage tissue slices causes marked changes in the metabolism and physiological properties of the component cells, and these may be associated both with a wound response (Lipetz, 1970) and with some degree of cell differentiation (Van Steveninck, 1975). It is therefore of interest to compare the structural and cytochemical changes associated with this washing phenomenon with the processes of root differentiation and abscission which have been discussed earlier. Peroxidase activity, for example, is known to increase in response to cutting injury in both tobacco stem pith and storage roots (Birecka and Miller, 1974) and this may involve the *de novo* synthesis of peroxidase isoenzymes (Shannon *et al.*, 1971). In this study, peroxidase localization has been investigated in fresh and washed beetroot

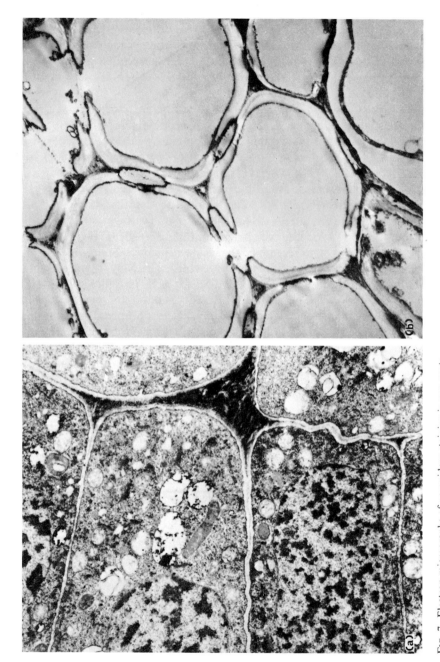

FIG. 7. Electron micrographs of peroxidase staining in pea roots.
(a) Portions of epidermal cells showing heavy staining in the cell walls and vacuoles ×6500.
(b) Developing xylem fibres. 90–95 mm behind the tip, showing heavy staining in the cell walls ×4500.

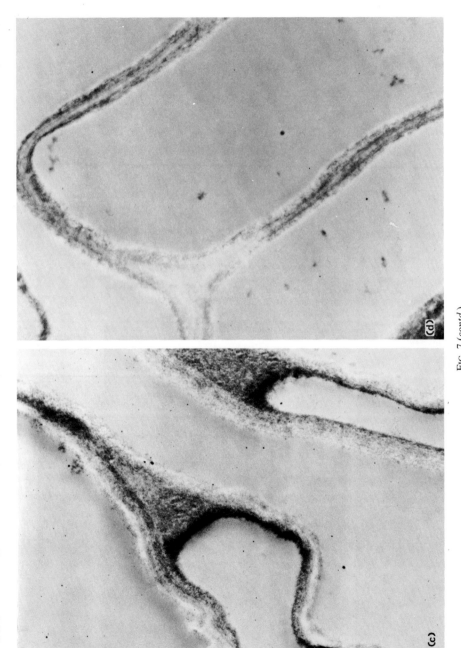

Fig. 7 (*contd.*)

(c) Peroxidase staining of isolated cell walls ×30 000.
(d) Peroxidase staining of isolated cell walls after washing with 0·5 M NaCl × 22 500.

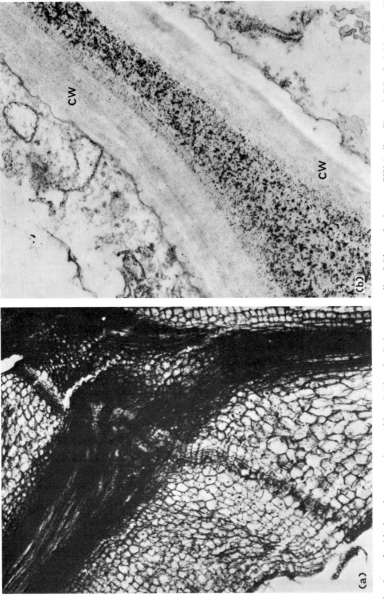

FIG. 8. Localization of acid phosphatase and peroxidase in the abscission zone cells of *Phaseolus* leaves. CW (cell wall) and GB (Golgi body). (From Hall and Sexton, 1974.)

(a) Longitudinal frozen section through the base of the primary leaf petiole cut three days after cold induction and stained for acid phosphatase. Note the high activity in the developing abscission zone and vascular tissue ×30.

(b) Electron micrograph of two cells at the proximal fracture face which appear to be separating and showing intense staining for acid phosphatase in the middle lamella ×25 500.

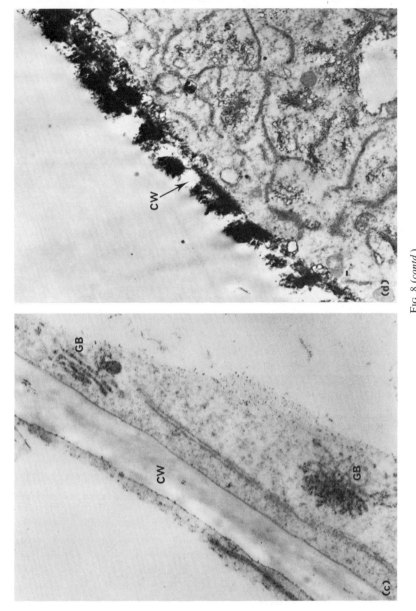

Fig. 8 (contd.)

(c) As in (b) showing staining for acid phosphatase in the Golgi bodies × 27 750.
(d) As in (b) but stained for peroxidase and showing intense staining in the cell wall and Golgi bodies × 9500.

discs by biochemical and cytochemical methods. Glucose-6-phosphatase activity has also been studied since this enzyme is considered to be a marker for the ER in animal cells (see Leskes *et al.*, 1971), and the ER has been shown to increase in amount in washed storage tissue slices (Van Steveninck, 1975).

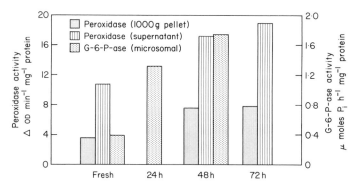

FIG. 9. Development of peroxidase and glucose-6-phosphatase activity in washed beetroot discs. 1 mm discs were cut and washed under sterile conditions for the periods shown. Peroxidase was assayed at pH6 as in Fig. 3, and glucose-6-phosphatase by the release of inorganic phosphate at pH 5·5.

Both peroxidase and glucose-6-phosphatase activities were present in freshly cut discs of beetroot, although the levels of cell wall and supernatant peroxidase, and glucose-6-phosphatase in a microsomal fraction, increased with washing for periods up to 72 h (Fig. 9). When sections of washed discs were stained for peroxidase and examined by light microscopy, heavy staining was observed in the cell walls, particularly in the outer few layers of cells (Fig. 10a). This is consistent with a biochemical study of peroxidase enhancement in sweet potato slices where highest activity was observed in the layer closest to the wound (Lipetz, 1970). When cell division was observed, very little peroxidase staining was found in the new cell walls (Fig. 10b). Electron microscopy showed that the staining was associated largely with the cell walls, Golgi bodies and ER; staining was found throughout the cell wall and did not appear to be heavier in the regions bordering the intercellular spaces (Fig. 10c). This pattern of staining was similar to that described earlier for peroxidase in the abscission zone of bean leaves, and to that reported in a study of peroxidase activity in relation to injury and infection of tobacco leaf tissue (Birecka *et al.*, 1975). In both these studies, it was speculated that the Golgi bodies and ER might be involved in the movement of increased peroxidase activity to the cell wall by reverse pinocytosis. A similar process may operate in washed storage tissue discs and in all cases may be associated with the formation of a protective layer. The lack of activity in new cells walls in this tissue may be compared with the low cell wall activity reported earlier for the root meristem region.

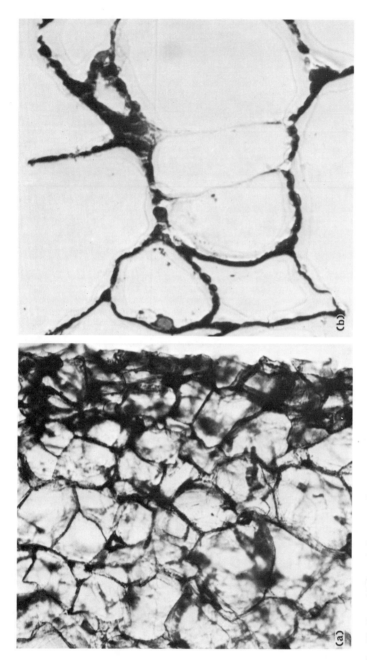

Fig. 10. Peroxidase and glucose-6-phosphatase localization in washed storage tissue discs. CW (cell wall), GB (Golgi body), V (vacuole).

(a) Light micrograph of Vibratome-cut section of the surface of a washed artichoke disc showing intense peroxidase staining, particularly in the cells nearest the surface ×160.

(b) 2 μm epon section of washed beetroot disc stained for peroxidase. Intense staining is associated with most cell walls, except those formed by new cell divisions ×675.

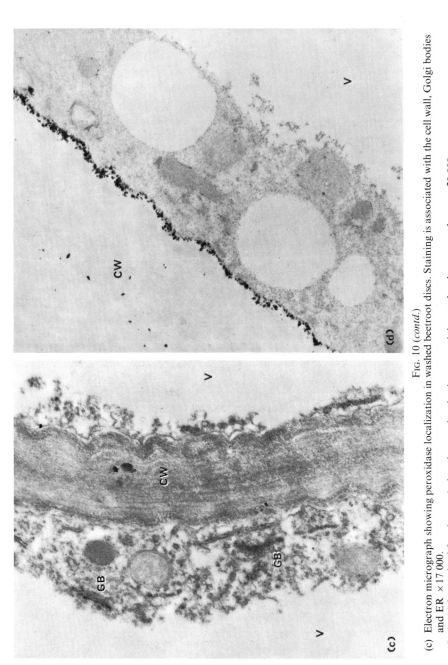

Fig. 10 (*contd.*)

(c) Electron micrograph showing peroxidase localization in washed beetroot discs. Staining is associated with the cell wall, Golgi bodies and ER ×17 000.

(d) As (c) but stained for glucose 6-phosphatase showing intense staining at the plasma membrane ×20 000.

Staining for glucose-6-phosphatase activity was found largely at the cell wall/plasma membrane interface (Fig. 10d); some activity was found associated with the tonoplast. Staining was very much heavier in washed discs than in the freshly cut tissue. This observation was consistent with the biochemical finding of increased microsomal activity with washing, since the plasma membrane and tonoplast would presumably be sedimented in this fraction, together with other cellular membranes. There was no evidence of activity associated with the ER as found with animal cells, and so glucose-6-phosphatase may not be a reliable marker for the ER in higher plants.

REFERENCES

Avers, C. J. (1958). *Am. J. Bot.* **45**, 609.

Al-Azzawi, M. J. and Hall, J. L. (1976). *Pl. Sci. Let.* **6**, 285.

Bal, A. K., Verma, D. P. S., Byrne, H. and MacLachlan, G. A. (1976). *J. Cell Biol.* **69**, 97.

Beck, F., Lloyd, J. B. and Squier, C. A. (1972). *In* "Lysosomes" (Dingle, J. T., ed.), pp. 200–239. North Holland, Amsterdam and London.

Birecka, H. and Miller, A. (1974). *Pl. Physiol. Lancaster*, **53**, 569.

Birecka, H., Catalfamo, J. L. and Urban, P. (1975). *Pl. Physiol., Lancaster* **55**, 611.

Brown, R. and Broadbent, D. (1950). *J. exp. Bot.* **1**, 247.

Burstone, M. S. (1962). *In* "Enzyme Histochemistry and its Application to the Study of Neoplasms" Academic Press, New York and London.

Chayen, J., Bitensky, L. and Butcher, R. G. (1973). *In* "Practical Histochemistry" John Wiley and Sons, Chichester.

Christie, K. N. and Stoward, P. J. (1974). *Proc. R. Soc. Lond.* B **186**, 137.

Cornelisse, C. J. and van Duijn, P. (1973a). *J. Histochem. Cytochem.* **21**, 607.

Cornelisse, C. J. and van Duijn, P. (1973b). *J. Histochem. Cytochem.* **21**, 614.

Danielli, J. F. (1953). *In* "Cytochemistry. A Critical Approach" John Wiley and Sons, Chichester.

Daoust, R. (1965). *Int. Rev. Cytol.* **18**, 191.

de Duve, C. and Wattiaux, R. (1966). *A. Rev. Physiol.* **28**, 435.

Edwards, M. L. and Hall, J. L. (1973). *Protoplasma* **78**, 321.

Ernst, S. (1972a). *J. Histochem. Cytochem.* **20**, 13.

Ernst, S. (1972b). *J. Histochem. Cytochem.* **20**, 23.

Essner, E. (1973). *In* "Electron Microscopy of Enzymes. Principles and Methods" (Hayat, M. A., ed.), Vol. I, pp. 44–76. Van Nostrand Reinhold, New York.

Fahimi, H. D. (1973). *J. Histochem. Cytochem.* **21**, 999.

Fahimi, H. D. and Drochmans, P. (1968). *J. Histochem. Cytochem.* **16**, 199.

Farquhar, M. G. and Palade, G. E. (1966). *J. Cell Biol.* **30**, 359.

Farquhar, M. G., Bergron, J. J. M. and Palade. G. E. (1974). *J. Cell Biol.* **60**, 8.

Fisher, D. B. (1968). *J. Cell Biol.* **39**, 745.

Fowler, M. W. (1975). *New Phytol.* **75**, 461.

Gahan, P. B. and Maple, A. J. (1966). *J. exp. Bot.* **17**, 151.

Gahan, P. B., McLean, J., Kalina, M. and Sharma, W. (1967). *J. exp. Bot.* **18**, 151.

Graham, R. C. and Karnovsky, M. J. (1966). *J. Histochem. Cytochem.* **14**, 291.

Goff, C. W. (1975). *Am. J. Bot.* **62**, 280.

Goff, C. W. and Klohs, W. D. (1974). *J. Histochem. Cytochem.* **22**, 945.

Gomori, G. (1939). *Proc. Soc. exp. Biol. Med.* **42**, 23.

Gruber, P. J., Becker, W. M. and Newcomb, E. H. (1973). *J. Cell Biol.* **56**, 500.

Hall, J. L. (1969a). *Planta* **89**, 254.

Hall, J. L. (1969b). *Ann. Bot.* **33**, 339.

Hall, J. L. (1971). *J. Microsc.* **93**, 219.

Hall, J. L. (1973). *In* "Ion Transport in Plants" (Anderson, W. P., ed.), pp. 11–24. Academic Press, London and New York.

Hall, J. L. and Davie, C. A. M. (1971). *Ann. Bot.* **35**, 849.

Hall, J. L. and Sexton, R. (1972). *Planta* **108**, 103.

Hall, J. L. and Sexton, R. (1974). *Ann. Bot.* **38**, 855.

Halperin, W. (1969). *Planta* **88**, 91.

Harkin, J. M. and Obst, J. R. (1973). *Science* **180**, 296.

Hayat, M. A. (1973). *In* "Electron Microscopy of Enzymes. Principles and Methods" (Hayat, M. A., ed.), Vol. I, pp. 1–43. Van Nostrand Reinhold, New York.

Henry, E. W., Valdovinos, J. G. and Jensen, T. E. (1974). *Pl. Physiol., Lancaster* **54**, 192.

Hepler, P. K., Rice, R. M. and Terranova, W. A. (1972). *Can. J. Bot.* **50**, 977.

Herzog, V. and Fahimi, H. D. (1974). *J. Cell Biol.* **60**, 303.

Holt, S. J. and O'Sullivan, D. G. (1958). *Proc. R. Soc. Lond.* B **148**, 465.

Jacobsen, J. V. and Knox, R. B. (1973). *Planta* **112**, 213.

Jensen, W. A. (1955). *Pl. Physiol. Lancaster* **30**, 426.

Jensen, W. A. (1962). *In* "Botanical Histochemistry". W. H. Freeman, London.

de Jong, D. W. (1967), *J. Histochem. Cytochem.* **15**, 335.

Kyte, J. (1976). *J. Cell Biol.* **68**, 287.

Leskes, A., Siekevitz, P. and Palade, G. E. (1971). *J. Cell Biol.* **49**, 264.

Lipetz, J. (1970). *Int. Rev. Cytol.* **27**, 1.

Marchesi, V. T. and Palade, G. E. (1967). *J. Cell Biol.* **35**, 385.

Mayersbach, H. V. (1967). *J. R. Microsc. Soc.* **87**, 295.

Morimoto, T., Matsuura, S., Sasaki, S., Tashiro, Y. and Omura, T. (1976). *J. Cell Biol.* **68**, 189.

Morre, D. J. (1968). *Pl. Physiol., Lancaster* **43**, 1545.

Pearse, A. G. E. (1968). *In* "Histochemistry. Theoretical and Applied" Churchill London, 3rd ed.

Pitt, D. (1975). *In* "Lysosomes and Cell Function" Longman, London and New York.

Poovaiah, B. W., Rasmussen, H. P. and Bukovac, M. J. (1973). *J. Am. Soc. Hort. Sci.* **98**, 16.

Popham, R. A. (1955). *Am. J. Bot.* **42**, 529.

Poux, N. (1970). *J. Microsc.* **9**, 407.

Quicho, F. A. and Richards, F. M. (1966). *Biochemistry,* **5**, 4062.

Schejter, A. and Bar-Eli, A. (1970). *Archs Biochem. Biophys.* **133**, 325.

Sexton, R. and Hall, J. L. (1974). *Ann. Bot.* **38**, 849.

Sexton, R., Cronshaw, J. and Hall, J. L. (1971). *Protoplasma* **73**, 417.

Shannon, L. M., Uritani, I. and Imaseki, H. (1971). *Pl. Physiol., Lancaster* **47**, 493.

Shnitka, T. K. and Seligman, A. M. (1971). *A. Rev. Biochem.* **40**, 375.

Silverberg, B. A. (1975). *Protoplasma* **83**, 269.

Sternberger, L. A. (1973). *In* "Electron Microscopy of Enzymes. Principles and Methods" (Hayat, M. A., ed.), Vol. 1, pp. 150–191. Van Nostrand Reinhold, New York.

Strauss, W. (1967). *In* "Enzyme Cytology" (Roodyn, D. B., ed.), pp. 253–258. Academic Press, London and New York.

Sutcliffe, J. F. and Sexton, R. (1969). *In* "Root Growth" (Whittington, W. J., ed.), pp. 80–102. Butterworth, London.

Takamatsu, H. (1939). *Trans. Soc. Pathol., Japan* **29**, 492.
Trelease, R. N., Becker, W. M. and Burke, J. J. (1974). *J. Cell Biol.* **60**, 483.
Van Steveninck, R. F. M. (1975). *A. Rev. Pl. Physiol.* **26**, 237.
Vigil, E. L. (1970). *J. Cell Biol.* **46**, 435.
Vigil, E. L. (1973). *J. Histochem. Cytochem.* **21**, 958.
Villiers, T. A. (1971). *Nature new Biol.* **233**, 57.

Author index

Numbers in italic are those pages on which references are listed in full

A

Aach, H. J., 35, 36, *38*
Aarnes, H., 30, 36, *38*, *39*, 257, *267*
Acton, G. J., 98, 106, 111, *125*, 236, 239, 241, *242*
Adams, M. J., 280, 284, *287*
Adams, P. A., 89, *91*
Agarwal, K. L., 272, *287*
Ainslie, R. E., 156, *174*
Aitchison, J. M., 73, 74, 75, 77, *80*
Aitchison, P. A., 72, 74, 75, 76, *80*, *81*
Akazawa, T., 87, *92*, 221, *223*
Akeson, A., 284, *286*
Aki, K., 4, *19*
Al Azzawi, M. J., 331, *361*
Albersheim, P., 84, *92*
Albertsson, P. A., 300, 312, 313, *323*, *325*
Albrecht, J., 249, *269*
Aliotta, G., 9, *21*
Allen, R. H., 272, *286*
Amiraian, K., 207, *222*
Amrhein, N., 103, *125*
Andersen, R. A., 251, *267*
Anderson, J. H., 60, *61*
Anderson, J. M., 294, *323*
Anderson, J. W., 247, 250, 251, 253, 254, 255, *267*
Anderson, L., 284, *286*
Anderton, B. A., 284, *286*
Andrews, P., 263, *267*
Ap Rees, T., 25, *38*, 48, 51, *61*
Ap Rees, T., 248, *269*
Armstrong, J. E., 87, *91*
Arnon, R., 198, 207, *222*
Ashford, A. E., 84, 87, *91*
Ashworth, J. M., 18, *19*
Astell, C. R., 272, *286*
Atkins, C. A., 36, *39*
Atkinson, A., 284, *286*
Atkinson, D. E., 23, *39*
Attridge, T. H., 99, 102, 104, 106, 111, *125*, *126*, 232, 234, 235, 236, 237, 238, 240, 241, *242*, *243*

Avers, C. J., 332, *361*
Azou, Y., 88, *92*

B

Baccari, V., 52, *61*
Bahn, E., 139, *153*
Bain, J., 314, 316, *323*
Bal, A. K., *361*
Baldry, C., 39, 254, 262, *267*
Baldry, C. W., 292, 301, *324*, *326*
Ball, E., 255, *268*
Baltscheffsky, H., 300, 312, *323*
Banno, Y., 179, *194*
Barash, I., 2, *19*
Bar-Eli, A., 340, *362*
Barett, J. T., 207, *222*
Barnard, R. A., 38, *40*
Barrett, H. C. M., 294, *326*
Barry, S., 272, 275, 276, 284, *286*, *288*
Bartels, H., 52, *61*
Barthe, F., 99, *127*
Bassham, J. A., 295, 301, 311, 320, *323*, *325*
Bate-Smith, E. G., 247, 249, *267*, *269*
Battaile, J., 250, 251, *000*
Bauohuin, P., 308, *324*
Bayley, J. M., 2, *19*
Beck, F., 346, *361*
Becker, W. M., 335, 347, *362*, *363*
Beevers, H., 26, *39*, 51, 55, *61*, 292, 293, 294, 295, 297, 298, 299, 300, 303, 305, 306, 307, 308, 309, 322, *323*, *324*, *325*, *326*
Beevers, L., 2, *19*
Behrens, M., 310, *323*
Bell, E., 118, *126*
Belser, W. L., 4, *20*
Bender, R. A., 5, 14, *21*
Bennett, P. A., 84, *91*
Bennett, V., 98, *125*
Ben-Tal, Y., 86, *91*
Berger, E., 207, *222*
Berglund, O., 276, *286*

Bergron, J. J. M., 346, *361*
Bernardi, G., 265, *267*
Berlin, C. M., 7, *20*
Betz, H., 18, *20*, 178, 187, 188, 189, *193 194*
Bianchetti, R., 3, *20*
Bidwell, R. G. S., 1, 4, 6, *21*
Bielski, R. L., 3, *20*, *21*
Bilderbach, D. E., 87, *91*
Bird, I. F., 299, 310, 311, *323*, *324*
Birecka, H., 353, 358, *361*
Birnie, G. D., 230, *243*
Birt, L. M., 248, *267*
Bishop, R., 79, *80*
Bitensky, L., 335, 340, 345, *361*
Black, C. C., 292, *324*
Blackwood, G. C., 24, *39*
Blair, G. E., 31, *39*
Boardman, N. K., 294, *323*
Bock, R. M., 100, *126*, 226, 228, 229, 230, *243*
Boisard, J., 99, 115, *125*, *126*
Bond, H. E., 232, *243*
Bonner, W. D., *324*
Borriss, H., 2, *20*, 90, *91*
Borthwich, H. A., 96, *125*
Bosmann, H. B., 76, *80*
Bottomley, R. C., 276, 284, *288*
Bottomley, W., 98, *125*
Boudet, A., 236, *242*
Boulter, D., 267, *267*
Bowman, D. E., 207, *222*
Bowman, H. G., 264, *267*
Bowman, P. D., 322, *324*
Box, V., 79, *80*
Boyer, P. D., 187, *193*
Brawerman, G., 117, *125*
Breidenbach, R. W., 308, *324*
Brenchley, J. E., 5, 14, *21*
Brent, T. P., 69, *80*
Brewin, N. J., 111, *125*
Briggs, D. E., 84, 85, *91*
Briggs, T. M., 251, *267*
Briggs, W. R., 97, 115, *125*, *126*, 183, 184, 185, *194*
Brinckmann, E., 255, *268*
Broadbent, D., 347, *361*
Brodelius, P., 284, *286*, *288*
Broue, P., 142, *153*
Brown, B., 26, *40*
Brown, C. M., 11, *20*
Brown, J. E., 292, *324*
Brown, J. R., 202, *223*

Brown, R., 63, 77, *80*, 347, *361*
Brüning, K., 98, *125*
Brunk, C. F., 233, *242*
Brunner, C. E., 28, 30, *39*
Bryan, J. K., 28, 30, 36, 38, *39*
Bryan, P. A., 28, 30, *39*
Bryce, G. F., 184, 193
Buc, J., 156, 157, 158, 159, 160, 161, 162, 163, 164, 166, *175*
Bucci, E., 207, *222*
Bücher, T.h., 42, 46, *61*
Bucke, C., 254, 262, *267*, 292, *324*
Bukovac, M. J., 349, *362*
Buller, D. C., 86, *91*
Bullock, G., 314, *324*
Bunning, J., 311, *326*
Burbolt, A. J., 253, *267*
Burke, J. J., 335, *363*
Burns, J. A., 42, 44, 45, *61*
Burr, G. O., 292, *325*
Burr, H. E., 232, *243*
Burstone, M. S., 335, *361*
Butcher, R. G., 335, 340, 345, *361*
Butler, J. A. V., 69, *80*
Butler, L. G., 98, *125*
Byrne, H., 89, *91*, *92*, *361*

C

Camm, E. E., 102, 103, *125*
Canvin, D. T., 36, *39*
Carbonara, A. O., 199, *223*
Cardiff, I. D., 64, *80*
Carfantan, N., 202, 203, *222*
Carlin, L., 133, *153*
Carlson, P. S., 5, *20*
Carminatti, H., 281, *286*
Carr, D. J., 85, *91*
Carr, N. G., 7, *20*
Carsiotis, M., 5, *20*, 207, *223*
Cashion, P. J., 272, *287*
Cass, K. H., 284, *288*
Cataldo, D. A., 192, 193, *194*, 248, *269*
Catalfamo, J. L., 358, *361*
Cave, P. J., 27, 29, *39*
Cave, P. R., 4, 5, *21*
Cawley, R. D., 28, 30, *39*
Chamberlain, C. J., 64, *80*
Chambers, J. F., 309, 310, *326*
Chan, P. H., 247, *267*, 284, *286*
Chan, W. W. C., 272, *286*
Chance, B., 46, 47, *61*
Chang, H., 292, 298, 300, 307, *324*
Changeux, J. P., 155, 168, *175*, 272, *288*

Chapman, E. A., 47, *61*
Charney, J., 185, *193*
Chatson, B., 65, 66, *80*
Chayen, J., 335, 340, 345, *361*
Chen, R. F., 87, *91*
Chen, Y. M., 88, *91*, 292, 298, 300, 307, *324*
Cheshire, R. M., 27, 28, 30, *39*
Chichibu, K., 179, *194*
Chin, C. K., 248, *267*
Chipchase, M. I., 307, *326*
Chrispeels, M. J., 84, 86, 87, *91*, 300, *324*
Christian, R. A., 314, *324*
Christie, K. N., 340, *361*
Christou, N. V., 89, *91*
Chroboczek-Kelker, H., 2, *20*
Chu, M., 4, *20*
Cinader, B., 198, *222*
Clarke, H. G. M., 200, *222*
Cleary, T. J., 5, *20*
Cleland, W., 256, *267*
Cleland, W. W., 156, *174*
Clowes, F. A. L., 65, 66, *80*
Clutterbuck, V. J., 84, 85, *91*
Cobb, A. H., 118, *125*, 306, *324*
Cockburn, W., 301, *324*, *327*
Cocking, E. C., 190, *193*
Collin, C., 312, *325*
Collins, G. G., 86, *91*
Comer, M., 284, *286*
Conn, E. E., 99, *126*, 261, *268*
Connelly, C. M., 46, *61*
Connolly, T. N., 248, *268*, 300, *324*
Constabel, F., 65, 66, 67, *80*
Coombe, B. G., 84, *91*
Coombs, J., *39*, 113, 114, *125*, 254, 262, *267*, 292, *324*
Coon, M. J., 261, *268*
Cooper, T. G., 308, *324*
Cornelius, M. J., 299, 310, *324*
Cornlisse, C. J., 345, *361*
Corvazier, P., 87, *91*, 218, 219, *222*
Coultate, T. P., 256, *267*
Covey, S. N., 117, 122, *125*
Cox, B. J., 70, 71, 73, 77, *80*
Cox, D. J., 202, *223*
Crabtree, B., 59, *62*
Craigie, J. S., 52, *62*
Crathorn, A. R., 69, *80*
Craven, D. B., 276, 282, 284, 286, *286*, *287*
Creanor, J., 74, *81*
Creasy, L. L., 107, 110, *125*

Criddle, R. A., 308, *324*
Crofts, A. R., 301, *327*
Cronshaw, J., 337, 338, 339, 340, 344, 346, *362*
Croteau, R., 253, *267*
Crumpton, M. J., 198, *222*
Cuatrecasas, P., 272, 279, *286*, *287*, *288*
Cybis, J., 6, 17, *20*

D

Dalgardo, L., 248, *267*
Dalziel, K., 156, *174*
Danielli, J. F., 344, 345, *361*
Daoust, R., 335, *361*
Das, U. S. R., 297, 298, 304, 305, *325*, *326*
Daussant, J., 87, *91*, 198, 201, 202, 203, 204, 207, 208, 210, 218, 219, 220, *222*, *223*
Davidson, A. W., 68, *80*, *81*
Davie, C. A. M., 347, *362*
Davies, D. D., 24, *39*, 53, 54, 56, 57, 59, 60, *61*, *62*, 89, *92*, 236, *242*
Davies, E., 118, *125*
Davies, J., 9, *22*
Davis, B. D., 6, *20*
Davis, K. A., 260, *267*
Davis, R. H., 6, 17, *20*
Dean, P. D. G., 274, 276, 277, 278, 280, 282, 284, 285, 286, *286*, *287*
Dedonder, A., 249, *269*
de Duve, C., 308, *324*, 346, *361*
De Flora, A., 284, *286*
De Leo, A. B., 5, 14, *21*
De Leo, P., 89, *91*
Delmer, D. P., 4, *20*
Dennis, D. T., 30, 36, *40*, 51, *61*, 256, *267*
Dennis, S., 254, *268*
De Pooter, H., 249, *269*
Desbuquois, B., 272, *287*
Dey, P., 256, *269*
Dicamelli, C. A., 38, *39*
Dijke, U., 298, 303, 307, *325*
Dilworth, M. F., 90, *91*
Di Marco, 258, *267*
Dixon, J. E., 277, 284, *287*
Djurtoft, J., 201, *223*
Doi, E., 190, *193*
Doley, S. G., 277, *286*
Donachie, W. D., 4, *20*
Doronton, W. J. S., 250, *267*
Douce, R., 309, *324*
Douglass, S. A., 308, *324*
Doyle, D., 2, *21*

Drochmans, P., 340, *361*
Drumm, H., 98, 111, *125*, 236, 239, *242*
Dubois, E., 5, 14, *20*
Durst, F., 99, *125*
Dyer, A. F., 63, *80*
Dyer, T. A., 299, 310, *324*

E

Eames, A. J., 64, *80*
Easterday, I. M., 284, *286*
Easterday, R. L., 284, *286*
Ebel, J., 103, *125*
Ebner, E., 18, *20*, 178, *194*
Ebner, K. E., 272, *287*
Eckstein, F., 276, *286*
Edelman, G. M., 272, *286*
Edwards, G. E., 294, *325*
Edwards, M. L., 341, 343, *361*
Eickhoff, F., 31, *39*
Elchinger, I., 98, *125*
Elion, G. B., 272, *287*
Ellis, R. J., 31, *39*, 291, 296, 304, 308, *324*, *325*
Elsässer, S., 178, 188, *194*
Endo, T., 133, *153*
Engelmann, T. W., 289, *324*
Engelsma, G., 102, 106, 107, *125*
Eppenberger, H. M., 130, *153*
Eriksson, T., 65, 66, *80*
Erickson, R. O., 64, 65, *80*
Ernst, S., 346, *361*
Essner, E., 345, *361*
Esterby, 159, *174*
Evans, P. K., 65, 66, 67, 71, *81*, 190, *193*
Evans, A., 306, *324*
Everse, J., 276, 277, 284, *287*
Evins, W. H., 86, *91*
Ewings, D., 89, *92*
Ewing, E. E., 108, *125*
Eye, J. D., 251, *267*

F

Fahimi, H. D., 340, 341, 344, *361*, *362*
Fahrney, D., 183, 184, *193*
Faiz-ur-Rahman, A. T. M., 53, 54, *61*
Fan, D. F., 88, *91*
Farquhar, M. G., 346, *361*
Fasold, H., 284, *286*
Faye, L., 202, 206, 208, 209, *222*
Feeley, J., 118, *126*
Feeney, R. E., 183, 184, *194*
Feirabend, J., 98, *125*

Felder, M. R., 133, 136, 141, 142, 143, 144, *153*
Feldman, K., 272, *287*
Felsted, R. L., 207, *222*
Ferber, E., 272, *288*
Fegruson, A. R., 2, 9, 10, *20*, 188, *193*
Ferrari, T. E., 9, *20*
Fields, M. A., 251, 255, *267*
Filmer, D., 155, 168, *175*
Filner, B., 98, *125*
Filner, P., 2, 3, 7, 9, 11, *20*, *22*, 29, *39*, 76, *81*, 100, *125*, 227, 228, 236, *243*
Finn, F. M., 272, *287*
Firn, R. D., 86, 87, *91*
Firenzuoli, A. M., 52, *61*
Fischer, H., 272, *288*
Fischer, M., 141, *153*
Fisher, D. B., 335, *361*
Fitzgerald, M. P., 295, 296, 313, 321, *325*
Flamm, W. G., 232, *243*
Flesher, D., 146, *153*
Fletcher, J. S., 26, *39*
Flodin, P., 263, *268*
Folkes, B. F., 79, *80*
Fogg, G. E., 7, *20*
Forti, G., 170, 171, *174*, *175*
Foster, P., 298, 306, *326*
Fouchier, F., 166, 167, 168, 169, 170, *175*
Foust, G. P., 170, *174*
Fowden, L., 256, *268*
Fowler, M. W., 52, *61*, 70, 71, 73, 77, *80*, 347, 348, *361*
Frankhauser, D. B., 5, *20*
Fraser, R. S. S., 65, *80*
Frederick, S. E., 308, *324*
Freeling, M., 220, *222*
Freeman, T., 200, *222*
French, C. J., 107, *125*
Fridkin, M., 272, *287*
Frieden, C., 59, *61*, 246, *267*
Fuchs, S., 272, *287*
Furuya, M., 102, *125*

G

Gage, L. P., 85, *92*
Gahan, P. B., 339, 347, *361*
Galliard, T., 254, *268*
Galston, A. W., 102, *125*
Gamborg, O. L., 2, *19*, *20*, 65, 66, 67, *80*
Gander, J. E., 272, *287*
Gandhi, A. P., 10, *20*, 183, *194*
Garg, G. K., 184, *193*
Garnier, V. R., 18, *20*

Gautheron, D. C., 284, *287*
Gawronski, T. H., 272, *287*
Geokas, M. C., 185, *194*
Gerbrandy, S., 256, *268*
Germershausen, J., 272, *288*
Gerstein, J. F., 207, *222*
Ghiron, C. A., 207, *222*
Gibbs, M., 18, *20*, 99, *126*, *127*, 300, 305, *326*
Gibson, R. A., 87, *91*
Giebel, W., 201, *222*
Gilham, P. T., 272, *288*
Ginsburg, A., 246, *268*
Giuliano, F., 284, *286*
Givol, D., 272, *288*
Gladstone, G. P., 26, *39*
Glasziov, K. T., 3, *20*, *21*
Goatly, M. B., 99, 112, 113, 114, *125*, 233, *243*
Godinot, C., 284, *287*
Goff, C. W., 332, 341, 347, *361*
Gold, A. M., 183, 184, *193*
Goldstein, J. L., 251, *268*
Gomori, G., 333, *361*
Good, N. E., 248, *268*, 300, *324*
Good, R. A., 272, *287*
Gooding, L. R., 221, *223*
Goodwin, P. B., 85, *91*
Goodwin, T. W., 98, *126*
Gorini, L., 4, *21*
Gove, D., 314, 316, *323*
Grabar, P., 87, *91*, 199/200, 218, *223*
Gracy, R. W., 184, *193*
Graham, D., 47, *61*, 98, *125*
Graham, R. C., 346, *361*
Gray, J. C., 221, *223*
Green, C. E., 27, 28, *39*
Green, D. E., 86, *92*
Green, T. R., 51, *61*
Greenshields, R. N., 202, *223*
Greenwood, D., 310, *326*
Grego, S., 258, *267*
Grenson, M., 5, 14, *20*
Greppin, H., 99, 115, *126*
Grierson, D., 117, 122, *125*
Grieve, A. M., 98, *125*
Griffiths, P., 280, 284, *287*
Grisebach, H., 99, 103, *125*, *126*
Gross, G. G., 251, *268*
Grosse, W., 31, *39*
Grover, A. K., 284, *287*
Gruber, P. J., 347, *362*
Guerzoni, M. E., 5, *20*

Guilbault, G. G., 272, *287*
Guilford, H., 284, *288*,
Guilfoyle, T. J., 88, *91*
Guilpoyle, T., 292, 298, 300, 307, *324*
Gulyaev, N. N., 284, *288*
Gundlach, H. G., 207, *223*
Güntermann, U., 230, *243*
Gurman, A. W., 36, 38, *39*

H

Haard, N. F., 250, *268*
Häcker, M., 241, *243*
Hageman, R. H., 2, 3, *19*, *20*, *21*, 31, *39*, 146, *153*, 311, *326*
Hahlbrock, K., 99, 101, 102, 103, 105, *125*, *126*, 264, *268*
Hahn, H., 90, *91*
Haissig, B. E., 251, 257, 258, *268*
Hall, D. O., 298, 300, 313, 320, *324*
Hall, J. L., 331, 332, 337, 338, 339, 340, 341, 343, 344, 345, 346, 347, 348, 349, 350, 351, 353, 356, *361*, *362*
Hallaway, M., 292, 295, 302, *324*
Halperin, W., 353, *362*
Halvorson, H. O., 226, 228, 229, 230, *243*
Halvorson, M. O., 100, *126*
Hamaguchi, Y., 179, *194*
Hamms, G. G., 284, *287*
Han, T., 272, *288*
Hanson, J. B., 300, *324*
Hanson, K. R., 102, *126*, 166, 167, *174*
Hanstein, W. G., 260, *268*
Hardie, D. G., 84, *91*
Harel, E., 37, *39*, 256, *268*
Harkin, J. M., 347, *362*
Harland, J., 71, 73, 74, *80*
Harley, J. L., 51, *61*
Harmey, M. A., 298, *325*
Harper, E., 183, *195*
Harris, J. I., 284, *287*
Hart, C. E., 292, *325*
Hart, G., 133, *153*
Hartmann, K. M., 96, *126*
Harvey, M. J., 276, 277, 278, 280, 282, 284, 285, 286, *286*, *287*
Hasilik, A., 190, 191, *193*
Hassid, W. Z., 284, *286*
Hatch, M. D., 18, *20*, 98, *125*, 292, 300, *324*, *325*
Hatefi, Y., 260, *267*, *268*
Hauck, R. A., 251, 267
Haverbach, B. J., 185, *194*
Havill, D. C., 3, *22*

Havir, E. A., 102, *126*, 166, 167, *174*
Hawke, J. C., 302, 321, *324*
Hawker, J. S., 250, *267*
Hayat, M. A., 335, 340, *362*
Hayem, A., 201, *223*
Haystead, A., 31, *39*
Heber, M., 311, *324*
Heber, U., 32, 35, 36, *38*, *39*, 301, 310, 311, 320, 321, *324*
Hegeman, G. D., 2, *22*
Heimer, Y. M., 3, 9, 13, *20*
Heinrich, R., 42, 44, 45, 83, *61*, *62*
Heldt, H. W., 33, 34, *40*, 301, 327
Helmreich, E., 272, *287*
Hendricks, S. B., 96, 102, *125*, *126*
Henk, D., 298, 303, 307, *325*
Henke, R. R., 28, 29, *39*
Henry, E. W., 349, *362*
Henshall, J. D., 98, *126*
Hepler, P. K., 347, *362*
Heremans, J. F., 199, *223*
Herriott, R. M., 207, *223*
Hersh, L. S., 272, *288*
Herzog, V., 341, *362*
Hew, C. S., 300, *326*
Hewitt, E. J., 2, 7, *20*
Heyns, W., *287*
Higgins, J., 42, 46, *61*
Higgins, T., 159, *174*
Higgins, T. J. V., 85, *91*, 117, *126*
Hill, A. C. R., 103, *126*, 260, 261, *268*
Hill, R., 289, 297, 320, *324*, *327*
Hill, R. J., 201, *223*
Hind, G., 33, 34, *39*
Hinde, R. W., 79, *80*
Hinze, H., 187, *193*
Hipkin, C. R., 9, 11, *22*
Hirji, R., 29, *39*
Hirvonen, A. P., 304, *326*
Hizukuri, S., 248, 256, *269*
Hjebten, S., 264, *269*
Ho, D., 233, *243*
Ho, D. H., 85, *91*, 183, *193*
Ho, D. H. T., 147, 148, 150, 151, *153*
Hocking, J. D., 284, *287*
Hofmann, K., 272, *287*
Hoglund, S., 272, *288*
Hohorst, H. J., 42, *61*
Holborrow, K., 291, *325*
Hollows, M. E., 284, *287*
Holmes, W., 46, *61*
Holt, S. J., 345, *362*
Holtz, R. B., 309, *324*

Holzer, H., 18, *20*, 177, 178, 187, 188, 189, 190, 191, *193*, *194*
Honda, S. I., 295, *324*, *327*
Hong, S.-C., 77, *81*
Hongladoram, T., 295, *324*, *327*
Honigman, W. A., 85, *92*
Horecker, B. L., 183, 184, *194*
Horinishi, H., 185, *194*
Hornby, W. E., 272, *287*
Horton, P., 306, *324*
Horvath, C., 272, *287*
Hotta, Y., 3, *20*, 69, 73, 77, *80*
Houlton, A., 280, 284, *287*
Hsiao, T. C., 87, *91*
Hu, A. S. L., 100, *126*, 226, 228, 229, 230, *243*
Huang, A. H., 308, 322, *324*
Huang, A. H. C., 299, 303, *324*
Huffaker, R. C., 10, 17, 18, *20*, *22*, 98, *126*, 183, 185, 187, *194*, 198, 218, 219, 220, *223*
Hulla, F. W., 284, *286*
Hulme, A. C., 250, 251, *268*, *269*, 300, *324*
Hultin, H. O., 250, *268*
Humphrey, T. J., 236, *242*
Huskins, C. L., 64, *80*
Hutter, R., 5, *21*
Hütterman, A., 230, *243*
Hyodo, H., 103, *126*

I

Ichihara, A., 4, *19*
Ikuma, H., 89, *91*
Illiano, G., 272, *286*
Imase Ki, H., 353, *362*
Indge, K. J., 190, *194*
Ingle, J., 3, *20*
Inman, D. J., 272, *287*
Inouye, H., 272, *287*
Ireland, S. E., 102, *126*
Izawa, S., 33, 34, *39*
Ito, T., 87, *92*
Izawa, S., 248, *268*
Izaway, S., 300, *324*

J

Jackson, J. F., 71, 73, 74, *80*
Jackson, R. J., 284, *287*
Jacobasch, G., 44, 45, 53, *62*
Jacobs, E. E., 260, *268*
Jacobsen, J. V., 84, 85, 86, 87, *91*, 117, *126*, 336, *362*

Jacobson, A. B., 300, 306, *324*
Jacoby, W. B., *288*
Jaffe, M. J., 121, *126*
Jagendorf, A. T., 221, *223*
Jakob, K. M., 65, 66, *80*
Jakobs, M., 241, *243*
Jakoby, W. B., 246, 258, *268*, 272
James, W. O., 297, 298, 304, *325*
Jay, E., 272, *287*
Jeffries, R. L., 24, *39*
Jenner, C. F., 86, *91*
Jensen, R. A., 30, *39*
Jensen, R. G., 295, 301, 311, 320, *323*, *325*
Jensen, T. E., 349, *362*
Jensen, W. A., 332, 335, 347, 349, *362*
Jetschmann, K., 237, *243*
Jimenez, de Asua, L., 281, *286*
Johansson, B. G., 201, *223*
Johnson, C. B., 98, 106, 111, 112, *125*, *126*, 232, 234, 235, 236, 238, 240, 241, *242*, *243*
Johnson, F. J., 26, 38, *40*
Johnson, H. S., 300, *325*
Johnson, K. D., 86, *91*
Johnston, F. B., 307, *325*
Jones, J. D., 250, *268*, 300, *324*
Jones, R. F., 5, *20*
Jones, R. L., 84, 85, 87, *91*, *92*
de Jong, D. W., 347, *361*
Jordon, W. R., 10, 17, *22*
Jörnvall, H., 131, *153*, 284, *286*
Jovanneau, J. P., 65, 66, 67, *80*, *81*
Jovin, T. M., 272, *287*
Joy, K. W., 3, *20*, 31, 33, 34, 35, *40*, 311, *326*
Julliard, J. H., 284, *287*

K

Kacser, H., 42, 44, 45, *61*
Kadam, S. S., 183, *194*
Kafatos, F. C., 184, *194*, 207, *222*
Kagawa, T., 295, 297, 309, *325*
Kahl, G., 52, *61*
Kahn, A., 313, 320, *325*
Kalghatigi, K. K., 167, 169, *175*
Kalina, M., 339, *361*
Kanai, R., 294, *325*
Kanamori, T., 2, *20*
Kaplan, D., 2, 9, *21*
Kaplan, N. O., 276, 277, 284, *287*
Karlstam, B., 312, *325*
Karnovsky, M. J., 346, *361*
Karya Kina, T. I., 2, *20*

Katsunuma, T., 178, 179, 188, *193*, *194*
Katunuma, N., 177, 179, 180, *194*, 256, *268*
Kato, J., 87, *92*
Katzen, H. M., 272, *287*, *288*
Kaufman, P. B., 89, *91*
Kavam, S. K., 10, *20*
Kays, E., 90, *91*
Keates, R. A. B., 88, *91*
Kekwick, R. G. O., 221, *223*, 296, 327
Keller, C. J., 98, *126*
Kende, H., 86, 90, *91*
Kende, H. A., 86, *91*
Kenrick, K. G., 203, *223*
Kenyon, A. J., 272, *287*
Kerfoot, M. A., 284, *287*
Key, J., 88, *92*
Key, J. L., 88, *91*, 118, 120, *126*, *127*, 292, 298, 300, 307, *324*
Keys, A. J., 33, 35, *39*, 299, 310, *324*, *325*, *326*
Khandker, R., 156, *175*
Khorana, H. G., 272, *287*
Kim, Y. S., 7, 8, *21*
King, J., 2, *19*, 29, *39*
King, P. J., 65, 66, 67, 70, 71, 73, 77, *80*, *81*
Kinghorn, J. R., 14, *21*
Kinze, G., 255, *268*
Kirk, P. R., 31, 32, *39*, 304, 305, *325*
Kito, T., 177, *194*
Klein, A. O., 98, 118, 123, *125*, *126*, *127*
Klein, S., 306, *325*
Kleinkopf, G. E., 199, 218, 219, 220, *223*
Klohs, W. D., 341, *361*
Kneen, E., 210, *223*
Knight, R. H., 118, *125*
Knobloch, K., 99, 102, 103, 105, *125*, *126*
Knobloch, K. H., 264, *268*
Knox, R. B., 336, *362*
Knypl, J. S., 2, 9, 10, *20*
Kobashi, K., 183, *194*
Kobayashi, K., 179, *194*
Kobr, M. J., 55, *61*
Kochetkov, S. N., 284, *288*
Koehler, D. E., 86, *92*
Kohen, C., 45, *61*
Kohen, E., 45, *61*
Kojima, M., 103, *127*
Koller, B., 121, *126*
Kollöffel, C., 298, 303, 307, *325*
Komani, H., 272, *287*
Kominami, E., 177, 179, *194*, 256, *268*
Kominami, S., 177, *194*

Komoszynski, M., 260, *268*
Konishi, S., 2, *20*
Kornberg, A., 272, *287*
Kortschak, H. P., 292, *325*
Koshland, D. E. Jr., 245, *268*
Koshland, D. E., 155, 168, *175*
Kosuge, T., 25, *40*
Kovacs, C. J., 65, 66, *80*
Kranz, B., 272, *288*
Krebs, H. A., 41, 42, 59, 60, *61*
Krentz, F. H., 42, *61*
Kretovich, V. L., 2, *20*
Kreuzaler, J. R., 99, 102, 105, *126*
Krøll, J., 203, *223*
Krueger, W. A., 300, 324
Krug, E., 272, *287*
Kruger, J. E., 210, 212, 219, *223*
Kuhlen, E., 103, *126*
Kula, M. R., 189, *195*
Kung, S., 293, *325*
Kung, S.-d., 247, *268*
Kurz, W. G. W., 65, 66, 67, *80*
Kyte, J., 336, *362*
Kyuwa, K., 147, *153*

L

Lacy, A. M., 5, *20*
Laetsch, W. M., 292, 293, *325*, *326*
Laidman, D. L., 87, *92*
Lam, T. H., 251, *268*
Lamb, C. J., 103, *126*, 238, *243*
Lamed, R., 276, 278, 284, *287*, *288*
Lange, H., 52, *61*
Lappi, D. A., 276, 284, *287*
Lara, J. C., 5, *20*
Larkins, B. A., 118, *125*
Larsson, C., 312, 313, *325*
Larsson, P.-O., 276, 284, *286*, *287*
Laskowski, M., 185, *194*
Lata, M., 2, *21*
Latzko, E., 18, *20*, 300, *326*
Laurell, C. B., 200, 203, *223*
Laurière, C., 201, 208, *223*
Law, J. H., 184, *194*, 207, *222*
Laycock, D., 24, *39*
Lea, P. J., 2, 3, *21*, 31, 32, 37, *39*
Leaver, C. J., 118, *126*, 298, *325*
Lebas, J., 201, *223*
Leblová, S., 147, *153*
Lee, C.-L. T., 277, 284, *287*
Lee, C.-Y., 276, 277, 284, *287*
Lee, J. A., 3, *22*

Lee, Y. P., 185, *194*
Leech, R. M., 31, 32, *39*, 293, 298, 300, 302, 303, 304, 305, 306, 313, 315, 316, 320, 321, 322, *324*, *325*, *326*, (332)
Leese, B. M., 298, 300, 302, 304, 316, *325*
Lehman, E., 256, *268*
Lehrer, H. I., 207, *223*
Leick, V., 233, *242*
Lembi, C. A., 292, 295, 297, 299, 303, 309, 319, *326*, *327*
Lenney, J. F., 178, *194*
Leonard, R. T., 292, 296, 297, 299, 300, 302, 303, 309, 319, *325*
Lerner, N. H., 249, *267*
Leskes, A., 358, *362*
Levin, O., 264, *269*
Levin, Y., 276, 278, 298, *287*
Levine, L., 207, *222*
Lilley, R. McC., 291, 292, 295, 296, 298, 301, 313, 320, 321, *325*, *326*, *327*
Lin, C. Y., 88, *91*, 120, *127*, 292, 298, 300, 307, *324*
Lindberg, M., 276, 284, *287*
Lindl, T., 103, *126*
Lipetz, J., 353, 358, *362*
Lippin, A., 272, *288*
Lips, S. H., 2, 9, *21*
Littauer, U. Z., 272, *287*
Littlefield, J. W., 77, *81*
Lloyd, J. B., 346, *361*
Loening, U. E., 65, *80*
Long, S. P., 254, 262, *267*
Longo, C. P., 3, *21*
Longo, G. P., 3, *21*
Loomis, W. D., 248, 250, 251, 253, 254, 267, *268*
Lopez, C., 272, *287*
Lord, J. M., 295, 297, 309, *325*
Losada, M., 182, *194*
Lowe, C. R., 274, 276, 282, 284, 286, *286*, *287*, *288*
Lowe, M., 284, *288*
Lowenstein, J. M., 55, *62*
Lu, A. Y. H., 261, *268*
Luppis, B., 184, *194*
Lutstorf, U., 131, 133, *153*
Luttge, U., 255, *268*
Lyndon, R. F., 296, 307, *325*
Lynen, F., 246, *268*
Lyon, F. M., 64, *80*
Lyttleton, J. W., 307, *325*

M

MacDonald-Brown, D. S., 11, 20
Mackenzie, J. M., 115, 126
Maclachlan, G. A., 88, 89, 91, 92, 299, 309, 326, 361
MacLennan, D. H., 51, 61
Macleod, A. J., 72, 73, 80
Mc Adoo, M. H., 108, 125
McCarty, R. E., 260, 268
McClendon, J. H., 294, 325
McClure, J. W., 28, 39
McKersie, B., 309, 325
McNeil, M., 84, 92
McLean, J., 339, 361
Magalhaes, A. C., 31, 39
Magasanik, B., 5, 14, 21
Maheshwari, P., 64, 80
Majerus, P. W., 272, 286
Mancini, G., 199, 223
Mandelstam, J., 2, 6, 18, 21
Mander, L. N., 84, 91
Mannella, C. A., 324
Manning, R. F., 85, 92
Mansell, R. L., 251, 268
Mansfield, K. J., 65, 67, 70, 80
Maple, A. J., 347
March, S., 279, 288
Marchesi, V. T., 346, 362
Marcus, A., 118, 126
Margolis, J., 203, 223
Market, C. L., 130, 153
Marmé, D., 115, 125, 126
Marshall, D. L., 272, 287
Marshall, D. R., 142, 153
Martin, J. P., 201, 223
Martin, M. A., 272, 288
Mascarenhas, J. P., 118, 126
Maslowski, P., 260, 268
Masoner, M., 98, 127
Massey, V., 170, 174, 272, 287
Masters, M., 74, 81
Matern, H., 189, 194
Matheson, A., 199, 218, 223
Mathews, M. B., 117, 126
Matoba, T., 190, 193
Matsuda, Y., 189, 194
Matsuura, S., 336, 362
Matthews, B. F., 36, 38, 39
Mattingly, E., 65, 66, 81
Maw, S. L., 292, 324
Mawal, R., 272, 287
May, S. W., 286, 287

Mayer, A. M., 256, 268
Mayersbach, H. V., 337, 362
Mayhew, S. P., 170, 174
Mazelis, M., 31, 39
Means, G. E., 183, 184, 194
Meers, J. L., 11, 20
Megnet, R., 133, 153
Meijer, G., 102, 125
Mellor, G. E., 37, 39
Mense, R. M., 85, 87, 92
Merrill, A. L., 246, 269
Meselson, M., 228, 243
Messing, R. A., 272, 288
Meunier, J. C., 156, 157, 158, 159, 160, 161, 162, 163, 164, 166, 175, 272, 288
Michaelis, M., 45, 61
Miflin, B. J., 2, 3, 4, 5, 6, 21, 24, 25, 26, 27, 28, 29, 30, 31, 32, 36, 37, 39, 40, 297, 298, 300, 303, 305, 306, 307, 308, 325
Milborrow, B. V., 25, 39
Millard, D. L., 292, 306, 326
Miller, A., 353, 361
Millette, C. F., 272, 286
Milovancev, M., 33, 34, 40
Mills, S. E., 4, 5, 20
Miozzari, J., 5, 21
Miron, T., 278, 288
Mitchell, D. J., 38, 40
Mitchell, J. P., 65, 81
Mitchison, J. M., 74, 77, 79, 81
Miyasaka, K., 65, 66, 81
Mohr, H., 98, 99, 102, 104, 111, 114, 115, 121, 125, 126, 127, 236, 239, 241, 242, 243
Mollenhauer, H. H., 303, 309, 310, 326
Möller, J., 98, 125
Møller, F., 130, 153
Molloy, G. R., 76, 79, 81
Momotani, Y., 87, 92
Monod, J., 155, 168, 175
Montague, M. J., 89, 91
Montgomery, M. W., 254, 268
Mor, H., 2, 19
de Moor, P., 284, 287
Moore, G. L., 185/187, 194
Moore, T. S., 293, 294, 297, 303, 309, 325, 326
Morelli, A., 284, 286
Morgenthaler, J. J., 305, 326
Morimoto, T., 336, 362
Morré, D. J., 292, 295, 297, 299, 303, 309, 310, 319, 326, 327, 353, 362
Morré, H., 303, 326

Morris, I., 9, *21*
Morrison, J. F., 272, *287*
Morton, R. K., 260, *268*
Mosbach, K., 276, 277, 284, *286, 287, 288*
Moses, M. J., 64, *81*
Mouttet, C. H., 102, *126*, 156, 157, 158, 160, 166, 167, 168, 169, 170, *175*
Moyed, H. S., 6, *21*
Mudd, J. B., 315, *326*
Muller, B., 18, *21*
Müller, H., 190, 191, *193*
Murphy, G. P., 272, *288*
Murphy, J. B., 4, *20*

N

Nagao, R. T., 88, *91*
Nagata, T., 65, 66, *81*
Naguchi, M., 295, *326*
Naik, G. G., 65, 66, 75, *81*
Naik, M. S., 10, *20*, 183, *194*
Nakatini, H. Y., 33, 34, *39*
Nari, T., 102, *126*, 156, 157, 158, 160, 166, 167, 168, 169, 170, *175*
Nasatin, M., 307, *325*
Nascimento, K. H. D., 59, 60, *61*
Navarro, A., 156, 159, 161, 162, 163, 164, *175*
Neet, K. E., 156, *174, 175*, 245, *268*
Neidhardt, F. C., 236, *243*
Nemethy, G., 155, 168, *175*
Nesterova, M. V., 284, *288*
Neurath, A. R., 272, *288*
Neurath, H., 202, *223*
Nevins, B. I., 64, *81*
Newcomb, E. H., 308, *324*, 347, *362*
Newsholme, E. A., 59, *62*
Neyra, C. A., 31, *39*
Nielsen, G., 147, *153*
Niku-Paavola, M. L., 87, *91*, 204, *223*
Nilsson, B., 207, *222*
Nishi, A., 65, 66, *81*
Nishimura, H., 221, *223*
Niskich, J. T., 292, 306, *326*
Nobel, P. S., 33, 34, *40*
Nola, E., 272, *288*
Noltmann, E. A., 184, *193*
Norris, R. D., 256, *268*
Northcote, D. H., 65, 66, 67, *81*, 111, *125*
Northrop, J. H., 185, *194*
Novelli, G. D., 118, *127*
Nylen, V., 156, *175*

O

Oaks, A., 1, 3, 4, 6, *21*, 26, 37, 38, *40*
Obst, J. R., 347, *362*
O'Carra, P., 272, 275, 276, 284, *286, 288*
O'Donnell, I. J., 266, *268*
Oelze-Karow, H., 98, 114, 115, *126*
Ohlsson, R., *288*
Ohlsson, P., 284, *288*
Ohtsura, C., 190, *193*
Okamura, S., 65, 66, *81*
Oleinick, N. L., 74, *81*
Olered, R., 210, *223*
Olsen, L. C., 2, *22*
Olsen, R. W., 272, *288*
Omura, T., 336, *362*
O'Neal, D., 31, 33, 34, 35, *40*, 300, *326*
Ongun, A., 310, 315, *326*
Opik, H., 292, 295, 306, 309, *326*
Orebamjo, T. O., 3, 16, *21, 22*
Ornston, L. N., 2, *22*
Ortmann, R., 99, *126*
Osborne, D. J., 88, *92*
Osmond, C. B., 255, *268*
O'Sullivan, D. G., 345, *362*
Ouchterlony, O., 199, 203, *223*

P

Packer, 44
Paigen, K., 272, *288*
Palade, G. E., 346, 358, *361, 362*
Paleg, L. G., 84, 86, 87, *91, 92*
Palmer, G., 272, *287*
Palmer, J. K., 254, *268*
Pardee, A. B., 74, *81*
Parikh, I., 272, 279, *288*
Paris, C. G., 5, 14, *21*
Parker, M. W., 96, *125*
Parker, W., 86, *91*
Pate, J. S., 18, *21*
Pateman, J. A., 14, *21*
Patil, K. D., 56, 59, 60, *61*
Patterson, B., 117, *126*
Pe, I., 249, *269*
Pearse, A. G. E., 334, 335, *362*
Peaud-Lenöel, C., 67, *81*
Penefsky, H. S., 260, *268*
Penel, C., 99, 115, *126*
Penon, P., 88, *92*
Perez-Villasenor, J., 184, *195*
Perlmann, G. E., 207, *222*
Perrin, D., 207, *223*
Peter, K., 98, *125*

Peterson, D. M., 192, 193, *194*, 248, *269*
Peterson, E. A., 263, *268*
Peterson, L. V., 185, *194*
Peterson, L. W., 18, *20*, 183, 185, 187, *194*, 219, 220, *223*
Petterson, G., 156, *175*
Phillips, R. L., 27, 28, *39*
Pierpoint, W. S., 249, 250, 255, *268*
Pierson, D. L., 30, *39*
Pietruszko, R., 131, *153*
Pike, G. S., 183, 184, 185, *194*
Pine, K., 118, *126*
Pinna, M. H., 102, *126*, *175*
Pirson, A., 98, *125*
Pitt, D., 347, *362*
Plachy, C., 99, *126*
Plummer, T. H. Jr., 207, *222*
Pogell, B. M., 281, *288*
Poljakoff-Mayber, A., 306, *325*
Pomerantz, Y., 85, *92*
Pon, N. G., 311, *324*
Pontremoli, S., 184, *194*
Poovaiah, B. W., 349, *362*
Popham, R. A., 349, *362*
Porath, J., 263, *268*
Porter, H. K., 311, *323*
Potts, J. R. M., 261, *268*
Potts, R. M., 99, 102, 105, *126*
Poulik, M. D., 201, *223*
Poux, N., 344, *362*
Pratt, H. M., 31, *39*
Pressey, R., 180, *194*
Price, C. A., 247, *268*, 304, 305, *326*
Pridham, J. B., 256, *269*
Priess, J., 25, *40*
Prince, A. M., 272, *288*
Prival, J. M., 5, 14, *21*
Pryor, A. J., 142, *153*
Puca, G. A., 272, *288*

Q

Quail, P. H., 115, *126*, 148, *153*
Queiroz, O., 98, *126*
Quicho, F. A., 340, *362*

R

Rabin, B. R., 156, *175*, 184, *193*
Racker, E., 260, *268*
Radin, J. W., 3, 9, *21*
Raftery, M. A., 272, *288*
Ram Chandra, G., 228, 241, *243*
Ramponi, G., 52, *61*
Ramshaw, J. A. M., 267, *267*

Rapoport, S., 44, 45, 52, *61*, *62*
Rapoport, T. A., 42, 44, 45, 52, *62*
Rasmussen, H. P., 349, *362*
Rathnam, C. K. M., 305, *326*
Raval, D. M., 48, *62*
Ray, M. M., 310, *326*
Ray, P. M., 310, *326*
Raylè, D. L., 89, *91*
Reid, J. S. G., 86, *91*
Reid, M. S., 3, *21*
Reiland, J., 263, *269*
Renard, M., 210, *222*
Rendon, G. A., 3, 4, 6, 11, 13, 15, 19, *21*
Ressler, N., 200, *223*
Rho, J. H., 307, *326*
Rhodes, D., 3, 4, 6, 11, 13, 15, 16, 17, 18, 19, *21*, *22*
Rhodes, M. J. C., 103, *126*, 248, 254, 260, 261, 262, *268*, *269*
Ribbereau-Gayon, P., 249, *269*
Ricard, J., 88, *92*, 102, *126*, 156, 157, 158, 159, 160, 161, 162, 163, 164, 166, 167, 168, 169, 170, *175*
Ricardo, C. P. P., 248, *269*
Rice, R. M., 347, *362*
Richards, F. M., 340, *362*
Rickwood, D., 230, *243*
Ridge, I., 88, *92*
Ridley, S. M., 302, 322, *326*
Rienits, K. G., 295, 296, 313, 321, *325*
Rigano, C., 9, *21*
Rijven, A. H. G. C., 3, *21*
Rinderknecht, H., 185, *194*
Rippa, M., 184, *194*
Ritenour, G. L., 311, *326*
Robbins, K. C., 207, *223*
Roberts, B., 117, *126*
Roberts, G. R., 310, *326*
Roberts, J. B., 254, *268*
Roberts, K., 65, 66, 67, *81*
Robertson, A. I., 75, *81*
Robertson, D., 293, *326*
Robertson, R. N., 292, 306, 309, *324*, *325*
Robinson, J. M., 305, 320, 321, *326*
Rocha, V., 297, 303, 305, 307, *326*
Rognes, S. E., 30, 36, *39*, 257, *267*, *269*
Roland, S. C., 309, 319, *326*
Rolleston, F. S., 47, *62*
Rollin, P., 98, 116, *127*
Rose, C., 265, *269*
Rosenstock, G., 52, *61*
Ross, C. W., 120, *127*
Roth-Bejerano, N., 2, 9, *21*

Roughen, P. G., 294, *326*
Rowan, K. S., 254
Roy, H., 221, *223*
Rozengurt, E., 281, *286*
Rubery, P. H., 238, *243*
Rubsamen, H., 156, *175*
Rumbsy, M. G., 293, 302, 321, *324, 325*
Russell, D. W., 99, *126*, 261, *269*
Rüssmann, W., 46, *61*
Rutishauser, U., 272, *286*

S

Sacher, J. A., 3, *21*, 89, *91, 92*
Sadon, T., 2, *19*
Saechtling, H., 201, *222*
Saheki, T., 178, 187, 189, *194*
Sahulka, J., 2, *21*
Sakano, K., 247, *267*
Salomon, Y., 284, *288*
Sampson, D., 272, *288*
Sanadi, D. R., 260, *268*
Santarius, K. A., 31, *39, 40*, 311, *324, 326*
Sanwal, B. D., 2, *21*
Sarawek, S., 53, *62*
Sardi, A., 272, *287*
Sartirana, M. L., 3, *20*
Sasaki, S., 336, *362*
Savageau, M. A., 4, *21*, 45, 48, *62*
Sawkney, S. K., 10, *20*, 183, *194*
Scandalios, J. G., 87, *91*, 130, 133, 136, 138, 141, 142, 143, 144, 147, 148, 150, 151, *153*, 183, *193*, 208, *223*, 233, *243*, 245, *269*
Schaller-Hekeler, B., 103, *125*
Schechter, B., 207, *222*
Schejter, A., 340, *362*
Schenker, T. M., 133, *153*
Schill, J. P., 156, *174, 175*
Schimke, R. T., 2, 7, *20, 21*, 111, *126*
Schipper, A. L. Jr., 251, 257, 258, *268*
Schlamowitz, M., 185, *194*
Schmidt, J., 272, *288*
Schmidt, R. R., 75, 76, 79, *80*
Schoellmann, G., 184, *194*
Schopfer, P., 98, 99, 105, 106, 121, *125, 126, 127*, 236, 239, 241, *242*
Schött, E., 178, 188, *194*
Schött, E. H., 178
Schrader, L. E., 3, *21*, 192, 193, *194*, 248, *269*
Schröder, J., 101, 105, *125*

Schurch, A., 5, *21*
Schwartz, D., 133, 139, 141, *153*, 206, *223*
Schwenn, J. D., 291, 292, 296, *325, 326*
Scott, K. J., 52, *62*
Scott, M., 284, *288*
Scouten, W. H., 286, *288*
Scrutton, M. C., 47, *62*
Seastone, C. V., 207, *223*
Segal, H. L., 7, 8, *21*
Seifter, S., 183, *195*
Sela, M., 272, *287*
Seligman, A. M., 334, 335, 336, 340, *362*
Selniger, Z., 284, *288*
Sercarz, E. E., 4, *21*
Severin, E. S., 284, *288*
Sexton, R., 331, 332, 337, 338, 339, 340, 344, 345, 346, 347, 348, 349, 350, 351, 353, 356, *361, 362*
Sgarbieri, V. C., 254, *268*
Shankar, V., 256, *268*
Shannon, L. M., 208, *223*, 353, *362*
Shapiro, B. M., 17, *21*
Shapiro, S. S., 207, *223*
Shargool, P. D., 4, 6, *21*
Sharma, W., 339, *361*
Shaw, C. R., 130, *153*
Shaw, E., 183, 184, *194, 195*
Shaw, M., 251, *268*
Shen, T. C., 2, *21*, 90, *91*
Shepard, D. V., 2, *21*
Shewry, P. R., 30, 31, 36, *40*
Shih, H. Y., 272, *288*
Shininger, T. L., 310, *326*
Shiota, T., 284, *287*
Shiotani, T., 179, *194*
Shivaram, K. N., 256, *268*
Shnitka, T. K., 334, 335, 336, 340, *362*
Shore, G., 299, 309, *326*
Shumway, L. K., 311, *326*
Shuster, L., 265, *269*
Shyluk, J. P., 2, *20*
Sica, V., 272, *288*
Siegelman, H. W., 102, *126*
Siekevitz, P., 358, *362*
Siepen, D., 189, *195*
Silverberg, B. A., 347, *362*
Silverman, P., 185, *194*
Sims, A. P., 17, *21*, 24, *39*, 79, *80*
Singh, S., 247, *267*
Singh, R. M. M., 248, *268*, 300, *324*
Sitz, T. O., 76, *81*
Skakoun, A., 87, *91*, 201, 202, 204, 208, 220, *222, 223*

Slack, C. R., 18, *20*, *21*, 98, *125*, 250, 254, *269*, 292, 294, *324*, 326
Slaughter, C. J., 50, 51, *62*
Sluyterman, L. A. E., 272, *288*
Small, D. A. P., 276, 284, *288*
Smillie, R. M., 52, *62*, 98, 125, 311, *326*
Smith, F. W., 3, *22*
Smith, H., 99, 102, 104, 106, 107, 112, 113, 114, 118, 119, 120, 121, 123, *125*, *126*, 232, 233, 234, 235, 236, 237, 238, 240, 241, *242*, *243*, 306, *324*
Smith, M., 272, *286*
Smith, S. G., 64, *80*
Smithers, M. J., 272, *287*
Sober, H. A., 263, *268*
Soderman, D. D., 272, *287*, *288*
Solomonson, L. P., 237, *243*
Sorensen, D. B., 284, *288*
Sorger, G. J., 9, *22*
Soulen, T. K., 2, *22*
Sowers, J. A., 251, *267*
Sparrow, A. H., 64, *81*
Spencer, D., 294, 313, 320, *323*, *326*
Spotswood, T. M., 86, *92*
Spurr, A. R., *326*
Squier, C. A., 346, *361*
Stadtman, E. R., 17, *21*, 246, *268*
Stahl, C. A., 88, *92*
Stanier, R. Y., 2, *22*
Starkenstein, E., 271, *288*
Steele, R., 64, *81*
Steer, B. T., 17, *22*, 99, *126*, *127*, 295, *326*
Stegemann, H., 256, *268*
Stellwagen, E., 284, *288*
Stenflo, J., 201, *223*
Stern, H., 3, *20*, 65, 69, 73, 77, *80*, *81*, 307, *325*
Sternberger, L. A., 336, *362*
Stewart, G. R., 2, 3, 4, 6, 11, 13, 14, 15, 16, 17, 18, 19, *21*, *22*, 24, *39*, 102, *125*
Sthal, F. W., 228, *243*
Stiborová, M., 147, *153*
Stocking, C. R., 31, *40*, 293, 310, 311, 320, 321, *323*, *326*
Stolzenbach, F. E., 277, 284, *287*
Stoward, P. J., 340, *361*
Stowe, B. B., 102, *125*
Strating, M., 284, *288*
Strauss, W., 346, *362*
Street, H. E., 65, 66, 67, 70, 71, 73, 77, *80*, *81*

Streicher, S. L., 5, 14, *21*
Sturani, E., 170, *174*
Subba Rao, P. V., 167, 169, *175*
Summaria, L., 207, *223*
Sund, H., 131, *153*
Sundberg, L., 272, *288*
Suskind, S. R., 207, *223*
Sutcliffe, J. F., 332, 347, 348, *362*
Sutter, A., 99, *126*
Suzuki, K., 179, *194*
Suzuki, Y., 147, *153*
Svendsen, P. J., 201, *223*, 265, *269*
Swain, T., 249, 251, 252, *269*
Swanson, E. S., 315, *326*
Swanson, H. R., 300, *324*
Synge, R. L. M., 249, *269*
Syrett, P. J., 9, 11, *21*, *22*

T

Tabahoff, B., 284, *288*
Taiz, L., 84, *92*
Takahashi, E., 2, *20*
Takahashi, T., 185, *194*
Takamatsu, H., 333, *362*
Takebe, I., 65, 66, *81*
Takeda, Y., 248, 256, *269*
Tamaki, E., 295, *326*
Tanaka, R., 74, *81*
Tanaka, Y., 87, *92*, 103, *127*
Tartakoff, A. M., 184, *194*
Tashiro, Y., 336, *362*
Tautvydas, K. J., 292, 307, *326*
Taylor, S. S., 277, 284, *287*
Teeple, L. J., 133, *153*
Teissere, M., 88, *92*
Tenner, A. J., 86, *91*
Tepler, M., 89, *91*
Terranova, W. A., 347, *362*
Tezuka, T., 98, *127*
Thalacker, R., 310, *323*
Theorell, H., 131, *153*
Thien, W., 121, *127*
Thompson, J. E., 309, *325*
Thompson, J. R., 3, *22*
Thompson, S. T., 284, *288*
Thomson, W. W., 293, 298, 300, 302, 304, 306, 315, 316, *325*, *326*
Thorell, B., 45, *61*
Thurman, D. A., 2, *21*
Ting, I. P., 297, 303, 305, 307, *326*
Tiselius, A., 264, 265, *269*
Tkemaladze, G. S., 2, *20*
Tobin, E. M., 123, *127*

Tolbert, N. E., 246, *269*, 299, 308, *326*, *327*
Tomarelli, R. M., 185, *193*
Tomino, S., 272, *288*
Toole, E. H., 96, *125*
Toole, V. K., 96, *125*
Tornheim, K., 55, *62*
Towers, G. H. N., 102, 103, *125*
Traniello, S., 184, *194*
Travis, R. L., 10, 17, *22*, 120, *127*
Trayer, I. P., 272, 276, 284, *288*
Trayer, H. R., 272, 276, 284, *288*
Treffry, T., 312, *327*
Tregunna, E. B., 37, *39*
Trelease, R. M., 335, *363*
Trewavas, A. J., 53, 54, *61*
Trosko, J. E., 65, 66, *80*
Truffa-Bachi, P., 272, *288*
Tsai, H., 189, *195*
Turner, D. H., 51, 55, 57, *62*
Turner, J. F., 51, 55, 57, *62*
Turner, J. V., 84, *91*
Tyson, C. A., 86, *92*
Tyson, H., 251, 255, *267*
Tyler, B. M., 5, 14, *21*
Tzagoloff, A., 260, *268*

U

Umbarger, H. E., 23, 26, 29, *40*
Unser, G., 98, *127*
Unt, H., 313, 320, *326*
Urban, P., 358, *361*
Uriel, J., 198, 199, *223*
Uritani, I., 103, *127*, 353, *362*
Ursprung, H., 133, *153*
Utter, M. F., 47, *62*

V

Valdovinos, J. G., 349, *362*
Van Bruggen, J. M. H., 107, *125*
Vanderhoef, L. N., 88, *92*
Van der Woude, W. J., 292, 295, 296, 297, 299, 300, 302, 303, 309, 319, *325*, *327*
Van Duijn, P., 345, *361*
Van Huystee, R. B., 88, *92*
Vanni, P., 52, *61*
Van Poucke, M., 99, *127*
Van Steveninck, R. F. M., 353, 358, *363*
Van Sumere, C. F., 249, *269*
Van't Hof, J., 65, 66, 80
Van Vunakis, H., 207, *222*, *223*

Varner, J. E., 2, 9, *20*, 84, 85, 86, 87, *91*, *92*, 100, *125*, 227, 228, 241, *243*
Varty, K., 87, *92*
Vennesland, B., 237, *243*
Verma, D. P. S., 89, *91*, *92*, *361*
Vesterberg, O., 266, *269*
Vigil, E. L., 309, *327*, 347, *363*
Villiers, T. A., 347, *363*
Vinograd, J., 228, *243*
Violante, U., 9, *21*
Virupaksha, T. K., 184, *193*
Visser, J., 284, *288*
Von Wartburg, J. P., 131, 133, *153*, 284, *288*
Vutz, H., 276, 284, *288*

W

Wade, R. N., 202, *223*
Waldron, J. C., 3, *20*
Walker, J. R. L., 251, *269*
Wallace, W., 10, *22*, 37, *40*, 90, *92*, 111, *127*, 180, 182, 192, 193, *195*, 237, *243*, 248, 257, *269*
Wallach, D. F. H., 272, *288*
Walker, D. A., 33, *40*, 291, 292, 295, 296, 298, 301, 302, 313, 320, 321, 323, *324*, *325*, *326*, *327*
Walls, D., 79, *80*
Walt, B. K., 246, *269*
Wang, P., 284, *288*
Wattiaux, R., 346, *361*
Weiare, P. J., 296, *327*
Weier, T. E., 310, 320, *326*
Weintraub, B., 272, *288*
Weith, H. L., 272, *288*
Weklych, R., 261, *268*
Wellburn, A. R., 118, *125*, 298, 306, 312, *326*
Wellburn, F. A. M., 298, 306, 312, *324*, *326*
Wellmann, E., 99, 102, 103, 105, *125*, *126*
Werdan, K., 33, 34, *40*, 301, *327*
Wermuth, B., 276, 284, *287*
West, K. R., 305, 321, *327*
Weston, G. D., 248, *267*
Wettstein, D., 313, 320, *325*
Whitaker, J. R., 184, *195*
White, H. A., 284, *286*
Whittingham, C. P., 310, *326*
Wiame, J. M., 5, 14, *20*
Wickham, M. L., 207, *223*
Widholm, J. M., 4, *20*, *22*, 33, *40*

Wiebers, J. L., 272, *288*
Wijdenes, J., 272, *288*
Wilchek, M., 272, 276, 278, 284, *287*, *288*
Wildman, S. G., 247, *267*, 295, *324*, *327*
Willenbrink, J., 310, *324*
Williams, C. A., 200, *223*
Williams, G. R., 118, *127*, *326*
Williams, L. S., 236, *243*
Wilson, C. M., 300, *324*
Wilson, K. G., 28, 29, *39*
Wilson, M., 64, *81*
Wilson, S. B., 65, 66, *81*
Winger, D., 248, *268*
Winget, G. D., 300, *324*
Winter, W., 248, *268*, 300, *324*
Wiskich, J. T., 305, 321, *327*
Witzel, H., 156, *175*
Woenckhaus, C., 276, 284, *288*
Wofsy, L., 272, *288*
Wolcott, R. M., 284, *287*
Wold, F., 272, *287*
Wolfe, R. G., 48, *62*
Wong, K., 56, *62*
Wong, K. F., 30, 36, *40*
Wood, A., 86, *92*
Woods, E. F., 266, *268*
Woods, J. S., 272, *287*
Wooltorton, L. S. C., 103, *126*, 248, 250, 254, 262, *268*, *269*, 300, *324*
Worthington, C. C., 272, *288*

Wray, J. L., 2, *20*
Wu, R., 44, *62*
Wyman, J., 155, 168, *175*

Y

Yagil, G., 7, 8, *22*
Yamamoto, Y., 98, *127*
Yamasaki, M., 202, *223*
Yang, N. S., 85, *92*, 148, *153*
Yang, S. F., 103, *126*
Yeoman, M. M., 65, 66, 67, 68, 71, 72, 73, 74, 75, 76, 77, *80*, *81*
Yoder, O. C., 9, *20*
Yomo, H., 84, 85, 87, *92*
Yonezawa, Y., 74, *81*
Yu, P. H., 189, *195*

Z

Zaborsky, O. R., 286, *287*
Zande, H. V., 86, *92*
Zanetti, G., 171, *175*
Zeisel, H., 272, *287*
Zenk, M. H., 103, *125*, 251, *268*
Ziegler, H., 18, *21*
Ziegler, I., 18, *21*
Zielke, H. R., 236, *243*
Zielke, R. H., 7, *22*, 76, 77, *81*
Zucker, M., 102, 105, 106, *127*
Zouaghi, M., 98, 116, *127*
Zwar, J. A., 85, *91*, 117, *126*

Subject Index

A

Abscisic acid   84, 85, 89
Acalypha   64
Acer pseudoplatanus   66, 67, 73
  suspension culture   70–71, 78
Acetabularia   313
Acetate   26
Acetohydroxyacid synthetase   4, 5, 6
Acetolactate synthase   26, 27, 29, 30, 31, 36
Acetylornithine-glutamate acety trans-
  ferase   5, 6
Acid phosphatase   84, 89, 113, 233, 235, 240
  localization   344, 346, 353, 356–357
Adenine nucleotides   53–56
Adenylate deaminase   55
Adenylate kinase   55
Adenyl cyclase
  localization   331
Adenylosuccinate synthetase   55
Agrostemma githago   83, 90
Alcohol dehydrogenase   129–153, 203, 205, 206, 220, 233
  characteristics   131, 143–146
  genetic control   138–141
  heterogeneity   131–133
  inhibitor   148–153, 183
  purification and properties   141–146
  regulation of activity   146–153
  spatial distribution   134–138, 191
  temporal control of expression   133–134, 150
  zymogram   132, 133, 134, 135, 137, 138, 140, 141
Aldolase   44, 48, 49, 51–53
  inhibition of   53
Algae   29
Allosteric control   6, 23, 29, 32, 43, 155–174
  and pH desensitization   36, 48
  of malic enzyme   59–61
α-Amanitin   88
Amaranthus   292

Amides   13
Amino acid
  aminotransferase   4, 5
  biosynthetic enzymes for aromatic   4, 5, 31
  biosynthetic enzymes for branched-
    chain   4, 26–29, 30, 31
  effects on enzymes   2, 3, 13, 26
Ammonia   2, 3, 11, 13, 15, 16, 17, 32, 37, 48
α-amylase   84, 85, 110, 117, 208, 210–217, 336
  density labelling   227–228
β-amylase   208, 213, 219
Anacystis nidulans   30
Anthranilate synthase   33, 35
Antibodies   *see* Immunochemistry
Antigens   *see* Immunochemistry
Arabinoxylan   84
Arginine   4, 5
  biosynthesis in *Lemna*   6
  in *Neurospora*   17
Arginosuccinate lyase   5, 6
Ascorbic acid oxidase   111, 239, 240
Aspartate kinase   27, 28, 29, 30, 31, 35, 36, 257
Aspartate semialdehyde   28
Aspartate transcarbamylase (ATC)   71, 73, 75
Aspergillus nidulans   14
ATPase
  localization   330, 336, 341–343, 344, 345, 346, 347
ATP-glucokinase   72, 73
Autoradiography   123, 335
Auxin   88
  2,4-dichlorophenoxyacetic acid   88
  pea seedling response to   88–89
  Rhoeo leaf section response to   89
  soyabean hypocotyl response to   88
Auxotrophs   4, 5
Avena
  coleoptile   88
  response to gibberellins   89

Azocoll   178, 183
  measuring protease activity   185, 186,
    187

B

Bacteria   24, 25, 26
  blue-green   29, 30
  enzyme regulation in   3, 4, 5, 6, 17,
    19, 37, 38
Barley
  ADH   149, 203, 205
  aleurone layer   84, 87
    response to gibberellins   84–88,
      227, 228
  aspartate kinase   31
  nitrate degrading enzymes   10, 183
Beetroot
  disc   359, 360
Benzyladenine   90
Botrychium   64
Buckwheat
  seedlings   103

C

C 3 plants   113, 308
C 4 plants   37, 113, 294, 308
Carboxypeptidase A   202, 207
  pro-   202
Carotenoids   97
Carrot   35
  discs   46, 47, 54
Cascade control   18
Castor bean
  endosperm   51, 55
Catalase   341
Cell cycle
  synchronous systems   64–68
    natural   64–65, 78–79
    synchronized   65–68, 78–79
Cell wall
  degradation   84
Cellulase   89, 349, 353
Cellulose   88, 89
Cereal grain
  aleurone   83–88
Chloramphenicol   2
Chlorella pyrenoidosa   30, 75, 79
Chlorophyll   97
Chloroplast
  EM examination   314–322
  isolation   296–301, 304–306, 311, 312
  isolation of proplastids   306
  nitrogen metabolism pathway in   31–37

Chromatography
  affinity   271–286
  elution procedures   281–286
  gel permeation   263
  ion exchange   263–264
  principles   271–274
  selection of ligand   274–277
  synthesis of adsorbents   277–280
Cinnamic acid 4-hydroxylase (CA4H)
    102, 103, 104, 105, 110
Cocoonase
  pro-   207
Compartmentation
  in cells   31, 33
  of ADH   137
  of enzymes   246–247, 289–323, 329–
    361
  of enzyme synthesis   221
  of glycolytic intermediates   51
  of inactivating enzymes   190–191
  of ions   85
  of PAL inhibitor   110
Continuous-flow culture   9, 12
Control strength   see Sensitivity
Cordycepin   85, 118, 119, 121
Cotyledon   2
  cucumber   10
  di-   38
  mono-   38
  Phaseolus   2
  radish   107, 110
p-Coumaroyl: Coenzyme A ligase   102,
    103, 105
Crossover theorem   46–47
Cross-pathway regulation   5
Cucumis sativus   30
Cucurbita pepo   99, 115
Cycloheximide   8, 26, 76, 89, 106, 107,
    116, 147, 185, 237
Cytochrome   97
  C oxidase   305, 309
  C reductase   110, 182, 336
Cytokinin   2, 90

D

Daucus carota   66
Density labelling   75, 100, 106, 111, 148,
    149, 225–242
  detection of de novo protein synthesis
    233–236
  labels   226–228
    incorporation into amino acid pools
      240–242

measurement of enzyme activity   238–242

measurement of enzyme turnover   236–238

principles   225–226

solutes for centrifugation   228–233

Development
  photoregulation of   93–95
  pollen   69
  programmed   77–78

Dextrinase   84

Diaminopimelate decarboxylase   31

Dictyosomes
  isolation   309–310

Digitaria   292

Dihydrodipicolinate synthase   28, 29

Dihydroxyacetone phosphate   48

Dioon   64

Dixon plot   109

DNA
  histones   74
  polymerase   71, 73
  synthesis   71, 74, 77–78
    inhibition of   65

Drosophila melanogaster
  ADH   133

E

Effector strength   *see*   Elasticity coefficient

Elasticity coefficient   43

Electrophoresis   265–266 (*see also* Immunochemistry)
  micro-   45

Energy charge   23, 33

Endoplasmic reticulum   86, 87

Enolase   52

Enzyme
  assays   48–51
  conformation changes   155–174
  distribution studies   289–323
  effect of light   97–118, 208–210, 242
    table of enzymes   98–99
  end product repressible   1–10, 23–38
  in nitrogen assimilation   11–17
  extraction and purification of   245–
    267, 322–323
    concentration of extracts   261–262
    use of low molecular wt protective
      agents   253–255
    use of polymeric protective agents
      250–253
  induced-fit   155, 156, 157, 166, 245

isoenzyme   33, 207, 208, 245
  alcohol dehydrogenase   129–153
  of α-amylase   87, 210–219
  PEPC   113
  terminology   130

levels during development   2, 95, 133–
  134, 150, 293–295

metalloenzyme   131

monomeric   156–174
  molecular memory   156–165

ping-pong mechanism   171

proenzyme   202

regulation by specific inhibitors   90,
  111, 148–153

substrate-inducible   1–10
  in animal systems   2
  in bacteria   2

Escherichia coli
  inducible enzymes in   2

Esterase   178
  activity assay   187
  EM localization   334

Ethylene   84, 85, 86, 88

Etioplast
  isolation   306

Euglena gracilis   5

F

Fatty acid synthetase   246

Ferredoxin   11, 36, 170
  hydride transfer   170–174

Fixation   337–340
  aldehyde   340–343, 348

Frittilaria   64

Fructose diphosphate   48, 49, 53

G

α-galactosidase   77, 256

β-galactosidase   110, 207

Gene dosage   76–77

Gene expression
  photocontrol of   93–124
  sites for regulation   95, 96

Gherkin
  hypocotyl   104, 106, 237
    PAL inhibitor from   107–110

Gibberellins   84–89

Gingko   64

β-glucanase   84

Glucose-6-phosphate   45, 46, 113, 159,
  162, 163, 165
  dehydrogenase   70, 71, 72, 73, 77,
    257, 279, 280

Glucose-6-phosphate—*continued*
  isomerase 44
  localization 358–361
α-glucosidase 84
β-glucosidase 110
Glutamate dehydrogenase 2, 11, 12, 13, 14, 304
  in tulip tepals 203, 204, 215
  isoenzymes in oat leaves 2
Glutamate synthase 12, 13, 16, 31, 36, 37
Glutamine synthetase 3, 5, 11, 12, 13, 14, 16, 31, 32, 33, 37
  ATP requirement 11, 16, 33
  effect of light 16, 17, 18, 19, 33, 34, 35, 37
  effect of pH 33, 34
  inactivation of 16, 33
  in higher plants 3, 11, 15
  in wheat embryos 3
Glutathione reductase 279, 280
Glyceraldehyde-3-phosphate 47
  dehydrogenase 52, 257, 258, 281, 285
Glycolysis
  and control of pH 41–61
  effect of anaerobiosis on intermediates 46–47
  effect of anoxia on intermediates 54–55
α-glycerophosphate 48
α-glycerophosphate dehydrogenase 48, 49
β-glycerophosphatase
  localization 330, 338
Glycine max 66, 67
Golgi apparatus
  isolation 309–310

H

Haemanthus 64
Haplopappus gracilis 65, 66, 74
Helianthus annuus 30
Helianthus tuberosus 66, 73
Helminthosporic acid 84
Helminthosporin
  analogues 84
Helminthosporol 84
Hexokinase 44, 45, 52
  wheat germ 159ff
    desensitization by urea and SDS 164
    glucose binding 159, 161, 162, 163, 165

Hexose-phosphate-isomerase 52
Hill equation 45, 58
Histochemistry 329–361
Histones 74
Homoserine dehydrogenase 28, 29, 31, 35, 36, 38
Hordeum vulgare 30
Hydrolases 87, 89
Hydroxylamine 28
Hypoxanthine guanine phosphoribosyl-transferase 77

I

Immunochemistry 197–222, 322, 336–337
  absorption techniques 201–202, 208–210, 217
  α-amylase 210–222
  double diffusion 203
  electrophoresis 87, 200
    crossed 200, 201, 203, 204, 208, 216
    line 203, 204, 205, 216
    two-dimensional 205, 216
  principles 198–202
Immunofluorescence 79, 336
Immunology 100, 101, 107, 117
Invertase 3
  in radish 115–116
Ionophore 68
Isocitrate lyase 3
Isocitric dehydrogenase 305
Isoetes 64
Isopropyl malate synthase 26
Isozyme (*see* Enzyme)
ITP-ase 330

J

Jerusalem artichoke
  enzyme levels 71–75, 78
  tuber 67

K

Kalanchoe blossfeldiana 98
Klebsiella aerogenes 5, 14
Klotz plots 159, 161

L

Lactate dehydrogenase 49, 275, 278, 284
Lemna gibba 123
Lemna minor 4, 5, 6, 9, 11, 14, 15, 16, 17, 18, 19, 29
  growth in continuous-flow culture 9

levels of aldolase 53
nitrogen assimilation pathways in 12
Lignification 349
Liliaceae 64
Lilium 64, 69, 73, 77, 78
Lineweaver–Burke plot 109, 159, 163, 167
Lipoxygenase 113–115
Lupinus alba 98

M

Maize 56
ADH 129–153, 206
biosynthesis of aspartate amino acids in 29
nitrate degrading enzymes in 10, 37, 90, 180–183, 187–193
root tips 26
localization of enzymes 330–331, 341–343
seedlings 27, 38
Malate dehydrogenase 24, 49, 149, 188, 208, 220, 257, 278, 284
Malate synthetase 3
Malic enzyme 56–61
Membrane
-bound organelles 17
-bound PAL 110
cell free preparations 86, 308–309
EM examination 319
model 86
Mesophyll enzymes 37
solubilization 259–261
Metabolism
carbohydrate 25, 41
nitrogen 25, 32–27, 38, 41
Metal
ion effects on enzymes 24
6-methyl purine 8
Michaelis–Menten kinetics 45, 57, 156, 158, 165
constants for ADH 144
negative cooperativity 156, 160, 167, 169
positive cooperativity 156, 160, 169
Microbiology
isolation 308
Microscopic localization
autoradiography 335
capture mechanisms 334–335
electron 305, 306, 314–319
histo-chemistry 329–361
immunochemical 336

list of enzymes 333
·substrate film 335
Microscopy
light 313, 353
Microsomes 110, 343
Mimulus cardinalis 28
Mitochondria 343
purification 306–307
Mnium 64
Modification control 23–39
Mustard
cotyledon 104, 106, 110, 112, 239, 240, 241
seedlings 114, 240, 241

N

NAD 2
oxidation reduction transients 4, 5
structure 277
NADH 12
NADP 2, 5
-linked dehydrogenases 60
reductase 170, 171
hydride ion transfer 170–174
NADPH 17
Neurospora crassa 5, 9
arginine biosynthesis in 17
histidine starvation 5
yeast inhibitor complex 189
Nicotiana 5, 66, 67, 76, 353
Nitella 190
Nitrate reductase 2, 3, 11, 12, 13, 14, 15, 37, 111, 112, 242, 257
cytokinin dependent increases 90
specific inhibitor 90, 110, 177, 178, 180–183, 185, 186, 187–193, 248
substrate induction 7, 8, 9
Nitrite reductase 3, 11, 31, 36, 90
Nitrogen assimilation 1–22
Nuclei
isolation 307

O

Oat
internode
response to gibberellins 83, 89
protease 183, 185
Orebranche majus 3
Organelles
isolation 289–323
Ornithine
aminotransferase 256, 257

Ornithine—*continued*
  transaminase
    proteolysis   179–180
Ornithine carbamyl transferase   5, 6

P

Parsley
  cell suspension cultures   103, 104, 105
Pentose phosphate pathway   46
Pepsin   207
Peroxidase   115
  localization   331, 336, 338, 339, 341,
      343, 346, 348, 349, 350, 351, 352,
      354–361
Petroselinum hortense   99
pH control   56–61, 88
Phaseolus
  aureus   122
  vulgaris   2, 98, 99, 118, 123
Phenylalanine ammonia-lyase (PAL)
      102, 111, 206
  immunoabsorption   208–210
  inhibitor   107–110
  isopycnic centrifugation   231, 234,
      237
  kinetics and binding   166–170
o-phenanthroline   85
Phenylpropanoids   102
  biosynthetic pathway of secondary
      products   101
    photoregulation of enzymes in
        102–110
Phosphatase   3
  alkaline   3, 260
  localization   334
Phosphoenol-pyruvate carboxylase
      (PEPC)   48, 53, 56, 57, 112, 113,
      114, 233
Phosphofructokinase   44, 45, 47, 48, 49,
      52, 256
  activation   53
  effect of aldolase   51
6-phosphogluconate dehydrogenase
      279, 280
Phosphoglucose
  isomerase   279
  mutase   279
Phosphoglyceraldehyde   48
3-phosphoglycerate   47
  dehydrogenase   50
    inhibition by pyridoxamine phos-
        phate   50, 51

kinase   77
mutase   52
Phosphorylcholine cytidyl transferase
      86
Phosphorylcholine glyceride transferase
      86
Photomorphogenesis   94, 95, 96
Photoreceptor   96, 97
Photosynthetic enzymes   97
  reversible inactivation of   18
Physarum polycephalum   74
Phytate
  hydrolysis   85
Phytochrome   96, 97, 115, 116, 118, 120,
      183
  regulation of transcription   121, 242
Pisum sativum   30, 57, 66, 98, 99, 292,
      348, 349
  ADH   147, 149
  apical buds   110
  epicotyls   89
  seedling   83, 88, 89, 103, 104
Placenta
  extract   275
Plasmalemma   87
Plasmin
  pro-   207
Plastid   (*see* Chloroplast)
Platymonas tetrathele   9, 11
Polysomes   87, 89
  photocontrol of   118–121
Potato
  sweet   103, 358
  tubers   56, 103, 106
Post-transcriptional control   7, 89
Post-translational control
  of ADH   147–153
Prolyl *t*-RNA synthetase   256
Protease
  active sites   183–184
  as ADH inhibitor   151
  control during extraction   192–193
  inactivation of enzymes   177–193
    FAD dependent   178
    NAD dependent   178
    pyridoxal dependent   178, 179, 180
Protein   7, 87
  fingerprint   106
  radiolabelling   100, 101, 105, 106, 148
  synthesis
    inhibition of   7, 8, 89, 90, 100, 106,
        107, 148
    photocontrol of   117–118, 220

Pseudomonas
  inducible enzymes 2
Purine nucleotide cycle 55
Puromycin 106
Pyridoxal phosphate 41, 42, 50, 51
Pyridoxamine phosphate 41, 42, 50, 51
  absorption at 340 nm 51
Pyruvate kinase 49, 50, 52, 57

R

Raphanus sativus 30, 98
Rhizoctonia solani 167, 169
Rhodotorula glutinus 102, 166
Rhoeo 83, 89
Ribosome
  isolation 307–308
Ribonuclease 71, 72, 73, 84, 89, 118
Ribulose diphosphate
  carboxylase 33, 75, 183, 185, 213,
      218, 219, 220, 221, 247, 301, 311
  oxygenase 221, 247
Rice
  α-amylase 87
  nitrate degrading enzymes 10
RNA 7
  capping 85
  degradation of mRNA 85, 95
  poly A 85, 89
    photocontrol of 122, 123
  ribosomal 121
  synthesis 75, 85, 88, 111, 117, 118–
      124
    inhibition of 7, 8, 85, 89, 90, 147,
      148
RNA polymerase
  I 88
  II 88

S

Saccharomyces cerevisiae 5
Saprophytes 3
Scatchard plot 159, 161
Scenedesmus obliquus 280
Schiff's base 42
Schizosaccharomyces pombe 76, 78
Secale cereale 98
Selaginella 64
Sensitivity 43
Sepharose 312
  (6-aminohexyl)-oxamate 275
  NADP$^+$ 278, 280, 286
  N^6-(6-aminohexyl)5′-AMP 277, 278,
    281, 283, 284, 285

NMN 279
P^2-ADP 283, 284, 285
Sinapis alba 30, 98, 99
Soyabean 294
  hypocotyl 83
    response to auxin and cytokinins
      88
  suspension culture 67, 103
Spherocarpos 64
Spinacia oleracea 292
Spirodela oligorrhiza 3
Stroma 33, 34
Succinate dehydrogenase (SDH) 70, 71,
    73, 77, 191, 260
Sucrose
  phosphate synthetase 311
  synthetase 311
Sugar cane 112, 113
Sunflower
  leaf extracts 107
  PAL inhibitor 110
Suspension cultures
  Acer pseudoplatanus 70–71
Swede
  root 103
Sycamore 74, 77

T

Tannins 248–250
Thrombin 207
  pro- 207
Thymidine kinase (Td RK) 3, 69, 70–
    74, 77, 78
  dTMPK 69, 70–74
Transaminases 31
Transcription 95
  linear 76–77
  photocontrol 116–117
Translation 95, 116
  in vitro 117, 122–123, 221
    E. coli 241
  photocontrol 116–117
Trillium 64, 69
Triose phosphate dehydrogenase 44,
    305, 311
Triose phosphate isomerase 49, 305
Triticum aestivum 30
Trypsin 202, 207
  inhibitor 256
  trypsinogen 202
Tryptophan synthase 35, 178, 209
  inactivation 178

V

Vicia faba
  peroxidase   347
Vitamin B$_6$ 180

W

Wheat
  α-amylase   87, 210–222
  β-amylase   219
Wheat germ
  hexokinases   (*see* Hexokinase)
  protein synthesis system   117, 122–123

X

Xanthium   105, 110

Y

Yeast
  ADH   149
  protease A   188, 189
  protease B   188, 189
  synchronous cultures   79
  tryptophan synthase   178, 187–191

Z

Zamia   64
Zea mays   30, 66, 98, 130, 260